METHODS IN MOLECULAR BIOLOGY

Series Editor
John M. Walker
School of Life and Medical Sciences
University of Hertfordshire
Hatfield, Hertfordshire, AL10 9AB, UK

For further volumes:
http://www.springer.com/series/7651

Organoids

Stem Cells, Structure, and Function

Edited by

Kursad Turksen

Ottawa Hospital Research Institute, Ottawa, ON, Canada

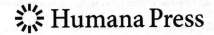

Editor
Kursad Turksen
Ottawa Hospital Research Institute
Ottawa, ON, Canada

ISSN 1064-3745   ISSN 1940-6029 (electronic)
Methods in Molecular Biology
ISBN 978-1-4939-7616-4   ISBN 978-1-4939-7617-1 (eBook)
https://doi.org/10.1007/978-1-4939-7617-1

Library of Congress Control Number: 2018965202

© Springer Science+Business Media, LLC, part of Springer Nature 2019
Open Access Chapter "Drug Sensitivity Assays of Human Cancer Organoid Cultures" is licensed under the terms of the Creative Commons Attribution 4.0 International License (http://creativecommons.org/licenses/by/4.0/). For further details see licence information in the chapter.
This work is subject to copyright. All rights are reserved by the Publisher, whether the whole or part of the material is concerned, specifically the rights of translation, reprinting, reuse of illustrations, recitation, broadcasting, reproduction on microfilms or in any other physical way, and transmission or information storage and retrieval, electronic adaptation, computer software, or by similar or dissimilar methodology now known or hereafter developed.
The use of general descriptive names, registered names, trademarks, service marks, etc. in this publication does not imply, even in the absence of a specific statement, that such names are exempt from the relevant protective laws and regulations and therefore free for general use.
The publisher, the authors, and the editors are safe to assume that the advice and information in this book are believed to be true and accurate at the date of publication. Neither the publisher nor the authors or the editors give a warranty, express or implied, with respect to the material contained herein or for any errors or omissions that may have been made. The publisher remains neutral with regard to jurisdictional claims in published maps and institutional affiliations.

This Humana Press imprint is published by the registered company Springer Science+Business Media, LLC, part of Springer Nature.
The registered company address is: 233 Spring Street, New York, NY 10013, U.S.A.

Preface

One of the outstanding challenges in regenerative medicine is how to instruct stem/early progenitor cells to progress through the appropriate steps to generate functional 3-dimensional organs. The field of organoids is geared towards defining and demonstrating the in vitro conditions that achieve this goal. I have attempted to collect representative protocols that show the exciting advances that have been made over the last several years in this area. I am very grateful to the contributors who have shared their hard-won successes in this volume.

Once again, I would like to acknowledge Dr. John Walker, Editor in Chief of the *Methods in Molecular Biology* series, for his support and leadership from the beginning of this project. In addition, I thank Patrick Marton, Senior Editor of the *Methods in Molecular Biology* series, for his encouragement on this project. I also remain very grateful to David Casey, the Editor of the Series, for keeping me on track and pointing out critical issues along the way so that this volume lives up the high standards of the series.

Ottawa, ON, Canada *Kursad Turksen*

Contents

Preface .... *v*
Contributors .... *xi*

A Simple Method of Generating 3D Brain Organoids Using Standard
Laboratory Equipment 1
Magdalena Sutcliffe and Madeline A. Lancaster

Clinically Amendable, Defined, and Rapid Induction of Human Brain
Organoids from Induced Pluripotent Stem Cells 13
Eva Tomaskovic-Crook and Jeremy M. Crook

Organoid Culture of Human Cancer Stem Cells 23
*Yohei Shimono, Junko Mukohyama, Taichi Isobe, Darius M. Johnston,
Piero Dalerba, and Akira Suzuki*

Construction of Thymus Organoids from Decellularized Thymus Scaffolds 33
*Asako Tajima, Isha Pradhan, Xuehui Geng, Massimo Trucco,
and Yong Fan*

Expansion of Human Airway Basal Stem Cells and Their Differentiation
as 3D Tracheospheres 43
Robert E. Hynds, Colin R. Butler, Sam M. Janes, and Adam Giangreco

Human Pluripotent Stem Cells (iPSC) Generation, Culture,
and Differentiation to Lung Progenitor Cells 55
Mahboobe Ghaedi and Laura E. Niklason

Organoid Culture of Lingual Epithelial Cells in a Three-Dimensional
Matrix 93
Hiroko Hisha and Hiroo Ueno

Generation of Functional Kidney Organoids In Vivo Starting
from a Single-Cell Suspension 101
Valentina Benedetti, Valerio Brizi, and Christodoulos Xinaris

Efficient Culture of Intestinal Organoids with Blebbistatin 113
Zhen Qi and Ye-Guang Chen

Isolation and Culture of Adult Intestinal, Gastric, and Liver Organoids
for Cre-recombinase-Mediated Gene Deletion 123
*Dustin J. Flanagan, Renate H. M. Schwab, Bang M. Tran,
Toby J. Phesse, and Elizabeth Vincan*

The Three-Dimensional Culture of Epithelial Organoids Derived
from Embryonic Chicken Intestine 135
*Malgorzata Pierzchalska, Malgorzata Panek, Malgorzata Czyrnek,
and Maja Grabacka*

New Trends and Perspectives in the Function of Non-neuronal Acetylcholine
in Crypt–Villus Organoids in Mice 145
Toshio Takahashi

Derivation of Intestinal Organoids from Human Induced Pluripotent
Stem Cells for Use as an Infection System 157
*Jessica L. Forbester, Nicholas Hannan, Ludovic Vallier,
and Gordon Dougan*

Murine Colonic Organoid Culture System and Downstream
Assay Applications 171
Yang-Yi Fan, Laurie A. Davidson, and Robert S. Chapkin

Intestinal Organoids as a Novel Tool to Study Microbes–Epithelium
Interactions 183
*Giulia Nigro, Melissa Hanson, Cindy Fevre, Marc Lecuit,
and Philippe J. Sansonetti*

The Isolation, Culture, and Propagation of Murine Intestinal Enteroids
for the Study of Dietary Lipid Metabolism 195
Diana Li, Hongli Dong, and Alison B. Kohan

Oncogenic Transformation of Human-Derived Gastric Organoids 205
*Nina Bertaux-Skeirik, Jomaris Centeno, Jian Gao, Joel Gabre,
and Yana Zavros*

Intestinal Crypt Organoid: Isolation of Intestinal Stem Cells, In Vitro Culture,
and Optical Observation 215
Yun Chen, Chuan Li, Ya-Hui Tsai, and Sheng-Hong Tseng

Human Intestinal Enteroids: New Models to Study Gastrointestinal
Virus Infections 229
*Winnie Y. Zou, Sarah E. Blutt, Sue E. Crawford, Khalil Ettayebi,
Xi-Lei Zeng, Kapil Saxena, Sasirekha Ramani, Umesh C. Karandikar,
Nicholas C. Zachos, and Mary K. Estes*

Study Bacteria–Host Interactions Using Intestinal Organoids 249
Yong-guo Zhang and Jun Sun

Disaggregation and Reaggregation of Zebrafish Retinal Cells for the Analysis
of Neuronal Layering 255
Megan K. Eldred, Leila Muresan, and William A. Harris

Antibody Uptake Assay in the Embryonic Zebrafish Forebrain to Study
Notch Signaling Dynamics in Neural Progenitor Cells In Vivo 273
Kai Tong, Mahendra Wagle, and Su Guo

Scaffold-Based and Scaffold-Free Testicular Organoids from Primary
Human Testicular Cells 283
Yoni Baert, Charlotte Rombaut, and Ellen Goossens

Use of a Super-hydrophobic Microbioreactor to Generate and Boost
Pancreatic Mini-organoids 291
*Tiziana A. L. Brevini, Elena F. M. Manzoni, Sergio Ledda,
and Fulvio Gandolfi*

Tissue Engineering of 3D Organotypic Microtissues by Acoustic Assembly 301
*Yuqing Zhu, Vahid Serpooshan, Sean Wu, Utkan Demirci, Pu Chen,
and Sinan Güven*

Cell Microencapsulation in Polyethylene Glycol Hydrogel Microspheres
Using Electrohydrodynamic Spraying 313
Mozhdeh Imaninezhad, Era Jain, and Silviya Petrova Zustiak

Gastrointestinal Epithelial Organoid Cultures from Postsurgical Tissues 327
Soojung Hahn and Jongman Yoo

Drug Sensitivity Assays of Human Cancer Organoid Cultures 339
*Hayley E. Francies, Andrew Barthorpe, Anne McLaren-Douglas,
William J. Barendt, and Mathew J. Garnett*

Erratum to: Drug Sensitivity Assays of Human Cancer Organoid Cultures 353
*Hayley E. Francies, Andrew Barthorpe, Anne McLaren-Douglas,
William J. Barendt, and Mathew J. Garnett*

Correction to: The Three-Dimensional Culture of Epithelial Organoids
Derived from Embryonic Chicken Intestine 355
*Malgorzata Pierzchalska, Malgorzata Panek, Malgorzata Czyrnek,
and Maja Grabacka*

Index *357*

Contributors

YONI BAERT • *Biology of the Testis, Research Laboratory for Reproduction, Genetics and Regenerative Medicine, Vrije Universiteit Brussel (VUB), Brussels, Belgium*

WILLIAM J. BARENDT • *Wellcome Trust Sanger Institute, Cambridge, UK*

ANDREW BARTHORPE • *Wellcome Trust Sanger Institute, Cambridge, UK*

VALENTINA BENEDETTI • *IRCCS-Istituto di Ricerche Farmacologiche 'Mario Negri', Centro Anna Maria Astori, Science and Technology Park Kilometro Rosso, Bergamo, Italy*

NINA BERTAUX-SKEIRIK • *Department of Molecular and Cellular Physiology, University of Cincinnati College of Medicine, Cincinnati, OH, USA*

SARAH E. BLUTT • *Department of Molecular Virology and Microbiology, Baylor College of Medicine, Houston, TX, USA*

TIZIANA A. L. BREVINI • *Laboratory of Biomedical Embryology, Centre for Stem Cell Research, Università degli Studi di Milano, Milan, Italy*

VALERIO BRIZI • *IRCCS-Istituto di Ricerche Farmacologiche 'Mario Negri', Centro Anna Maria Astori, Science and Technology Park Kilometro Rosso, Bergamo, Italy*

COLIN R. BUTLER • *Lungs for Living Research Centre, UCL Respiratory, University College London, London, UK*

JOMARIS CENTENO • *Biology Department, University of Puerto Rico-Río Piedras, San Juan, Puerto Rico*

ROBERT S. CHAPKIN • *Program in Integrative Nutrition and Complex Diseases, Department of Nutrition & Food Science, Texas A&M University, College Station, TX, USA; Center for Translational Environmental Health Research, Texas A&M University, College Station, TX, USA*

PU CHEN • *Department of Biomedical Engineering, School of Basic Medical Sciences, Wuhan University, Wuhan, China; Institute of Model Animals of Wuhan University, Wuhan, China*

YE-GUANG CHEN • *The State Key Laboratory of Membrane Biology, Tsinghua-Peking Center for Life Sciences, School of Life Sciences, Tsinghua University, Beijing, China*

YUN CHEN • *Department of Surgery, Far Eastern Memorial Hospital, Pan-Chiao, New Taipei, Taiwan; Department of Chemical Engineering and Materials Science, Yuan Ze University, Taoyuan, Taiwan*

SUE E. CRAWFORD • *Department of Molecular Virology and Microbiology, Baylor College of Medicine, Houston, TX, USA*

JEREMY M. CROOK • *ARC Centre of Excellence for Electromaterials Science, Intelligent Polymer Research Institute, AIIM Facility, Innovation Campus, University of Wollongong, Wollongong, NSW, Australia; Illawarra Health and Medical Research Institute, University of Wollongong, Wollongong, NSW, Australia; Department of Surgery, St Vincent's Hospital, The University of Melbourne, Fitzroy, VIC, Australia*

MALGORZATA CZYRNEK • *Department of Food Biotechnology, Faculty of Food Technology, The University of Agriculture in Kraków, Kraków, Poland*

PIERO DALERBA • *Department of Pathology and Cell Biology, Columbia University, New York, NY, USA*

LAURIE A. DAVIDSON • *Program in Integrative Nutrition and Complex Diseases, Department of Nutrition & Food Science, Texas A&M University, College Station, TX,*

USA; Center for Translational Environmental Health Research, Texas A&M University, College Station, TX, USA

UTKAN DEMIRCI • Bio-Acoustic MEMS in Medicine (BAMM) Lab, Department of Radiology, Canary Center for Early Cancer Detection, Stanford University School of Medicine, Stanford, CA, USA

HONGLI DONG • Department of Nutritional Sciences, University of Connecticut, Storrs, CT, USA

GORDON DOUGAN • Wellcome Trust Sanger Institute, Cambridge, UK

MEGAN K. ELDRED • Department of Physiology, Development and Neuroscience, Cambridge University, Cambridge, UK

MARY K. ESTES • Department of Molecular Virology and Microbiology, Baylor College of Medicine, Houston, TX, USA; Department of Medicine, Baylor College of Medicine, Houston, TX, USA

KHALIL ETTAYEBI • Department of Molecular Virology and Microbiology, Baylor College of Medicine, Houston, TX, USA

YANG-YI FAN • Program in Integrative Nutrition and Complex Diseases, Department of Nutrition & Food Science, Texas A&M University, College Station, TX, USA

YONG FAN • Institute of Cellular Therapeutics, Allegheny Health Network, Pittsburgh, PA, USA; Department of Biological Sciences, Carnegie Mellon University, Pittsburgh, PA, USA; Department of Microbiology and Immunology, Medical College of Drexel University, Philadelphia, PA, USA

CINDY FEVRE • Biology of Infection Unit, Inserm U1117, Institut Pasteur, Paris, France

DUSTIN J. FLANAGAN • Molecular Oncology Laboratory, University of Melbourne, Melbourne, VIC, Australia; Victorian Infectious Diseases Reference Laboratory, Doherty Institute, Melbourne, VIC, Australia

JESSICA L. FORBESTER • Wellcome Trust Sanger Institute, Cambridge, UK

HAYLEY E. FRANCIES • Wellcome Trust Sanger Institute, Cambridge, UK

JOEL GABRE • Department of Internal Medicine, University of Cincinnati, Cincinnati, OH, USA

FULVIO GANDOLFI • Laboratory of Biomedical Embryology, Centre for Stem Cell Research, Università degli Studi di Milano, Milan, Italy

JIAN GAO • Department of Pediatrics, W.F. Maternal & Child Health Hospital, Weifang, People's Republic of China

MATHEW J. GARNETT • Wellcome Trust Sanger Institute, Cambridge, UK

XUEHUI GENG • Department of Dermatology, University of Pittsburgh School of Medicine, Pittsburgh, PA, USA

MAHBOOBE GHAEDI • Departments of Anesthesia and Biomedical Engineering, Yale University, New Haven, CT, USA; Department of Biomedical Engineering, Yale University, New Haven, CT, USA

ADAM GIANGRECO • Lungs for Living Research Centre, UCL Respiratory, University College London, London, UK

ELLEN GOOSSENS • Biology of the Testis, Research Laboratory for Reproduction, Genetics and Regenerative Medicine, Vrije Universiteit Brussel (VUB), Brussels, Belgium

MAJA GRABACKA • Department of Food Biotechnology, Faculty of Food Technology, The University of Agriculture in Kraków, Kraków, Poland

SU GUO • State Key Laboratory of Genetic Engineering, Department of Genetics, School of Life Sciences, Fudan University, Shanghai, China; Department of Bioengineering and Therapeutic Sciences, Programs in Human Genetics and Biological Sciences, ELi and

Edythe Broad Center of Regeneration Medicine and Stem Cell Research, University of California, San Francisco, CA, USA

SINAN GÜVEN • *Izmir International Biomedicine and Genome Institute, Dokuz Eylul University, Izmir, Turkey; Department of Medical Biology, Faculty of Medicine, Dokuz Eylul University, Izmir, Turkey*

SOOJUNG HAHN • *Department of Microbiology and Institute of Basic Medical Sciences, School of Medicine, CHA University, Seongnam-si, Gyeonggi-do, South Korea*

NICHOLAS HANNAN • *University of Nottingham, Nottingham, UK*

MELISSA HANSON • *Biology of Infection Unit, Inserm U1117, Institut Pasteur, Paris, France*

WILLIAM A. HARRIS • *Department of Physiology, Development and Neuroscience, Cambridge University, Cambridge, UK*

HIROKO HISHA • *Kansai Medical University, Osaka, Japan*

ROBERT E. HYNDS • *Lungs for Living Research Centre, UCL Respiratory, University College London, London, UK*

MOZHDEH IMANINEZHAD • *Department of Biomedical Engineering, Saint Louis University, St. Louis, MO, USA*

TAICHI ISOBE • *Institute for Stem Cell Biology and Regenerative Medicine, Stanford University, Stanford, CA, USA*

ERA JAIN • *Department of Biomedical Engineering, Washington University in Saint Louis, St. Louis, MO, USA*

SAM M. JANES • *Lungs for Living Research Centre, UCL Respiratory, University College London, London, UK*

DARIUS M. JOHNSTON • *Department of Molecular and Cellular Physiology, Stanford University, Stanford, CA, USA*

UMESH C. KARANDIKAR • *Department of Molecular Virology and Microbiology, Baylor College of Medicine, Houston, TX, USA*

ALISON B. KOHAN • *Department of Nutritional Sciences, University of Connecticut, Storrs, CT, USA*

MADELINE A. LANCASTER • *MRC Laboratory of Molecular Biology, Cambridge Biomedical Campus, Cambridge, UK*

MARC LECUIT • *Biology of Infection Unit, Inserm U1117, Institut Pasteur, Paris, France*

SERGIO LEDDA • *Department of Veterinary Medicine, Università degli Studi di Sassari, Sassari, Italy*

CHUAN LI • *Department of Biomedical Engineering, National Yang-Ming University, Taipei, Taiwan*

DIANA LI • *Department of Nutritional Sciences, University of Connecticut, Storrs, CT, USA*

ELENA F. M. MANZONI • *Laboratory of Biomedical Embryology, Centre for Stem Cell Research, Università degli Studi di Milano, Milan, Italy*

ANNE MCLAREN-DOUGLAS • *Wellcome Trust Sanger Institute, Cambridge, UK*

JUNKO MUKOHYAMA • *Division of Molecular and Cellular Biology, Kobe University Graduate School of Medicine, Kobe, Hyogo, Japan*

LEILA MURESAN • *Department of Physiology, Development and Neuroscience, Cambridge University, Cambridge, UK*

GIULIA NIGRO • *Molecular Microbial Pathogenesis Unit, Inserm U1202, Institut Pasteur, Paris, France*

LAURA E. NIKLASON • *Departments of Anesthesia and Biomedical Engineering, Yale University, New Haven, CT, USA; Department of Biomedical Engineering, Yale University, New Haven, CT, USA*

MALGORZATA PANEK • *Department of Food Biotechnology, Faculty of Food Technology, The University of Agriculture in Kraków, Kraków, Poland*

TOBY J. PHESSE • *Molecular Oncology Laboratory, University of Melbourne, Melbourne, VIC, Australia; Victorian Infectious Diseases Reference Laboratory, Doherty Institute, Melbourne, VIC, Australia*

MALGORZATA PIERZCHALSKA • *Department of Food Biotechnology, Faculty of Food Technology, The University of Agriculture in Kraków, Kraków, Poland*

ISHA PRADHAN • *Institute of Cellular Therapeutics, Allegheny Health Network, Pittsburgh, PA, USA*

ZHEN QI • *The State Key Laboratory of Membrane Biology, Tsinghua-Peking Center for Life Sciences, School of Life Sciences, Tsinghua University, Beijing, China*

SASIREKHA RAMANI • *Department of Molecular Virology and Microbiology, Baylor College of Medicine, Houston, TX, USA*

CHARLOTTE ROMBAUT • *Biology of the Testis, Research Laboratory for Reproduction, Genetics and Regenerative Medicine, Vrije Universiteit Brussel (VUB), Brussels, Belgium*

PHILIPPE J. SANSONETTI • *Molecular Microbial Pathogenesis Unit, Inserm U1202, Institut Pasteur, Paris, France*

KAPIL SAXENA • *Department of Molecular Virology and Microbiology, Baylor College of Medicine, Houston, TX, USA*

RENATE H. M. SCHWAB • *Molecular Oncology Laboratory, University of Melbourne, Melbourne, VIC, Australia; Victorian Infectious Diseases Reference Laboratory, Doherty Institute, Melbourne, VIC, Australia*

VAHID SERPOOSHAN • *Stanford Cardiovascular Institute, Stanford University School of Medicine, Stanford, CA, USA*

YOHEI SHIMONO • *Division of Molecular and Cellular Biology, Kobe University Graduate School of Medicine, Kobe, Hyogo, Japan*

JUN SUN • *Division of Gastroenterology and Hepatology, Department of Medicine, University of Illinois at Chicago, Chicago, IL, USA*

MAGDALENA SUTCLIFFE • *MRC Laboratory of Molecular Biology, Cambridge Biomedical Campus, Cambridge, UK*

AKIRA SUZUKI • *Division of Molecular and Cellular Biology, Kobe University Graduate School of Medicine, Kobe, Hyogo, Japan*

ASAKO TAJIMA • *Institute of Cellular Therapeutics, Allegheny Health Network, Pittsburgh, PA, USA*

TOSHIO TAKAHASHI • *Suntory Foundation for Life Sciences, Bioorganic Research Institute, Kyoto, Japan*

EVA TOMASKOVIC-CROOK • *ARC Centre of Excellence for Electromaterials Science, Intelligent Polymer Research Institute, AIIM Facility, Innovation Campus, University of Wollongong, Wollongong, NSW, Australia; Illawarra Health and Medical Research Institute, University of Wollongong, Wollongong, NSW, Australia*

KAI TONG • *State Key Laboratory of Genetic Engineering, Department of Genetics, School of Life Sciences, Fudan University, Shanghai, China; Department of Bioengineering and Therapeutic Sciences, Programs in Human Genetics and Biological Sciences, ELi and Edythe Broad Center of Regeneration Medicine and Stem Cell Research, University of California, San Francisco, CA, USA*

BANG M. TRAN • *Molecular Oncology Laboratory, University of Melbourne, Melbourne, VIC, Australia; Victorian Infectious Diseases Reference Laboratory, Doherty Institute, Melbourne, VIC, Australia*

MASSIMO TRUCCO • *Institute of Cellular Therapeutics, Allegheny Health Network, Pittsburgh, PA, USA; Department of Biological Sciences, Carnegie Mellon University, Pittsburgh, PA, USA; Department of Microbiology and Immunology, Medical College of Drexel University, Philadelphia, PA, USA*

YA-HUI TSAI • *Department of Surgery, Far Eastern Memorial Hospital, New Taipei, Taiwan; Department of Chemical Engineering and Materials Science, Yuan Ze University, Taoyuan, Taiwan*

SHENG-HONG TSENG • *Department of Surgery, National Taiwan University Hospital, National Taiwan University College of Medicine, Taipei, Taiwan*

HIROO UENO • *Kansai Medical University, Osaka, Japan*

LUDOVIC VALLIER • *Wellcome Trust Sanger Institute, Cambridge, UK; Wellcome Trust-Medical Research Council Stem Cell Institute, Anne McLaren Laboratory, Department of Surgery, University of Cambridge, Cambridge, UK*

ELIZABETH VINCAN • *Molecular Oncology Laboratory, University of Melbourne, Melbourne, VIC, Australia; Victorian Infectious Diseases Reference Laboratory, Doherty Institute, Melbourne, VIC, Australia; School of Biomedical Sciences, Curtin University, Perth, WA, Australia*

MAHENDRA WAGLE • *Department of Bioengineering and Therapeutic Sciences, Programs in Human Genetics and Biological Sciences, ELi and Edythe Broad Center of Regeneration Medicine and Stem Cell Research, University of California, San Francisco, CA, USA*

SEAN WU • *Stanford Cardiovascular Institute, Stanford University School of Medicine, Stanford, CA, USA*

CHRISTODOULOS XINARIS • *IRCCS-Istituto di Ricerche Farmacologiche 'Mario Negri', Centro Anna Maria Astori, Science and Technology Park Kilometro Rosso, Bergamo, Italy*

JONGMAN YOO • *Department of Microbiology and Institute of Basic Medical Sciences, School of Medicine, CHA University, Seongnam-si, Gyeonggi-do, South Korea; CHA Biocomplex, Seongnam-si, Gyeonggi-do, South Korea*

NICHOLAS C. ZACHOS • *Division of Gastroenterology and Hepatology, Department of Medicine, Johns Hopkins University School of Medicine, Baltimore, MD, USA*

YANA ZAVROS • *Department of Molecular and Cellular Physiology, University of Cincinnati College of Medicine, Cincinnati, OH, USA*

XI-LEI ZENG • *Department of Molecular Virology and Microbiology, Baylor College of Medicine, Houston, TX, USA*

YONG-GUO ZHANG • *Division of Gastroenterology and Hepatology, Department of Medicine, University of Illinois at Chicago, Chicago, IL, USA*

YUQING ZHU • *Department of Biomedical Engineering, School of Basic Medical Sciences, Wuhan University, Wuhan, China; Institute of Model Animals of Wuhan University, Wuhan, China*

WINNIE Y. ZOU • *Department of Molecular Virology and Microbiology, Baylor College of Medicine, Houston, TX, USA*

SILVIYA PETROVA ZUSTIAK • *Department of Biomedical Engineering, Saint Louis University, St. Louis, MO, USA; Parks College of Engineering, Aviation and Technology, Saint Louis University, St. Louis, MO, USA*

A Simple Method of Generating 3D Brain Organoids Using Standard Laboratory Equipment

Magdalena Sutcliffe and Madeline A. Lancaster

Abstract

3D brain organoids are a powerful tool with prospective application for the study of neural development and disease. Here we describe the growth factor-free method of generating cerebral organoids from feeder-dependent or feeder-free human pluripotent stem cells using standard laboratory equipment. The protocol outlined below allows generation of 3D tissues, which replicate human early in vivo brain development up to the end of the first trimester, both in terms of morphology and gene expression pattern.

Keywords: Brain, In vitro, Neural differentiation, Neurobiology, Neurodevelopment, Organoids, Stem cells, Three-dimensional

1 Introduction

Recent development of pluripotent stem cell technology has enabled generation of 3D structures that recapitulate development and function of tissue in vivo under in vitro conditions. Currently available in vitro 3D models of human brain include, in order of increasing complexity, SFEBq (cortical tissues obtained through suspension aggregation culture) [1], cortical spheroids [2], forebrain organoids [3], and cerebral organoids [4]. Generation of these neural organoids in a dish allows for the study and manipulation of human brain tissue in vitro and provides a powerful tool to gain insight into fetal neocortical development, neurodevelopmental, and neurological disorders, such as autism and neurodegeneration, and even the potential to screen candidate drugs [5].

The method of generation of cerebral organoids as previously established [4, 6] and described here builds upon the intrinsic ability of pluripotent stem cells (PSCs) to self-organize upon precisely timed manipulation of culture conditions, but without addition of external growth factors or small molecules. PSCs in vitro can form aggregates, called embryoid bodies (EBs), which display differentiation of three germ layers. Upon transfer of EBs to minimal media, their outermost layer, ectoderm, acquires more neural properties, and differentiates into neuroepithelium, which later becomes the source of neural progenitors. Provision of a Matrigel scaffold supports the self-organization of the neuroepithelium and elicits the correct polarity signal to form a large, apicobasally polarized neuroepithelial bud [4, 6]. The neuroepithelia further develop into structures that bear remarkable similarity to developing human neocortex, both in terms of morphology and gene expression patterns [7].

Tissues developed with this protocol replicate human early in vivo brain development up to a stage equivalent to gestation week 13 [7]; however, for other studies, cerebral organoids can be cultured for up to 1 year and potentially longer [6].

2 Materials

All liquids and plasticware used for cell culture must be sterile. PBS used for all reagents does not contain calcium or magnesium. All reagents are to be stored at 4 °C unless indicated otherwise. When using reagents stored at −20 °C, avoid repeated freeze-thaw cycles by aliquoting into smaller volumes.

2.1 Feeder-Dependent hPSC Culture

1. Growth-arrested irradiated mouse embryonic fibroblasts (MEF, feeder cells).
2. MEF media: 10 % (v/v) ES-quality fetal bovine serum, 1 % (v/v) GlutaMAX in DMEM.
3. Gelatin-coated cell culture plates: 2 mg gelatin per 10 cm^2 using 1 mg/ml solution in water.
4. hESC medium: 20 % (v/v) KnockOut Serum Replacement, 3 % (v/v) ES-quality fetal bovine serum, 1 % (v/v) GlutaMAX supplement, 1 % (v/v) MEM-NEAA with 7 μl/l of neat 2-mercaptoethanol (*see* **Note 1**) in DMEM/F12.
5. Basic fibroblast growth factor (bFGF): 10 μg/ml in PBS with 0.1 % BSA (w/v) (store at −20 °C).
6. Complete hESC medium: hESC supplemented with 20 ng/ml of bFGF immediately before use (*see* **Note 2**).
7. Collagenase IV solution: 1 mg/ml in DMEM/F12 (store at −20 °C).
8. Cell culture dishes or plates.
9. Cell lifter.

2.2 Feeder-Independent hPSC Culture

1. Matrigel, growth factor reduced (store at −20 °C).
2. mTeSR1 medium.
3. DEMEM/F12.
4. 0.5 mM EDTA in PBS.
5. Parafilm.

2.3 Generation of Embryoid Bodies from Feeder-Dependent hPSC

1. Dispase solution: 1 mg/ml in DMEM/F12 (store at −20 °C).
2. 0.05 % trypsin-EDTA (store at −20 °C).
3. Soybean trypsin inhibitor. Prepare 1 ml trypsin inhibitor solution at 1 mg/ml in DMEM/F12 for every ml of trypsin solution used for dissociation just before use. Sterile-filter before use.
4. Low bFGF hESC medium: hESC supplemented with 4 ng/ml of bFGF immediately before use (*see* **Note 3**).
5. 0.4 % trypan blue solution in PBS or other isotonic solution.
6. U-bottomed ultra-low attachment 96-well plate.
7. Y-27623 ROCK inhibitor, 5 mM solution in water (store at −20 °C).

2.4 Generation of Embryoid Bodies from Feeder-Independent hPSC

1. Accutase (Store at −20 °C).

2.5 Germ Layer Differentiation and Induction of Primitive Neuroepithelia

1. Ultra-low attachment 24-well plate.
2. Neural induction (NI) medium: 1 % (v/v) N2 supplement, 1 % (v/v) GlutaMAX supplement, 1 % (v/v) MEM-NEAA solution in DMEM/F12 with 1 µg/ml heparin.

2.6 Expansion of Neuroepithelial Tissue

1. Matrigel, normal growth factors (store at −20 °C).
2. Parafilm.
3. 1:1 mixture of DMEM/F12 and Neurobasal medium.
4. Differentiation medium without vitamin A (DM-A): 0.5 % (v/v) N2 supplement, 1 % (v/v) B-27 supplement without vitamin A, 1 % (v/v) GlutaMAX supplement, 0.5 % (v/v) MEM-NEAA solution in 1:1 DMEM/F12:Neurobasal medium. Supplement with insulin 2.5 µg/ml, 3.5 µl/l of neat 2-mercaptoethanol, and add 100 U/ml penicillin and 100 µg/ml streptomycin.
5. 6 cm cell culture dishes.

2.7 Growth of Cerebral Tissue

1. Differentiation medium with vitamin A (DM + A): 0.5 % (v/v) N2 supplement, 1 % (v/v) B-27 supplement with vitamin A, 1 % (v/v) GlutaMAX supplement, 0.5 % (v/v) MEM-NEAA solution in 1:1 DMEM/F12:Neurobasal medium. Supplement with insulin 2.5 µg/ml, 3.5 µl/l of neat 2-mercaptoethanol, and add 100 U/ml penicillin and 100 µg/ml streptomycin.
2. 6 cm cell culture dishes.

3 Methods

This protocol is suitable for generation of cerebral organoids both from feeder-dependent and feeder-independent cultures. For feeder-dependent cells follow Sections 3.1 and 3.3, then 3.5, for feeder-independent follow Sections 3.2 and 3.4 then 3.5.

All liquids used for cell culture should be equilibrated at least to room temperature just before use.

3.1 Feeder-Dependent hPSC Culture and Passaging

1. Coat cell culture plates with gelatin solution and incubate at 37 °C for 30–60 min (see **Note 4**).
2. Plate 1.7×10^5 MEFs per 10 cm^2 in MEF media and incubate at 37 °C 5 % CO_2 overnight (see **Note 5**).
3. To passage hPSC wash 80 % confluent hPSC culture with PBS (see **Note 6**).
4. Remove PBS and replace with 1 ml collagenase IV solution, incubate for 5–10 min at 37 °C 5 % CO_2, then dislodge the colonies with a cell lifter and triturate with 1000 μl pipette tip to break colonies up and transfer to a 15 ml tube (see **Note 7**).
5. Wash the well with 1 ml of complete hESC medium and add to the 15 ml tube from step 4.
6. Pellet the colonies at 200 g for 2 min.
7. During the spin wash the MEF plate with PBS and add appropriate amount of complete hESC media to the well (see **Note 8**).
8. Remove supernatant from over the hPSC pellet and gently resuspend the colonies in appropriate volume of complete hESC medium. For routine maintenance hPSC are split at the ratio of 1:3–1:6 based on the growth area.
9. Transfer the colonies onto the MEF plate (see **Note 9**) and incubate at 37 °C with 5 % CO_2.
10. Change media daily to fresh complete hESC media.

3.2 Feeder-Independent hPSC Culture and Passaging

1. Coat a cell culture dish with 83 μg Matrigel, growth factors reduced, in 1 ml DMEM/F12 per 10 cm^2. Incubate overnight at 4 °C (see **Notes 10** and **11**).
2. Before use, pre-warm the plate at 37 °C for 20 min.
3. Passage cells when about 80 % confluent. Aspirate the spent medium and wash cells twice with EDTA.
4. After the second wash, add 600 μl of EDTA per 10 cm^2 and incubate at 37 °C for 4 min.
5. During incubation, aspirate the Matrigel solution from the pre-warmed plate and replace with an appropriate volume of mTeSR1 (see **Note 8**).

6. Aspirate the EDTA and spray off (*see* **Note 12**) the colonies from the required area of the well using a small volume (for example 500 μl) of mTeSR1. Routinely feeder-independent hPSC are split at the ratio of 1:3–1:6 based on the growth area.
7. Transfer the colonies to the plate prepared in steps 1 and 2, and triturate with a 1000 μl pipette tip to break the colonies up (*see* **Note 7**).
8. Incubate at 37 °C with 5 % CO_2.
9. Change media daily to fresh mTeSR1.

3.3 Generation of Embryoid Bodies from Feeder-Dependent hPSC

1. Use a culture no more than 80 % confluent (*see* **Note 13**).
2. Wash the colonies with PBS and add 1 ml dispase solution per 10 cm^2.
3. When the colony edges begin to curl off plate, remove the dispase solution and wash with 1 ml PBS. Remove the colonies by spraying (*see* **Note 12**) with 1 ml hES media using a P1000 tip, three times (3 ml total), and transfer to a 15 ml conical tube being careful to limit the disruption of colonies.
4. Allow colony clumps to settle for 3 min and aspirate supernatant gently with a P1000 tip.
5. Resuspend colonies in 1 ml Trypsin/EDTA and incubate 2 min at 37 °C. Add 1 ml trypsin inhibitor and triturate using a P1000 tip to obtain a single cell suspension (*see* **Note 14**). Take a small aliquot for cell counting, and then add 8 ml complete hES media.
6. Centrifuge cells at 270 × g for 5 min.
7. During the spin mix the aliquot of the cell suspension with an equal volume of trypan blue solution and count live cells (*see* **Note 15**). Use the average of two replicates for calculations in subsequent steps.
11. Resuspend the cell pellet in 1 ml low bFGF hESC medium with 1:100 ROCK inhibitor (*see* **Note 16**).
12. Calculate the volume of cell suspension needed. For 1 well use 150 μl and 9000 live cells (*see* **Note 17**). Prepare the suspension in complete hES medium with ROCK inhibitor.
13. Plate 150 μl per well of a 96-well ultra-low attachment plate and incubate the plate for 3 days (*see* **Note 18**).

3.4 Generation of Embryoid Bodies from Feeder-Independent hPSC

1. Use a culture no more than 80 % confluent (*see* **Note 13**).
2. Aspirate spent media from culture and wash cells twice with EDTA.
3. After the second wash, add 600 μl of EDTA to 10 cm^2 and incubate at 37 °C for 4 min.

4. Aspirate the EDTA solution and replace with 500 μl Accutase per 10 cm². Incubate for another 4 min at 37 °C.

5. Tap the culture dish vigorously to detach cells. Neutralize Accutase with 1 ml mTeSR1 and triturate to obtain a single cell suspension (*see* **Note 14**).

6. Transfer the cell suspension to a 15 ml conical tube, take an aliquot for counting, and centrifuge at $200 \times g$ for 4 min.

7. During the spin mix the aliquot of cell suspension with an equal volume of trypan blue solution and count live cells (*see* **Note 15**). Use the average of two replicates for calculations in subsequent steps.

8. Resuspend the cell pellet in 1 ml low bFGF hESC medium with 1:100 ROCK inhibitor (*see* **Note 16**).

9. Calculate the volume of cell suspension needed. For 1 well use 150 μl and 9000 live cells (*see* **Note 19**). Prepare the suspension in low bFGF medium with ROCK inhibitor.

10. Plate 150 μl per well of a 96-well ultra-low attachment plate and incubate the plate for 3 days (*see* **Note 18**).

3.5 Germ Layer Differentiation and Induction of Primitive Neuroepithelia

1. On day 3 gently aspirate about half of the medium from each well, and add 150 μl of fresh hESC medium without bFGF or ROCK inhibitor (*see* **Note 20**).

2. Typically on day 6, when the EBs grow to 500–600 μm in size and display smooth, bright edges under a cell culture microscope (Fig. 1a), transfer individual aggregates to a 24-well ultra-low attachment plate containing 500 μl of neural induction medium per well. To avoid damaging the EBs, use a cut P200 pipette tip (*see* **Notes 21 and 22**).

3. Feed the EBs every other day. Aspirate about half of the spent media and add 500 μl of fresh NI medium.

3.6 Expansion of Neuroepithelial Tissue

After 4–6 days, typically on days 10–12, the EBs should become brighter on the outside with a sharper distinction between outer bright tissue and inner dark tissue. The outer ectoderm should show signs of radial organization, which is a sign of neuroepithelium development. The edges of a healthy neuroepithelium should be smooth (see Fig. 1b):

1. Thaw appropriate volume of Matrigel on ice (*see* **Notes 10 and 23**).

2. Prepare a dimpled Parafilm sheet. Cut a 5 × 5 cm square of Parafilm and place on empty 10 μl pipette tip tray with the protective paper sheet facing up (Fig. 2a). With a gloved finger, press down over a hole to form a dimple. Make a grid of 4 × 4 dimples (Fig. 2b), remove the protective paper sheet, and trim

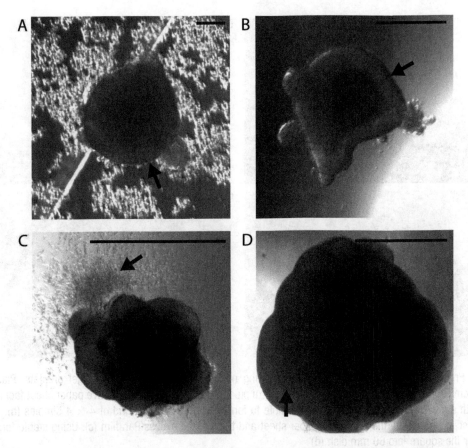

Fig. 1 Progression of cerebral organoid development from human PSCs. EB on day 6, showing ectodermal differentiation visible as clearing on the outside indicated with the *arrow*, scale bar 100 μm (**a**). A well-defined, pseudostratified neuroepithelium in EB on day 11, scale bar 500 μm (**b**). A budded organoid 8 days after Matrigel embedding, day 19. The *arrow* indicates some nonneural cells migrating out of the main mass of the organoid, scale bar 1 mm (**c**). Day 37 organoid with expanded cortical tissue and clearly visible ventricle-like structures, indicated with an *arrow*, scale bar 1 mm (**d**)

off the excess Parafilm (Fig. 2c). Then transfer the square into a 60 mm dish (Fig. 2d, *see* **Note 24**).

3. With a cut P200 pipette tip (*see* **Note 25**), transfer individual tissues into the dimples in the Parafilm sheet.
4. Remove excess media using an uncut P200 pipette tip, being careful not to suck up the tissue into the uncut tip, which will damage it (*see* **Note 26**).
5. Add about 30 μl of Matrigel to each dimple to cover the tissue.
6. Position each aggregate in the middle of the droplet by pushing the tissue with a small tip such as a P10 pipette tip (*see* **Note 27**).

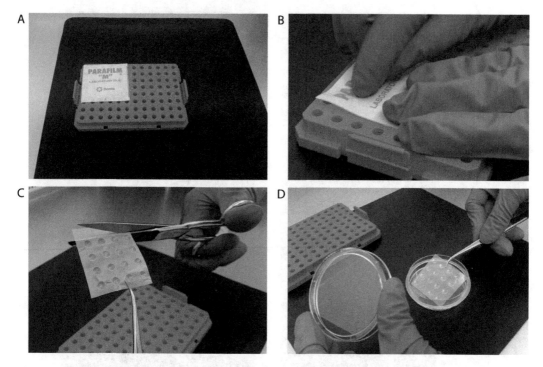

Fig. 2 Preparation of Parafilm sheet for embedding neuroepithelial tissues in Matrigel droplets. Place a 5 × 5 cm square of Parafilm on empty, sterile 10 μl pipette tip tray with the protective paper sheet facing up (**a**). With a gloved finger, press down over a hole to form a dimple. Make a grid of 4 × 4 dimples (**b**). With sterile scissors remove the protective paper sheet and trim off the excess Parafilm (**c**). Using sterile forceps transfer the square into 60 mm dish (**d**)

7. Place the 60 mm dish with sheet containing droplets in a 37 °C incubator for 20–30 min to allow the Matrigel to fully polymerize.
8. Add 5 ml of DM-A per dish.
9. Detach the Matrigel droplets containing neuroepithelial tissues from the Parafilm sheet. This can be done either by gently spraying off the droplets when adding media in the previous step or by holding the Parafilm sheet with forceps and agitating while immersed in the media (*see* **Note 28**).
10. Change media every other day. Tilt the dish to one side and wait for the tissues to sediment to one side; then aspirate the spent medium without touching the neuroepithelia. Then replace with 5 ml DM-A.

After 24 h the tissues should form buds of neuroepithelium, and within the following 48 h, the fluid-filled cavities in the center of each bud should become visible (Fig. 1c). Often, other migratory cell types appear around the neuroepithelial buds and spread into the Matrigel matrix. These cells do not seem to contribute to the development of the cerebral tissue and their identity has not been investigated.

3.7 Growth of Cerebral Tissue

1. Three to four days after embedding or when the tissues show more complex neuroepithelia with budding outgrowth and radial processes in the Matrigel, change the feeding media to DM + A (*see* **Note 29**).

2. At this point, the organoids are sufficiently large to require agitation. This can easily be done by simply placing a standard orbital shaker within a CO_2 tissue culture incubator. Agitate the dishes at 85 rpm (*see* **Note 30**) at 37 °C with 5 % CO_2.

3. Change the culture media every 3–4 days.

Tissues can be grown under such conditions until large lobes of cerebral tissue develop (Fig. 1d) containing neural stem cells, progenitors, and neurons. These tissues can be maintained for up to 1 year (*see* **Note 31**).

4 Notes

1. For smaller volumes of media, a 1:100 dilution of 2-mercaptoethanol can be used for the ease of liquid handling.

2. To obtain a bFGF concentration of 20 ng/ml, use 2 µl of bFGF stock solution per 1 ml of medium.

3. To obtain a bFGF concentration of 4 ng/ml, use 0.4 µl of bFGF stock solution per 1 ml of medium.

4. To avoid drying the gelatin coat, either plate MEFs or add media immediately after removing gelatin solution.

5. Minimum attachment time is 4–6 h.

6. This protocol uses cells in an established culture. It is not recommended that cells be used for generation of cerebral organoids immediately after thawing.

7. Do not triturate too much as this will decrease cell viability.

8. The final volume of media should be 2 ml per 10 cm^2. For example, if the volume of cells to be plated is 0.5 ml, add 1.5 ml of fresh medium to the plate at this step.

9. To distribute colonies evenly, move plate back and forth and side to side or in the shape of figure of eight before placing in an incubator.

10. Matrigel polymerizes at room temperature. Medium for dissolving Matrigel must be at 2–8 °C, and if small quantities of Matrigel are handled, the pipette tips and tubes used should also be chilled.

11. Coated plates can be used up to 1 week after coating if stored at 4 °C. To prevent plates from drying out, they can be sealed with Parafilm.

12. Aim to lift the colonies from the surface of the plate using liquid expelled from the pipette tip.
13. Cells in the active phase of growth generate better EBs. For more confluent cultures, increasing culture medium volume to 3 or 4 ml per 10 cm^2 can help maintain cells in the active growth phase.
14. Pipetting up and down up to five times is usually enough. The cell suspension should appear uniformly cloudy.
15. Performing the count on hemocytometer tends to be more accurate than using a cell counter and verifies single cell suspension.
16. Final concentration of ROCK inhibitor is 50 μM.
17. For example, for 10 wells, one needs 10×150 μl $= 1.5$ ml and 9000×10 cells $= 90{,}000$ cells. If the live cell density in the sample from step 6 was $1 \times 10[6]$ cells/ml, then the total number of cells is 2 ml $\times 1 \times 10[6] = 2 \times 10[6]$ cells. Since this number of cells was then pelletted and resuspended in 1 ml in step 12, the required volume of cell suspension needed is 90,000 cells/2,000,000 cells/ml $= 0.045$ ml $= 45$ μl. Mix the cell suspension by pipetting up and down a couple of times before taking the required volume of cells.
18. The day of generation of EBs is day 0.
19. For example, for 10 wells one needs 10×150 μl $= 1.5$ ml and 9000×10 cells $= 90{,}000$ cells. If the live cell density in the sample from step 6 was $1 \times 10[6]$ cells/ml, then the total number of cells is 1.5 ml $\times 1 \times 10[6] = 1.5 \times 10[6]$ cells. Since this number of cells was then pelletted and resuspended in 1 ml in step 12, the required volume of cell suspension needed is 90,000 cells/1,500,000 cells/ml $= 0.06$ ml $= 60$ μl.
20. If instead of one large EB several small ones are obtained, or there is a lot of debris with no main large EB by day 3, it indicates too much cell death had occurred during EB preparation. To increase cell viability, good-quality PSC cultures should be used and pipetting should be limited to minimum. Do not use ROCK inhibitor and bFGF that have been repeatedly freeze-thawed.
21. A P200 pipette tip should be cut with sterile scissors to obtain an opening of about 1–1.5 mm in diameter. Other methods of transferring EBs such as using a spatula will result in damage of the delicate structures.
22. To avoid transferring larger volumes of hESC media with the EBs, let it drop to the bottom of the tip by gravity and release only a small amount of liquid containing the EB. Carrying over hESC medium can interfere with neural induction.

23. About 500 µl Matrigel is enough to embed 16 EBs. Such volume takes about 2 h to thaw on ice at 4 °C. At this stage use regular Matrigel, not the more expensive growth factor reduced version.

24. Parafilm can be sterilized by spraying with 70 % ethanol and drying in cell culture cabinet. When pressing dimples into the Parafilm sheet, make sure you wear fresh gloves, cleaned with 70 % ethanol. To trim and transfer Parafilm sheets within a cell culture cabinet, use sterile scissors and forceps. To minimize the risk of contamination, antibiotics are added to DM-A medium.

25. A P200 pipette should be cut with sterile scissors to obtain an opening of about 1.5–2 mm in diameter. Other methods of transferring EBs such as using a spatula will result in damage of the delicate structures.

26. To avoid sucking the EB into the pipette and damaging it, aim the opening of the p200 tip away from the tissue and try to collect a few smaller portions of liquid rather than aspirate everything at once.

27. Matrigel polymerizes quickly at room temperature and can be kept on ice or cold block inside a cell culture cabinet. Neuroepithelial tissue can only be positioned within Matrigel droplets before they solidify. Embedding 16 tissues at a time is usually manageable before the Matrigel polymerizes.

28. The most resistant droplets can be detached by inverting the Parafilm sheet and agitating. If all other methods fail, individual droplets can be gently sprayed off using a P1000 tip, but care has to be taken since this can result in Matrigel matrix damage.

29. Alternatively, neuroepithelia can be cultured in a 125 ml spinner flask in 75 ml of DM + A medium, 32 tissues per flask. To transfer tissues embedded in Matrigel into a spinner flask, use a cut P1000 pipette tip. Agitate at 25 rpm and change media every 7 days.

30. The agitation speeds given are applicable to shakers with shaking diameter (throw) of 10 mm. We observed that the development of properly organized brain tissue is sensitive to the centrifugal force (Fc) produced by shaking. To adjust shaking speed to different shaking diameter, calculate the appropriate speed using the following formula: $Fc = rpm^2 \times throw$ where $Fc = 72{,}250$ for 10 mm throw with agitation at 85 rpm. For example, in order to adjust rpm for an orbital shaker with throw of 15 mm, the following calculation will yield the correct speed: $rpm = \sqrt{(72{,}250/15)} = 69.4$. This simplified formula does not account for the dish size, and organoids in dishes

smaller than 60 mm will experience lower Fc, while organoids in larger dishes will experience greater Fc.

31. As the tissues grow bigger, the maximum number of tissues per 60 mm dish should be halved roughly every 20 days, until finally only 2 organoids remain per dish.

Acknowledgments

We would like to thank all members of the Lancaster Lab for helpful discussion. Research in the Lancaster Lab is supported by the MRC (MC_UP_1201/9) as well as a 3Rs prize from the NC3Rs.

References

1. Eiraku M, Watanabe K, Matsuo-Takasaki M et al (2008) Self-organized formation of polarized cortical tissues from ESCs and its active manipulation by extrinsic signals. Cell Stem Cell 3:519–532
2. Paşca AM, Sloan SA, Clarke LE et al (2015) Functional cortical neurons and astrocytes from human pluripotent stem cells in 3D culture. Nat Methods 12:671–678
3. Kadoshima T, Sakaguchi H, Nakano T et al (2013) Self-organization of axial polarity, inside-out layer pattern, and species-specific progenitor dynamics in human ES cell-derived neocortex. Proc Natl Acad Sci U S A 110:20284–20289
4. Lancaster MA, Renner M, Martin C et al (2013) Cerebral organoids model human brain development and microcephaly. Nature 501:373–379
5. Kelava I, Lancaster MA (2016) Dishing out mini-brains: current progress and future prospects in brain organoid research. Dev Biol 420(2):199–209
6. Lancaster MA, Knoblich JA (2014) Generation of cerebral organoids from human pluripotent stem cells. Nat Protoc 9:2329–2340
7. Camp JG, Badsha F, Florio M et al (2015) Human cerebral organoids recapitulate gene expression programs of fetal neocortex development. Proc Natl Acad Sci U S A 112:15672–15677

Clinically Amendable, Defined, and Rapid Induction of Human Brain Organoids from Induced Pluripotent Stem Cells

Eva Tomaskovic-Crook and Jeremy M. Crook

Abstract

Human brain organoids provide opportunities to produce three-dimensional (3D) brain-like tissues for biomedical research and translational drug discovery, toxicology, and tissue replacement. Here we describe a protocol for rapid and defined induction of brain organoids from human induced pluripotent stem cells (iPSCs), using commercially available culture and differentiation media and a cheap, easy to handle and clinically approved semisynthetic hydrogel. Importantly, the methodology is uncomplicated, well-defined, and reliable for reproducible and scalable organoid generation, and amendable to principles of current good laboratory practice (cGLP), with the potential for prospective adaptation to current good manufacturing practice (cGMP) toward clinical compliance.

Keywords Brain organoid, Clinical compliance, Defined, GelMA, Human induced pluripotent stem cells, Hydrogel

1 Introduction

The ability to generate three-dimensional (3D) organoids in vitro that recapitulate features of the developing prenatal human brain in utero represents a quantum leap from conventional 2D neural cell culture. Whether they incorporate multiple brain regions as reported by Lancaster and colleagues [1, 2] or are comparatively simpler [3], they exhibit a remarkable level of complexity unattainable with cells in 2D. While current methods are unable to precisely recapitulate the complicated patterning and diverse environmental cues found in and experienced by an intact embryo, they are nonetheless rapidly evolving from first-generation systems to more defined, standardized, reproducible, and clinically amendable protocols, vital for experimentation and translational application [4]. In addition, with the need to model development and disease pathogenesis later as well as early in fetal development, extended culture for organoid maturation is necessary.

Here we describe in detail a method to generate human brain organoids for protracted culture and maturation, under defined conditions, amendable to cGLP and adaptable to cGMP for clinical compliance. Similar to other protocols for producing organoids described in this volume, a desired level of quality assurance (QA) can be applied for all aspects of the method, including but not limited to the manufacture, procurement, receipt, storage and handling of materials, reagents and iPSCs, as well as cell and organoid culture. While other guidances specifically consider QA for many of the abovementioned processes [5–7], in any event, standardized and reliable protocols are necessary for basic science through applied research and clinical product development.

This chapter details methods that are suitable for routine application, high-quality research, and adaptable for clinical compliance. The culture media used include commercially available cGMP, defined feeder-free and serum-free maintenance medium for human iPSCs, and similarly defined off-the-shelf serum-free medium for initial neural induction and end-stage differentiation medium for organoid formation. Importantly, iPSCs are differentiated, and organoids are maintained on a soft hydrogel substrate produced from photocrosslinkable gelatin methacrylate (GelMA), a derivative of collagen widely used for biomedical applications [8]. Finally, the protocol described is scalable to generate larger numbers of organoids.

2 Materials

2.1 iPSC Culture and Passaging

1. iPSCs.
2. Corning® Matrigel® (hESC qualified matrix, LDEV-free; Falcon, In Vitro Technologies, Cat. no. FAL354277).
3. Dulbecco's Modified Eagle Medium/Nutrient Mixture F-12 (DMEM/F-12; Gibco, Life Technologies, Cat. no. 11330-057).
4. mTeSR1 complete kit for hES maintenance (STEMCELL Technologies, Cat. no. 05850).
5. Dulbecco's phosphate-buffered saline, without Ca^{2+} and Mg^{2+} (PBS; Sigma Aldrich, Cat. no. D8537).
6. Ethylenediaminetetraacetic acid (EDTA; Sigma Aldrich, Cat. no. E8008).

2.2 Preparation of Cast Hydrogel

1. Gelatin methacrylate (GelMA, bloom 300, 80% degree of substitution, Sigma-Aldrich, Cat. no. 900496) or prepared as described elsewhere (see **Note 6**).
2. Irgacure 2959 (2-hydroxy-1-[4-(2-hydroxyethoxy)phenyl]-2-methyl-1-propanone; CIBA Chemicals, Cat. no. 29891301PS04).
3. Dulbecco's phosphate-buffered saline, without Ca^{2+} and Mg^{2+} (PBS; Sigma Aldrich, Cat. no. D8537).

4. 100% ethanol (Chem-Supply, Cat. no. EA043).
5. DMEM/F-12 medium (Gibco, Life Technologies, Cat. no. 11330-057).

2.3 Neural Induction of iPSCs: Generation of Neural Progenitor Cells

1. STEMdiff Neural Induction Medium (STEMCELL Technologies, Cat. no. 05835).

2.4 Differentiation of iPSC-Derived Neural Progenitor Cells

1. Neurobasal Medium (Gibco, Life Technologies, Cat. no. 21103-049).
2. DMEM/F-12 medium (Gibco, Life Technologies, Cat. no. 11330-057).
3. NeuroCult SM1 neuronal supplement (STEMCELL Technologies, Cat. no. 05711).
4. N2 supplement-A (STEMCELL Technologies, Cat. no. 07152).
5. Brain-derived neurotrophic factor (BDNF; Cat. no. AF-450-02; PeproTech, Lonza).
6. Tris–HCl (Sigma-Aldrich, Cat. no. T5941).
7. Human serum albumin (Sigma-Aldrich, Cat. no. A9511).
8. L-Glutamine (200 mM; Gibco, Life Technologies, Cat. no. 25030-081).

2.5 General Equipment

1. 6-well and 24-well tissue culture plates (Costar, Sigma-Aldrich).
2. Pipetman (micropipette) and tips.
3. Pipet-Aid (motorized pipette) and serological pipettes.
4. 15 mL conical tubes (Corning, Cat. no. 430052).
5. Parafilm M laboratory film (Edwards Group, Cat. no. PM992).
6. Class 2 Biological Safety Cabinet (biosafety cabinet).
7. Humidified 5% CO_2 in air incubator maintained at 37 °C.
8. Dry 37 °C incubator.
9. Inverted phase-contrast microscope.
10. Low-speed centrifuge.
11. Liquid nitrogen storage tank.
12. −80 °C freezer.
13. Refrigerator.
14. UV light source (365 nm range).
15. Luminometer.

3 Methods

Reagent preparation and cell culture work should be performed in a biosafety cabinet unless otherwise specified. All media and reagents should remain sterile. Incubations and culturing should be performed in a 37 °C incubator with a humidified atmosphere of 5% CO_2 in air. Centrifuge steps are performed at room temperature (RT).

3.1 Routine iPSC Culture and Passaging

iPSCs are cultured as feeder-free cultures as in the method outlined below (see **Note 1**). Importantly, a good quality iPSC culture with low spontaneous differentiation is essential for efficient neural induction.

1. Prepare Corning® Matrigel® matrix as per the manufacturer's instructions and store frozen aliquots at -80 °C (see **Note 2**).
2. Thaw Matrigel aliquot at 4 °C in refrigerator, 2 h before use.
3. Transfer contents of thawed aliquot to 6 mL DMEM/F-12 media prechilled to 4 °C in a sterile 15 mL conical tube. Gently pipet up and down with a serological pipette to mix contents.
4. Immediately add 1 mL diluted Matrigel matrix per well of 6-well plate. Swirl and rock plate back and forth to evenly coat the surface of the well. Seal plate edges with Parafilm, wrap in foil, and transfer to a level shelf within a refrigerator to incubate overnight at 4 °C.
5. The following day, carefully remove Matrigel matrix with serological pipette (without scratching bottom of well) and add 1 mL per well DMEM/F-12 to rinse.
6. Remove and add 2 mL DMEM/F-12 to cover each coated well. Coated wells may be used immediately on the same day or prepared up to 2 weeks in advance. If to be stored at 4 °C in a refrigerator, seal plate edges with Parafilm and wrap in foil before storage.
7. Prepare iPSC culture medium (mTeSR1) as per the manufacturer's instructions and store at 4 °C in refrigerator (see **Note 3**).
8. Monitor iPSCs daily by viewing under phase-contrast microscope to ensure appropriate timing of passaging and removal of areas of differentiation prior to passage. iPSCs are passaged prior to colonies becoming confluent (approximately 60–75% confluency across well) and when center of each colony has become dense, but not overgrown. For routine passaging, cells undergo a 1:6 to 1:12 split ratio.
9. When iPSC colonies are ready to passage, prewarm iPSC culture media, DMEM/F-12 media, PBS, and EDTA solution to

37 °C in dry incubator 1 h before use. Prior to use, allow stored Matrigel-coated plates to equilibrate to room temperature for 30 min.

10. Just before dissociating cells for passaging, aspirate DMEM/F-12 from Matrigel-coated wells and add 2 mL prewarmed iPSC culture media to each well of 6-well plate and place in humidified 37 °C/5% CO_2 incubator until ready to seed.

11. Aspirate spent culture media from wells to be passaged. Wash each well briefly with 1 mL PBS and aspirate PBS solution.

12. To dissociate iPSCs, add 1 mL 0.02% EDTA (see **Note 4**), and incubate for 2 min in 37 °C/5% CO_2 incubator.

13. Once detachment of edge of colonies viewed under phase-contrast microscope has been confirmed, carefully aspirate EDTA solution.

14. With a 5 mL serological pipette, immediately add 3 mL prewarmed DMEM/F-12 media to a well. While slowly dispensing the contents, scrape the tip of the serological pipette over the surface of the well to dislodge the cells.

15. Repeat one to two times to dislodge the iPSC colonies, but do over-pipette cell suspension to ensure dissociation of cell aggregates (approximately 10–20 cells) rather than single cells. Transfer dissociated cell aggregates to a 15 mL conical tube.

16. For routine passaging, plate three wells at a range of seeding densities to enable optimal passaging approximately every 3–4 days in culture. For example, depending on the iPSC line, cells may be seeded at a 1:6, 1:10, and 1:20 split ratio. For instance, to achieve a 1:6 split ratio, add 0.5 mL of dissociated cell aggregates per well of a 6-well plate prepared earlier (see Sect. 3.1, Step 10), containing 2 mL prewarmed iPSC culture media, and stored in humidified 37 °C/5% CO_2 incubator.

3.2 Preparation of Cast GelMA Hydrogel

GelMA hydrogels are prepared by directly casting into wells of standard 24-multiwell culture plates (see **Note 5**).

1. Freshly prepare 5–10% (w/v) GelMA (see **Note 6**) with 0.5% Irgacure 2959 in PBS solution (see **Note 7**). Dissolve by heating to 37 °C for 1 h and mix well.

2. Add 0.2 mL prepolymerized solution per well of 24-well plate for casting of hydrogel. Ensure an even coating of the well by rocking plate back and forth.

3. Immediately crosslink the prepolymerized solution by exposure of each well to 100 mW/cm^2 UV light (365 nm) for 60 s via an optical fiber connected to a Dymax BlueWave 75 UV spot lamp (see **Note 8**).

4. Immediately following polymerization, add 1 mL PBS to each well.
5. Soak hydrogel for 2 h at 4 °C. Rinse hydrogel twice in PBS.
6. Continue soaking of hydrogel overnight at 4 °C.
7. The following day, replace PBS solution with 1 mL DMEM/F12 media per well.
8. Store pre-made cast hydrogels at 4 °C in refrigerator. Seal plate edges with Parafilm and wrap in foil before storage.

3.3 Surface Seeding of iPSCs Onto Hydrogel and Generation of Neural Progenitor Cells

iPSCs are seeded onto cast hydrogel as dissociated cells in mTeSR1 (see **Note 5**). Colonies are allowed to attach and expand, before neural induction using STEMdiff Neural Induction Medium (Fig. 1).

1. Prewarm DMEM/F-12 and mTeSR1 media for 1 h at 37 °C in dry incubator before use.
2. Pre-made cast hydrogels stored at 4 °C are equilibrated to room temperature for 1 h.
3. Carefully remove DMEM/F-12 media from wells without disturbing cast hydrogel from well.
4. Wash hydrogel in each well of 24-well plate with 1 mL prewarmed DMEM/F-12 media. Allow to incubate for at least 15 min in a 37 °C/5% CO_2 incubator.
5. Just before dissociating cells for passaging, aspirate DMEM/F-12 from wells, and add 1 mL prewarmed iPSC culture media to each well of 24-well plate containing cast hydrogel, and place in humidified 37 °C/5% CO_2 incubator until ready to seed.
6. iPSCs are dissociated with EDTA as described in Sect. 3.1 for routine iPSC cell culture and dissociation.
7. Following dissociation, using a 2 mL serological pipette, slowly pipette cell aggregate suspension up and down twice, and seed dissociated cell aggregates at one drop per well (approximately 50 μL).
8. Return 24-well plate to humidified 37 °C/5% CO_2 incubator. Distribute cells across the surface of the hydrogel by moving plate side to side and front and back (at right angles).
9. Leave plates undisturbed overnight before confirming attachment of cells the following day.

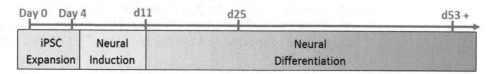

Fig. 1 Schematic for neural induction and differentiation of iPSCs seeded onto hydrogel

10. Replenish mTeSR1 media cells every 2 days. During media change, ensure the cells are minimally dislodged by slowly pipetting to the side of the well.

11. Before and after media change, check attachment of cells to hydrogel surface and cell clustering by phase-contrast microscopy. Return 24-well plate to humidified 37 °C/5% CO_2 incubator after media change.

12. On the fourth day after seeding, iPSC colonies are transitioned to Neural Induction Media (see **Note 9**). Prewarm STEMdiff Neural Induction Media for 1 h at 37 °C in dry incubator before use.

13. Carefully remove spent media and add 1 mL STEMdiff Neural Induction Media to each well. Observe cells with phase-contrast microscopy. Return 24-well plate to humidified 37 °C/5% CO_2 incubator.

14. To maintain cultures, perform a half media change every 2–3 days by removal of 0.5 mL of spent media and addition of 0.5 mL STEMdiff Neural Induction Media to each well. Observe cells with phase-contrast microscopy. Return 24-well plate to humidified 37 °C/5% CO_2 incubator.

3.4 Neural Differentiation of iPSC-Generated Neural Progenitor Cells

Following neural stem cell induction and proliferation, iPSC-neural progenitor cells are transitioned to undergo neural differentiation (Fig. 1).

1. Prepare Neural Differentiation Medium (e.g., 50 mL; see **Note 10**):
   - Neurobasal Medium: 25 mL
   - DMEM/F-12 medium: 25 mL
   - NeuroCult SM1 neuronal supplement: 500 µL (1% final conc.)
   - N2 supplement-A: 250 µL (0.5% final conc.)
   - L-Glutamine: 500 µL (1× final conc.)
   - BDNF (250 µg/mL stock; see **Note 11**): 10 µL (50 ng/mL final concentration).

2. Prewarm Neural Differentiation Medium at 37 °C in dry incubator, 1 h before use (see **Note 12**).

3. To transition iPSC-neural progenitor cells to undergo neural differentiation, remove spent Neural Induction Media from each well. Be particularly careful not to dislodge the cells or the hydrogel.

4. Slowly add 1 mL prewarmed Neural Differentiation Medium to each well. Return 24-well plate to humidified 37 °C/5% CO_2 incubator.

5. Feed cultures every 3–4 days with a half media change to replenish media. Perform a half media change by removal of 0.5 mL of spent media and addition of 0.5 mL Neural Differentiation Medium to each well. Due to evaporation of media over time, it may be necessary to remove a smaller amount of spent media (roughly half the residual volume of the media in well). Replace with 0.5 mL Neural Differentiation Medium.

6. Monitor cultures daily and inspect morphology (semitransparent microspikes on outer surface of neurosphere) and acidity of media (change media before reaching orange/yellow color). Continue culture until desired stage of development is reached.

4 Notes

1. For complete details of culture, adaptation, and expansion of pluripotent stem cells, refer to Brehm and Ludwig [9].

2. Refer to supplied Certificate of Analysis for Corning® Matrigel® matrix to prepare appropriate aliquot size using a dilution factor to achieve 6 mL total volume of diluted matrix.

3. The iPSC colonies are routinely propagated as colonies in mTeSR1 media. Thaw frozen mTeSR1 supplement overnight at 4 °C. Once thawed, prepare aliquots of mTeSR1 supplement in working volumes and store at −20 °C. Do not refreeze. To ensure stability of components, keep mTeSR1 medium for up to 2 weeks at 4 °C. Other feeder-independent culture media systems may also be employed.

4. Although EDTA (0.02% w/v in PBS; 0.5 mM) method of iPSC dissociation is described, iPSCs may also be dissociated using other reagents, such as Versene (0.02%, Lonza, Cat. no. 17-711E).

5. High-throughput production of organoids can be achieved by seeding and differentiating iPSCs onto either hydrogel cast into wells of standard plastic-multiwell culture plates or custom bioprinted hydrogel-multiwell plates.

6. A commercial supplier of GelMA is provided. However, depending on the required application, GelMA monomer can be produced from gelatin (Sigma-Aldrich) with tunable modification of the methacrylation degree and gel concentration; refer to Nichol et al. and O'Connel et al. [10, 11].

7. Alternatively, a 10% (w/v) Irgacure 2959 stock solution dissolved in 100% ethanol can be prepared and stored at RT, wrapped in foil to keep dark. Dilute to final concentration of 0.5% (w/v) in GelMA monomer solution before polymerization. Depending on the application, other photoinitiators, for

example, VA-086 (2, 2'-azobis[2-methyl-N-(2-hydroxyethyl) propionamide], Wako Pure Chemical Industries), can be used.

8. Apply UV light through a covered tissue culture plate to maintain sterility. To ensure consistent application of energy of UV light to polymer solution, measure intensity of light using a luminometer. Expose each well separately by positioning optical fiber directly over each well.

9. Thaw frozen STEMdiff Neural Induction Medium at room temperature or overnight at 4 °C. To ensure stability of components, keep STEMdiff Neural Induction Medium for up to 2 weeks at 4 °C. Once thawed, prepare aliquots in working volumes and store at −20 °C. Do not refreeze.

10. To ensure stability of components, prepare sufficient working volume of Neural Differentiation Medium for 2–3 weeks of neural differentiation at any one time. Store at 4 °C for 2–3 weeks.

11. To prepare 250 μg/mL BDNF stock solution, dissolve 100 μg vial of BDNF in 0.4 mL filter-sterilized 5 mM Tris–HCl containing 0.5% human serum albumin, pH 7.6. Store 10 μL aliquots of 250 μg/mL BDNF stock solution at −80 °C. Do not freeze/thaw vials.

12. Immediately after use, return tube of Neural Differentiation Medium to 4 °C to preserve stability of components.

Acknowledgment

The authors wish to acknowledge funding from the Australian Research Council (ARC) Centre of Excellence Scheme (CE140100012).

References

1. Lancaster MA, Knoblich JA (2014) Generation of cerebral organoids from human pluripotent stem cells. Nat Protoc 9:2329–2340
2. Lancaster MA, Renner M, Martin CA, Wenzel D, Bicknell LS, Hurles ME et al (2013) Cerebral organoids model human brain development and microcephaly. Nature 501:373–379
3. Li Y, Muffat J, Omer A, Bosch I, Lancaster MA, Sur M et al (2017) Induction of expansion and folding in human cerebral organoids. Cell Stem Cell 20:385–396 e3
4. Lindborg BA, Brekke JH, Vegoe AL, Ulrich CB, Haider KT, Subramaniam S et al (2016) Rapid induction of cerebral organoids from human induced pluripotent stem cells using a chemically defined hydrogel and defined cell culture medium. Stem Cells Transl Med 5:970–979
5. Crook JM, Ludwig TE (eds) (2017) Stem cell banking: concepts and protocols. methods in molecular biology, vol 1590. Springer, New York
6. Andrews PW, Baker D, Benvinisty N, Miranda B, Bruce K, Brustle O et al (2015) Points to consider in the development of seed stocks of pluripotent stem cells for clinical applications: international stem cell banking initiative (ISCBI). Regen Med 10:1–44
7. International Stem Cell Banking I (2009) Consensus guidance for banking and supply of human embryonic stem cell lines for research purposes. Stem Cell Rev 5:301–314

8. Yue K, Trujillo-de Santiago G, Alvarez MM, Tamayol A, Annabi N, Khademhosseini A (2015) Synthesis, properties, and biomedical applications of gelatin methacryloyl (GelMA) hydrogels. Biomaterials 73:254–271
9. Brehm JL, Ludwig TE (2017) Culture, adaptation, and expansion of pluripotent stem cells. In: Crook JM, Ludwig TE (eds) Stem cell banking: concepts and protocols. Methods in molecular biology, vol 1590. pp 139–150
10. Nichol JW, Koshy ST, Bae H, Hwang CM, Yamanlar S, Khademhosseini A (2010) Cell-laden microengineered gelatin methacrylate hydrogels. Biomaterials 31:5536–5544
11. O'Connell CD, Di Bella C, Thompson F, Augustine C, Beirne S, Cornock R et al (2016) Development of the Biopen: a handheld device for surgical printing of adipose stem cells at a chondral wound site. Biofabrication 8:015019

Organoid Culture of Human Cancer Stem Cells

Yohei Shimono, Junko Mukohyama, Taichi Isobe,
Darius M. Johnston, Piero Dalerba, and Akira Suzuki

Abstract

Organoid culture is a three-dimensional culture method that enables ex vivo analysis of stem cell behavior and differentiation. This method is also applicable to the studies on stem cell characters of human cancer stem cells. The components of organoid culture include Matrigel® and a culture medium containing growth factor cocktails that mimic the microenvironments of organ stem cell niches. Here, we describe the basic methods for the organoid culture of dissociated or FACS-sorted human cancer stem cells. Then, we introduce a method to dissociate the organoids for serial passage and propagation.

Keywords: Organoids, Cancer stem cells, Clonogenicity, Matrigel, Growth factor cocktails, Passage

1 Introduction

Adult tissue stem cells expand and generate lineage-restricted cells to meet the requirements of tissue maintenance and repair [1]. The behaviors of adult tissue stem cells are regulated by their interaction with niche cells and microenvironments. The culture of embryonic stem cells (ESCs) from an epiblast made it possible to propagate pluripotent cells and analyze somatic derivatives in vitro. However, the culture of adult tissue stem cells had been hindered by the lack of appropriate culture systems until the organoid culture and the air–liquid interface culture of murine intestine were developed [2–4].

The currently used method of organoid culture was first established for the culture of murine small intestinal epithelium [3]. And then, the organoid cultures of other tissue-derived stem cells, such as stomach, liver, and pancreas, are established through investigation of growth factor cocktails that mimic the various organ stem cell niches [5]. The essential components of this organoid culture include Matrigel ® (containing the mixture of basement membrane proteins secreted by the Engelbreth-Holm-Swarm mouse sarcoma cells) and culture medium containing the growth factor cocktails, including Wnt stimulators, such as Wnt3A and R-spondin-1, and activators of tyrosine kinase receptor signaling.

The cancer stem cell (CSC) hypothesis proposes that CSCs within a tumor have the ability to self-renew and generate differentiated progeny in a similar way as normal stem cells in the tissue [6–8]. The organoid culture method has been successfully applied to malignant tissues, such as primary colon, prostate, and pancreatic cancers [2, 5] and is proved to be an attractive method to analyze functions of CSCs in human cancer tissues [9].

The organoid culture method enables the in vitro propagation of CSCs by reflecting the complexity of tumor formation using primary cancer tissues and tumor xenografts. In addition, organoid culture allows functional analyses of CSCs including their genetic engineering using lentivirus constructs and/or CRISPR/Cas9-mediated genome editing [10, 11] (Fig. 1). Furthermore, patient-derived organoids can be applicable to predict drug response in a personalized fashion. Thus, combined with other

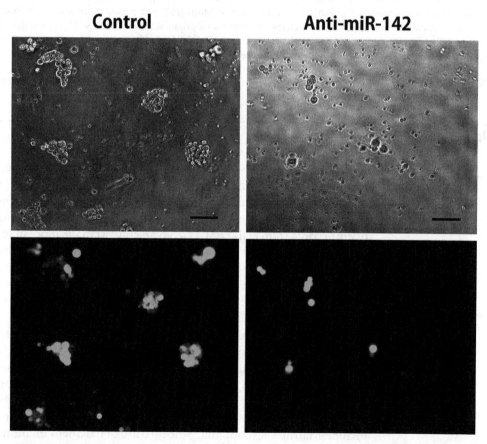

Fig. 1 Suppression of organoid formation by microRNA-expressing lentivirus. PDX-derived human breast cancer cells were infected with 20 moi of anti-miR-142 or control-expressing lentivirus. The *upper panels* are the phase-contrast images of the organoids, and the *lower panels* are the fluorescent microscopic images for the detection of GFP expressed from the infected lentivirus. The number of organoids was significantly reduced when the cells were infected with anti-miR-142-expressing lentivirus. Bars, 100 μm

in vivo experiments, such as xenotransplantation assays of CSCs, organoid cultures of human CSCs has a high potential to advance our understanding of human cancer biology.

2 Materials

Prepare all solutions using ultrapure water (attain a resistivity of 18 MΩ-cm at 25 °C) and tissue culture grade reagents. Diligently follow all waste disposal regulations when disposing of waste materials.

2.1 Plate Preparation and Seeding

Organoid base medium is premade and can be stored at 4 °C. Organoid culture medium is prepared freshly just before the experiments by adding the reagents listed in Table 1 to the organoid base medium.

1. Growth factor-reduced phenol red-free Matrigel® (GFRM).
2. Organoid base medium: Advanced DMEM/F12 medium, containing 4 mM glutamine (or GlutaMax®), 1 mM sodium

Table 1
List of reagents supplemented to the organoid culture medium

Reagent	Final concentration	Dilution	Stock solution
Y-27632 (ROCK inhibitor)	10 μM	×1,000	10 mM
EGF	10–50 ng/ml	×10,000	100 μg/ml
Noggin	100 ng/ml	×2,500	250 μg/ml
R-spondin 1	250 ng/ml	×400	100 μg/ml
Wnt3A	100 ng/ml	×2,000	200 μg/ml
B27 supplement	×1	×50	
Nicotinamide	10 mM		1 M
Jagged-1	1 μM		
ITES (Insulin–Transferrin–Ethanolamine–Selenium)	×1		
N-acetyl-L-cysteine	1 mM		500 mM
N2 supplement	×1	×100	
A83-01 (ALK5 inhibitor)	500 nM	×1,000	500 μM
SB202190 (p38 MAPK inhibitor)	10 μM	×3,000	30 mM
Gastrin I	10 nM	×10,000	100 μM

Stock solution: Dilute A83-01 and SB202190 with dimethyl sulfoxide (DMSO). Dilute other reagents with 0.15 % BSA/PBS

pyruvate, 10 mM HEPES, 2 % fetal bovine serum (FBS) (*see* **Note 1**) and 100 U/ml penicillin, 100 μg/ml streptomycin, 0.25 μg/ml amphotericin B (*see* **Note 2**).

To prepare the organoid base medium, add 10 ml of 200 mM glutamine (or GlutaMax®), 5 ml of 100 mM sodium pyruvate, 5 ml of 1 M HEPES, 5 ml of 10,000 U/ml penicillin–10,000 μg/ml streptomycin–25 μg/ml amphotericin B, and 10 ml FBS to 465 ml advanced DMEM/F12 medium.

3. Organoid culture medium: 10 μM Y-27632, 10 ng/ml EGF, 100 ng/ml Noggin, 250 ng/ml R-spondin 1, 100 ng/ml Wnt-3A (or 50 % Wnt3A-conditioned medium) (*see* **Note 3**).

    To prepare the organoid culture medium, add 5 μl Y-27632, 0.5 μl EGF, 2 μl Noggin, 12.5 μl R-spondin 1, and 2.5 μl Wnt3A to 5 ml organoid base medium (*see* **Note 4**).
    Other reagents listed in Table 1 can be added depending on the type of the cells for organoid culture.

4. Autoclaved or filtered phosphate buffed saline (PBS).

2.2 Seeding of Sorted Cells

1. 1.5 ml low attachment microcentrifuge tubes
2. Sorting solution: HBSS, containing 2 % FBS, 1 mM sodium pyruvate, 2 mM EDTA, and 10 mM HEPES.

2.3 Passage of the Organoids

1. Cell Recovery Solution (Corning, #354253)
2. 0.025 % trypsin-EDTA

3 Methods

3.1 Plate Preparation and Seeding

1. Add 250–500 μl of GFRM in each well of a 24-well plate (*see* **Notes 5–7**).
2. Place the plate in a tissue culture incubator at 37 °C for 30 min to allow gelation of GFRM (*see* **Note 8**).
3. Freshly prepare 5–10 ml of organoid culture media. Warm the medium in 37 °C water bath.
4. Resuspend the cells in pre-warmed culture media at the density of 10,000 cells/ml. Add 100 μl per well on the GFRM layer (*see* **Note 9**).
5. Place the plate in a tissue culture incubator at 5 % CO_2, 37 °C (*see* **Note 10**). Organoids will develop over 4–14 days (Fig. 2).
6. For the first week, do not change the media. Or feed colonies each day by carefully adding 25 μl of culture media to the top of the well (*see* **Note 11**).

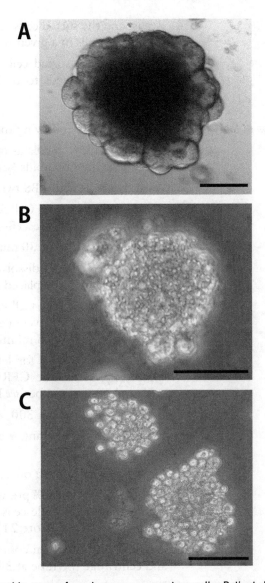

Fig. 2 Organoids grown from human cancer stem cells. Patient-derived tumor xenografts (PDX) were established by the xenotransplantation of surgical specimens from human cancer patients. The dissociated PDX cells were seeded on the GFRM layer and cultured for 14 days. Organoids grown from (**a**) colon, (**b**) pancreas, and (**c**) breast (invasive lobular carcinoma) cancers are shown. Bars, 50 μm

3.2 Seeding of Sorted Cells

1. Set up the plate following the **steps 1–3** in Section 3.1.
2. Add 100 μl per well of organoid culture media on the GFRM layer. Place the plate in the incubator for at least 30 min to equilibrate the GFRM matrix and the culture media before plating.
3. Sort cells into 1.5 ml low attachment microcentrifuge tubes that are preloaded with 100 μl of sorting solution.

4. Resuspend the sorted cells in culture media to approximate the number of cells in a given volume (*see* **Note 12**).

5. Gently add the sorted cells to each well (*see* **Note 13**). The volume added needs to be below the 10 % of liquid culture media volume.

3.3 Passage of the Organoids

1. Set up the plate following the **steps 1–3** in Section 3.1.
2. Carefully remove media as much as possible, paying attention not to lose the organoids (*see* **Note 14**).
3. Wash the wells with PBS twice (*see* **Note 14**).
4. Add 500 µl of ice-cold Cell Recovery Solution to each well of a 24-well plate (*see* **Notes 15–17**).
5. Incubate the plate for 30 min at 4 °C.
6. Transfer the cells and dissolved gel in a 1.5 ml low attachment microcentrifuge tube placed on ice (*see* **Note 18**).
7. Add 500 µl of ice-cold Cell Recovery Solution to each well of a 24-well plate to rinse wells (*see* **Note 19**). Collect and combine the solution in the 1.5 ml microcentrifuge tube on ice.
8. Place the tubes on ice for 1 h or until the GFRM is dissolved completely. When the GFRM is dissolved, the colonies will settle at bottom of tube (*see* **Note 20**).
9. Centrifuge the tube at $300 \times g$ for 5 min at 4 °C.
10. Remove the supernatant. Wash cell pellet by gentle resuspension in ice-cold PBS.
11. Repeat **steps 9** and **10** once.
12. Add 300 µl of 0.025 % pre-warmed trypsin–EDTA to dissociate colonies into single cells. Gently pipette and incubate for 5 min at 37 °C (*see* **Note 21**).
13. Add 1,000 µl of serum containing medium (HBSS + 2%FBS) and centrifuge the tube at $300 \times g$ for 5 min at 4 °C.
14. Remove the solution and resuspend the cell pellet in pre-warmed culture media at the density of 10,000 cells/ml.
15. Add 100 µl per well on the GFRM layer.

4 Notes

1. A higher concentration of FBS will be required for the growth of certain patient samples.
2. The concentration of antibiotics, such as penicillin and streptomycin, can be doubled especially when using primary cancer specimens potentially contaminated with microbes (e.g.

primary colorectal cancer). Take extra care for the first 2–3 days to prevent contamination into the organoid culture.

3. Prepare the Wnt3A-conditioned medium by culturing L-Wnt3A cells (ATCC catalog number CRL-2647) until the cells become confluent.

4. The reagents required for organoid culture of human cancer stem cells vary depending on the cancer cells [12]. Because of gene mutations and/or autocrine secretion, cancer cells may not require the supplementation of the factors that are normally required for the growth of human/murine normal counterparts. Success rates are variable depending on the tumor types. It is possible that addition of other tissue-specific factors not listed in Table 1 will increase success rates.

5. It is very important to cool plates and pipette tips beforehand. Keeping GFRM, plates, and equipment cool and applying GFRM from the center of each well will help to form level surface without forming bubbles. For example, when the surface of Matrigel layer is concave, cells tend to gather and aggregate around the center of a well. Bubbles will prevent observation when organoids have formed.

6. When using a 96-well plate, add 50–100 μl of GFRM in each well.

7. A trans-well plate can be used instead of a 24-well plate. Preparing the GFRM layer in a trans-well will make the culture and maintenance of organoids easier.

8. Add autoclaved or filtered PBS in several empty wells of a 24-well plate to prevent drying of GFRM surface.

9. Alternatively, cells will be directly suspended in GFRM at **step 1** in Section 3.1. This method will prevent too much aggregation of the cells during the organoid formation. However, microscopic observation will become a bit difficult because organoids will distribute more sparsely throughout the GFRM.

10. Some organoids will grow better in hypoxic conditions.

11. Do not pipette to mix the medium extensively, as you may dislodge the cells that are beginning to settle and grow on top of the GFRM matrix.

12. For example, if you sort 1,000 cells, resuspending in 100 μl will give you 100 cells per 10 μl aliquot. Alternatively, you can sort directly into the culture plate.

13. Take extra care to seed cells evenly on the Matrigel layer. When the surface of the GFRM layer is concave or the cells are not evenly distributed, the cells will easily aggregate during the culture.

14. Because the GFRM matrix is not rigid, the organoids can detach from it and be lost while pipetting.
15. It is very important to cool plates and pipette tips beforehand.
16. Alternatively, 1 ml of dispase (50 units/ml) and 500 μl collagenase III (200 units/ml) can be used. In this case, solution is incubated at 37 °C.
17. When using a 96-well plate, add 100 μl of ice-cold Cell Recovery Solution to each well.
18. Use the pre-cooled pipette tips. Do not warm the solution with dissolved gel to prevent the gelation of GFRM.
19. When using a 96-well plate, add 100 μl of ice-cold cell recovery solution to each well.
20. Do not pipet the gel to avoid the damage to the cells.
21. If cells are not fully dissociated, add 300 μl trypsin–EDTA again and incubate for 5-30 min at 37 °C. Periodically monitor cell dissociation under microscope.

Acknowledgments

We thank the divisions of Breast and Endocrine Surgery and of Gastrointestinal Surgery of Kobe University Graduate School of Medicine, and the division of Surgery of Nagoya University for contribution in the patient tumor collection.

This study was supported by Grants-in-Aid for Scientific Research from the Japan Society of the Promotion of Science and by the grants from Japan Foundation for Applied Enzymology and the Itoh-Chubei Foundation to Y.S.

References

1. Clevers H (2015) STEM CELLS. What is an adult stem cell? Science 350 (6266):1319–1320. doi:10.1126/science.aad7016
2. Huch M, Koo BK (2015) Modeling mouse and human development using organoid cultures. Development 142(18):3113–3125. doi:10.1242/dev.118570
3. Sato T, Vries RG, Snippert HJ, van de Wetering M, Barker N, Stange DE, van Es JH, Abo A, Kujala P, Peters PJ, Clevers H (2009) Single Lgr5 stem cells build crypt-villus structures in vitro without a mesenchymal niche. Nature 459(7244):262–265. doi:10.1038/nature07935, nature07935
4. Ootani A, Li X, Sangiorgi E, Ho QT, Ueno H, Toda S, Sugihara H, Fujimoto K, Weissman IL, Capecchi MR, Kuo CJ (2009) Sustained in vitro intestinal epithelial culture within a Wnt-dependent stem cell niche. Nat Med 15 (6):701–706. doi:10.1038/nm.1951
5. Clevers H (2016) Modeling development and disease with organoids. Cell 165 (7):1586–1597. doi:10.1016/j.cell.2016.05.082
6. Shimono Y, Mukohyama J, Nakamura S, Minami H (2015) MicroRNA regulation of human breast cancer stem cells. J Clin Med 5(1). doi:10.3390/jcm5010002
7. Clarke MF, Dick JE, Dirks PB, Eaves CJ, Jamieson CH, Jones DL, Visvader J, Weissman IL, Wahl GM (2006) Cancer stem cells—perspectives on current status and future directions: AACR workshop on cancer stem cells.

Cancer Res 66(19):9339–9344. doi:10.1158/0008-5472.CAN-06-3126

8. Lobo NA, Shimono Y, Qian D, Clarke MF (2007) The biology of cancer stem cells. Annu Rev Cell Dev Biol 23:675–699. doi:10.1146/annurev.cellbio.22.010305.104154

9. Sato T, Stange DE, Ferrante M, Vries RG, Van Es JH, Van den Brink S, Van Houdt WJ, Pronk A, Van Gorp J, Siersema PD, Clevers H (2011) Long-term expansion of epithelial organoids from human colon, adenoma, adenocarcinoma, and Barrett's epithelium. Gastroenterology 141(5):1762–1772. doi:10.1053/j.gastro.2011.07.050

10. Matano M, Date S, Shimokawa M, Takano A, Fujii M, Ohta Y, Watanabe T, Kanai T, Sato T (2015) Modeling colorectal cancer using CRISPR-Cas9-mediated engineering of human intestinal organoids. Nat Med 21(3):256–262. doi:10.1038/nm.3802

11. Isobe T, Hisamori S, Hogan DJ, Zabala M, Hendrickson DG, Dalerba P, Cai S, Scheeren F, Kuo AH, Sikandar SS, Lam JS, Qian D, Dirbas FM, Somlo G, Lao K, Brown PO, Clarke MF, Shimono Y (2014) miR-142 regulates the tumorigenicity of human breast cancer stem cells through the canonical WNT signaling pathway. elife 3. doi:10.7554/eLife.01977

12. Fujii M, Shimokawa M, Date S, Takano A, Matano M, Nanki K, Ohta Y, Toshimitsu K, Nakazato Y, Kawasaki K, Uraoka T, Watanabe T, Kanai T, Sato T (2016) A colorectal tumor organoid library demonstrates progressive loss of niche factor requirements during tumorigenesis. Cell Stem Cell 18(6):827–838. doi:10.1016/j.stem.2016.04.003

Construction of Thymus Organoids from Decellularized Thymus Scaffolds

Asako Tajima, Isha Pradhan, Xuehui Geng, Massimo Trucco, and Yong Fan

Abstract

One of the hallmarks of modern medicine is the development of therapeutics that can modulate immune responses, especially the adaptive arm of immunity, for disease intervention and prevention. While tremendous progress has been made in the past decades, manipulating the thymus, the primary lymphoid organ responsible for the development and education of T lymphocytes, remains a challenge. One of the major obstacles is the difficulty to reproduce its unique extracellular matrix (ECM) microenvironment that is essential for maintaining the function and survival of thymic epithelial cells (TECs), the predominant population of cells in the thymic stroma. Here, we describe the construction of functional thymus organoids from decellularized thymus scaffolds repopulated with isolated TECs. Thymus decellularization was achieved by freeze–thaw cycles to induce intracellular ice crystal formation, followed by detergent-induced cell lysis. Cellular debris was removed with extensive wash. The decellularized thymus scaffolds can largely retain the 3D extracellular matrix (ECM) microenvironment that can support the recolonization of TECs. When transplanted into athymic nude mice, the reconstructed thymus organoids can effectively promote the homing of bone marrow-derived lymphocyte progenitors and support the development of a diverse and functional T cell repertoire. Bioengineering of thymus organoids can be a promising approach to rejuvenate/modulate the function of T-cell mediated adaptive immunity in regenerative medicine.

Keywords: Thymus, Scaffold, Tissue engineering, Organoids, Decellularization

1 Introduction

As a pivotal immune organ in the adaptive immune system, the thymus is responsible for generating a diverse repertoire of T-cells that can effectively react to invading pathogens, while maintaining immune self-tolerance [1–3]. Paradoxically, the thymus glands in vertebrate animals begin to undergo a degenerative process termed "thymus involution," the progressive reduction of tissue mass and function, at extremely young ages [4]. As early as the first year after birth, the stromal compartment of the human thymus begins to shrink about 3–5 % per year until middle age and continues to decrease at an annual rate ~1 % in the years followed [5, 6]. As the consequence of thymus deterioration, the newly generated T cells can no longer effectively replenish those lost in the periphery, resulting in constriction of the naïve T cell repertoire and expansion of the memory T cell pool [7, 8]. In addition to age-related thymus senescence, other pathological and environmental factors can also contribute to thymus involution. Inflammation caused by infectious pathogens such as viruses and bacteria can perturb the organization of the thymus microenvironments and accelerate the degeneration of thymic function [9, 10]. Chemotherapy and other cancer related treatments could cause irreversible damages to the thymus stroma, impeding timely recovery from immune deficiency. Thymus regeneration could be an effective means to rejuvenate the adaptive immune system and would have broad impacts in medicine [11–13].

While making up less than 0.5 % of the total thymic cells, thymic epithelial cells (TECs) are the key population of residential stromal cells for the development of T-cells [14–16]. After homing to the thymus, bone marrow-derived lymphocyte progenitors follow a well-programmed, sequential order of steps (e.g., lineage restriction, somatic recombination of the T-cell receptor genes, and positive and negative selection), to differentiate into naïve T-cells [17]. TECs play essential regulatory roles in each of these steps. Proper cross talks between TECs and the developing thymocytes are critical to the development of a diverse and self-tolerant T-cell repertoire.

Unlike most of the epithelial cells making up the lining of tubular structures in organs and tissues, the endoderm-derived TECs are organized in three-dimensional (3D) configurations. TECs cultured under 2D conditions either undergo apoptosis or lose their molecular properties to support T cell development. Reproducing the 3D thymic microenvironment from artificial materials proves to be challenging as the composition and configuration of the extracellular matrix (ECM) provides not only the matrix support for the physical colonization of TECs but also the necessary physiochemical signals to maintain their function [18]. Alternative to the synthetic chemical engineering approach, biological scaffolds have been successfully prepared from various organs (e.g., the heart, the lung, the liver, and the kidney), and have been used to provide genuine ECM microenvironments for the parenchymal cells repopulated. Limited, but encouraging functional recoveries of the tissue-engineered organoids have been observed in preclinical studies [19, 20]. Taking advantage of the tissue engineering approach, we have reconstructed thymus organoids by repopulating decellularized thymus scaffolds with isolated TECs [21]. The thymus organoids can support T-cell development both in vitro and in vivo. When transplanted into athymic nude mice, the reconstructed thymus organoids enable the generation of a diverse T-cell repertoire that help to reestablish T-cell mediated adaptive immune responses. Thus, bioengineering of thymus organoids provide a novel means to genetically manipulate TECs and can be an effective strategy to rejuvenate the adaptive immune system.

2 Materials

- Equipment:
  - Dissecting scissors.
  - Curved forceps.
  - Cryotubes.
  - 5 ml or 12 ml round-bottom polystyrene tubes.
  - Pipets as needed.
  - Tube rocker.
  - 60 mm and 100 mm petri dish.
  - 50 ml conical tube.
  - 1.5 ml microcentrifuge tube.
  - 6-well tissue culture dish with transwell.
- Solutions to prepare: (**Note 1**).
  - Washing buffer: 1× PBS, 0.5 % BSA, 2 mM EDTA.
  - 0.5 % SDS: sodium dodecyl sulfate, ddH$_2$O.
  - 0.1 % SDS: sodium dodecyl sulfate, ddH$_2$O.
  - MgSO$_4$/CaCl$_2$/Triton X-100 buffer: dH$_2$O, 5 mM MgSO$_4$, 5 mM CaCl$_2$, 1 % Triton X-100.
  - 1× PBS.
  - RPMI-10: RPMI-1640, 10 % FBS, 1 % penicillin/streptomycin, 1 % L-glutamine, 1 % NEAA, 0.5 % HEPES, 1 % 2-mercaptoethanol.
  - 60 % w/v iodixanol.

  Digesting solution: RPMI-1640, 0.025 mg/ml Liberase TM Research Grade, 10 mM HEPES, 0.25 mg/ml DNaseI.

  RPMI-1640 with phenol red.

  Anti-CD16/CD32 antibody.

  Anti-CD45 antibody.

  Anti-EpCAM antibody.

3 Methods

3.1 Mouse Thymus Decellularization

1. Harvest thymi from euthanized, appropriately aged mice (**Notes 2–4**).
2. Freeze the thymi in cryotubes at −80 °C. Thymi can be stored at −80 °C until future use.
3. Thaw frozen thymi in 30 °C water bath for 20 min.
4. Freeze the thymi at −80 °C in a Styrofoam box for 20 min.

5. Repeat **steps 3** and **4** twice. It is possible to stop the procedure at this step and store the thymi at −80 °C until proceeding further.

6. Thaw thymi in 30 °C water bath for 20 min.

7. Under a sterile condition, transfer the thymic samples to 5 ml round-bottom polystyrene tubes with freshly prepared 3 ml 0.5 % SDS solution. One thymus per tube is recommended to ensure thorough permeation of reagents (**Note 5**).

8. Place the tubes on a tube rocker set at moderate speed (~12–20 cycles per minute) at room temperature. Check the clarities of the thymic samples every hour and change the SDS solution every 1.5–2 h for 2–3 times. The thymus will become transparent at the end of the procedure (**Notes 6** and **7**).

9. Transfer the thymus to a fresh 5 ml flow tube with 3 ml of 0.1 % SDS and rock at 4 °C overnight.

10. Transfer the decellularized thymus scaffolds to fresh tubes with 3 ml Triton buffer (**Note 8**). Rock the tubes on the rocker at 4 °C for 15 min. Repeat the washing step twice. Use fresh tubes each time.

11. Wash the thymus scaffolds in new tubes with 2 ml of 1× PBS at 4 °C for 30 min. Repeat the washing step two more times.

12. Store the thymus scaffolds in washing buffer at 4 °C until use (**Note 9**).

3.2 Thymic Cell Isolation

1. Harvest thymus from euthanized mice and place in flow tubes with 3 ml of washing buffer (**Note 10**).

2. Prepare the digesting solution and keep in a 37 °C water bath until use.

3. In a 60 mm petri dish with 6 ml RPMI-1640, tear the thymi into small fragments (about 1 mm) with 28G insulin syringes.

4. Rinse the thymic fragments briefly by pipetting up and down twice and discard the supernatant with a 5 ml glass pipet (**Note 11**).

5. Add 6–7 ml of fresh RPMI-1640 to the dish. Rinse the thymic fragments again with a glass pipet and discard the solution, leaving the fragments settled on the bottom of the dish (**Note 12**).

6. Transfer the thymic fragments into a 5 ml round-bottom polystyrene tube with 3 ml digesting solution (**Note 13**).

7. Incubate the thymic fragments on a rocker by gentle agitation at 37 °C for 6 min.

8. After the incubation, aspirate the supernatant and transfer to a 50 ml conical tube with 10 ml of washing buffer. Set aside on ice.

9. In the polystyrene tube with the thymic fragments, add another 3 ml of digesting solution and repeat the incubation at 37 °C for 6 min.

10. During the 2nd digesting step, centrifuge the 50 ml tube with the supernatant with from the 1st digestion. Discard the supernatant, resuspend the cells in 10 ml washing buffer and keep on ice.

11. After the 2nd digestion is over, let the fragments settle to the bottom by gravity and transfer the supernatant to the tube in **step 10**.

12. Add 3 ml digesting solution to the thymic fragments and incubate at 37 °C for 6 min. After the incubation, pipet up and down 5 times to further break down the thymic fragments (**Note 14**).

13. Centrifuge the 50 ml tube containing all of the digested thymic fragments and digesting solution. Discard the supernatant, resuspend in 10 ml of washing buffer and filter through 100 μm strainer. Count the cells as necessary. Keep the cells on ice until further use.

3.3 TEC Enrichment

1. Prepare 21 % gradient medium solution from 60 % w/v iodixanol.

2. Centrifuge the thymic cells and resuspend in 2.5 ml washing buffer.

3. Add 20 ml RPMI-1640 to the cell suspension (**Note 15**).

4. Place the tip of the pipette with 12 ml of 21 % gradient medium solution at the bottom of the tube. Let the density gradient solution drain from the pipette by gravity (**Note 16**).

5. Centrifuge at $600 \times g$ for 20 min at room temperature with decelerating brake off.

6. Transfer the top layer and the interface to a new 50 ml tube. These layers will include the enriched TECs (**Note 17**).

7. Wash the TECs by adding washing buffer to 40 ml and centrifuge at $400 \times g$ for 6 min.

8. Repeat the washing step twice (**Note 18**).

9. Resuspend the cells in 1 or 2 ml of washing buffer, depending on the size of the pellet. Count the cells.

3.4 TEC Isolation by FACS

1. Resuspend the cells in washing buffer at the concentration of 1×10^7 cells/100 μl.

2. Add 2 μl anti-CD16/CD32 antibody per 1×10^6 cells and incubate at 4 °C for 10 min.

3. Add anti-CD45 and anti-EpCAM antibodies and incubate at 4 °C for more than 20 min.

4. Wash the cells with 2 ml of washing solution and centrifuge.

5. Resuspend the cells in 2 ml 1× PBS (**Note 19**).

6. Filter the cells through 100 μm strainer.

7. Set the cell suspension for sorting. Select the CD45-G8.8+ population and sort into washing buffer by FACS. It is important to include not only the lymphocyte population but also bigger, more complicated cells in the SSC/FSC panel.

8. Keep the cells on ice until use.

3.5 Isolation of Progenitor Cells from the Bone Marrow

1. Harvest the bones from euthanized mice. Larger bones such as femur and tibia are more feasible to work with and the bone marrow can be collected more efficiently from these bones. Remove the muscles and connective tissues as much as possible with sharp scissors, and store in washing buffer.

2. Using two forceps and a sterile gauze, scrape off the remaining tissues from the bones on a sterile 100 mm petri dish. Transfer the bones into a 60 mm petri dish with washing buffer.

3. Hold a bone with the forceps. Fill a 28G insulin syringe with washing buffer, insert the needle into one end of the bone and flush out the bone marrow with washing buffer (**Note 18**). Work from both sides of the bone. Repeat the washing step with the washing buffer to ensure that most of the bone marrow cells are flushed (**Note 19**).

4. Repeat the procedure to collect bone marrow from all of the bones.

5. Break down the bone marrow by passing the clumps through a 21-gauge needle on a 5 ml syringe.

6. Pass the cells through a 40 μm strainer into a 50 ml conical tube.

7. Adjust the total volume with washing buffer to 30 ml and centrifuge.

8. Remove the supernatant, and resuspend the cell pellet with 5 ml of red blood cell lysis buffer. Incubate in the dark at room temperature for 5 min. Add 25 ml of washing buffer and centrifuge.

9. Resuspend the cells in 10 ml of washing buffer and count the cells.

10. Select the lineage negative cells with commercial kit (e.g., Miltenyi Biotec lineage cell depletion kit), following manufacturer's suggested protocol.

11. Collect the lineage negative cells. Resuspend the cells in washing buffer at 1×10^7/ml and keep on ice until use.

3.6 Construction of Thymus Organoids

1. Under sterile condition, transfer the thymus scaffolds from washing buffer to a complete medium at least 30 min prior to the cell injection. Keep at room temperature.
2. Mix the TECs and the lineage negative cells from the bone marrow at 1:1 ratio.
3. Centrifuge and collect the cells in a 1.5 ml microcentrifuge tube.
4. Centrifuge again and resuspend the cells in the complete medium at the concentration of 20 µl per 1×10^6 cells per scaffold (**Note 20**).
5. Fill the 28-gauge insulin needle with the cell suspension.
6. Take the scaffold out of the medium and place on a 12 mm petri dish. Under the dissection microscope, gently pinch the scaffold with fine forceps and puncture the scaffold with the syringe needle.
7. Gently infuse the cell suspension in the scaffold, and slowly pull out the needle. If there are any unnoticed ruptures in the scaffold, the cells will start coming out and this phenomenon can be observed under the dissection microscope.
8. In a 6-well transwell dish, add 2 ml of pre-warmed complete medium in the bottom well. Place the reconstructed thymus scaffold onto the upper dish of the 6-well transwell and add 50–100 µl of complete medium onto the scaffold (**Note 21**).
9. Incubate at 37 °C with 5 % CO_2 until use.
10. If it is to be cultured for a long term, change half of the medium in the bottom well every other day. Note that the reconstructed scaffolds will start to shrink after 2–4 days of incubation.

4 Notes

1. All solutions must be filtered sterile.
2. The size of the thymus gland varies with age. 2–6-week-old mice provide a thymus scaffold with a suitable size for injections.
3. When harvesting the thymus, it is critical that the thymus is not damaged. Any fissure or puncture will cause the thymus to collapse during the procedure, rendering it unusable for thymus organoid construction.
4. It is not necessary to remove all the connective tissue or blood clots from the thymus after harvesting. It will not affect the decellularization steps.
5. It is advisable to perform the procedures under the laminar flow hood when handling thymi, in order to keep the thymus scaffolds sterile for cell culture purposes.

6. Changing the solution after 1 h is recommended for the 1st cycle, in order to remove most of the cellular components that are released.

7. Depending on the size of thymus, decellularization with 0.5 % SDS solution can be switched to 0.1 % SDS once the thymus is clear, followed by an overnight incubation with 0.1 % SDS.

8. Thymi should be mostly transparent before replacing the solution to Triton buffer. It is fundamental to carefully verify the condition of the thymus to ensure the removal of debris from the thymus; in the next step, Triton X-100 in the washing buffer will renature the ECM proteins that are still left in the scaffold.

9. Successfully decellularized thymus scaffolds are bulbous. Scaffolds will retain this property for about 1-2 months at 4 °C. Scaffolds should be transferred to a culture medium for 30 min or more before cell injections.

10. It is recommended that three to four thymi are processed in one batch, to obtain enough number of TECs for thymus scaffolds.

11. Glass pipet is necessary to avoid the attachment of thymic fragments on the pipet wall. Fragments can easily stick to plastic pipets, causing the loss of materials.

12. Tilting the petri dish helps the fragments to sink to the bottom. When discarding the solution, it is better to aspirate slowly from the surface or from where fragments are not floating.

13. Use 1–1.5 ml of solution at a time to rinse off all the fragments from the surface of the petri dish.

14. Fat tissue and connective tissue may remain undigested. These are distinguishable because they will often float at the surface, and can be transferred together with the rest of the solution.

15. The red color in RPMI-1640 helps to distinguish the two layers.

16. Add the gradient density solution as gentle as possible. It is important to form two distinct layers at this step. Any vigorous mixing will result in an obscure boundary after the centrifugation.

17. The interface is visible as an opaque ring, if the cell number in the layer is high. In some cases, the "ring" may not be distinguishable first and the boundary between the top and the bottom layer might be blurry after the spin. However, it will become clearer when the interface is removed.

18. Cutting the tip of the bone will help for needle insertion into the bone, which is necessary to flush out the bone marrow. The practice will not negatively affect the amount progenitor cells obtained.

19. Bone marrow is clearly visible when it is pushed out of the bone; and the bones will appear more white than pink afterwards.
20. We typically limit the injection volume to 20 μl per scaffold, to prevent overloading.
21. Up to four scaffolds can be cultured together in the same 6-well insert.

Acknowledgements

This work was supported in part by the National Institutes of Health grant R01 AI123392 (YF) and by the generous support of Allegheny Health Network to the Institute of Cellular Therapeutics.

References

1. Klein L, Kyewski B, Allen PM, Hogquist KA (2014) Positive and negative selection of the T cell repertoire: what thymocytes see (and don't see). Nat Rev Immunol 14(6):377–391. doi:10.1038/nri3667
2. Fan Y, Rudert WA, Grupillo M, He J, Sisino G, Trucco M (2009) Thymus-specific deletion of insulin induces autoimmune diabetes. EMBO J 28(18):2812–2824. doi:10.1038/emboj.2009.212
3. Fan Y, Gualtierotti G, Tajima A, Grupillo M, Coppola A, He J et al (2014) Compromised central tolerance of ICA69 induces multiple organ autoimmunity. J Autoimmun 53:10–25. doi:10.1016/j.jaut.2014.07.001
4. Boehm T, Swann JB (2013) Thymus involution and regeneration: two sides of the same coin? Nat Rev Immunol 13(11):831–838. doi:10.1038/nri3534
5. Bodey B, Bodey B Jr, Siegel SE, Kaiser HE (1997) Involution of the mammalian thymus, one of the leading regulators of aging. In Vivo 11(5):421–440
6. Goronzy JJ, Fang F, Cavanagh MM, Qi Q, Weyand CM (2015) Naive T cell maintenance and function in human aging. J Immunol 194(9):4073–4080. doi:10.4049/jimmunol.1500046
7. Nikolich-Zugich J, Rudd BD (2010) Immune memory and aging: an infinite or finite resource? Curr Opin Immunol 22(4):535–540. doi:10.1016/j.coi.2010.06.011
8. Palmer DB (2013) The effect of age on thymic function. Front Immunol 4:316. doi:10.3389/fimmu.2013.00316
9. Borges M, Barreira-Silva P, Florido M, Jordan MB, Correia-Neves M, Appelberg R (2012) Molecular and cellular mechanisms of Mycobacterium avium-induced thymic atrophy. J Immunol 189(7):3600–3608. doi:10.4049/jimmunol.1201525
10. Ye P, Kirschner DE, Kourtis AP (2004) The thymus during HIV disease: role in pathogenesis and in immune recovery. Curr HIV Res 2(2):177–183
11. Black S, De Gregorio E, Rappuoli R (2015) Developing vaccines for an aging population. Sci Transl Med 7(281):281ps8. doi:10.1126/scitranslmed.aaa0722
12. Di Stefano B, Graf T (2014) Hi-TEC reprogramming for organ regeneration. Nat Cell Biol 16(9):824–825. doi:10.1038/ncb3032
13. Bredenkamp N, Nowell CS, Blackburn CC (2014) Regeneration of the aged thymus by a single transcription factor. Development 141(8):1627–1637. doi:10.1242/dev.103614
14. Takahama Y (2006) Journey through the thymus: stromal guides for T-cell development and selection. Nat Rev Immunol 6(2):127–135. doi:10.1038/nri1781
15. Ohigashi I, Kozai M, Takahama Y (2016) Development and developmental potential of cortical thymic epithelial cells. Immunol Rev 271(1):10–22. doi:10.1111/imr.12404
16. Anderson G, Takahama Y (2012) Thymic epithelial cells: working class heroes for T cell development and repertoire selection. Trends Immunol 33(6):256–263. doi:10.1016/j.it.2012.03.005

17. Starr TK, Jameson SC, Hogquist KA (2003) Positive and negative selection of T cells. Annu Rev Immunol 21:139–176. doi:10.1146/annurev.immunol.21.120601.141107
18. Seach N, Mattesich M, Abberton K, Matsuda K, Tilkorn DJ, Rophael J et al (2010) Vascularized tissue engineering mouse chamber model supports thymopoiesis of ectopic thymus tissue grafts. Tissue Eng Part C Methods 16(3):543–551. doi:10.1089/ten.TEC.2009.0135
19. Baptista PM, Orlando G, Mirmalek-Sani SH, Siddiqui M, Atala A, Soker S (2009) Whole organ decellularization—a tool for bioscaffold fabrication and organ bioengineering. Conf Proc IEEE Eng Med Biol Soc 2009:6526–9. doi:10.1109/IEMBS.2009.5333145
20. Booth C, Soker T, Baptista P, Ross CL, Soker S, Farooq U et al (2012) Liver bioengineering: current status and future perspectives. World J Gastroenterol 18(47):6926–6934. doi:10.3748/wjg.v18.i47.6926
21. Fan Y, Tajima A, Goh SK, Geng X, Gualtierotti G, Grupillo M et al (2015) Bioengineering thymus organoids to restore thymic function and induce donor-specific immune tolerance to allografts. Mol Ther 23(7):1262–77. doi:10.1038/mt.2015.77

Expansion of Human Airway Basal Stem Cells and Their Differentiation as 3D Tracheospheres

Robert E. Hynds, Colin R. Butler, Sam M. Janes, and Adam Giangreco

Abstract

Although basal cells function as human airway epithelial stem cells, analysis of these cells is limited by in vitro culture techniques that permit only minimal cell growth and differentiation. Here, we report a protocol that dramatically increases the long-term expansion of primary human airway basal cells while maintaining their genomic stability using 3T3-J2 fibroblast coculture and ROCK inhibition. We also describe techniques for the differentiation and imaging of these expanded airway stem cells as three-dimensional tracheospheres containing basal, ciliated, and mucosecretory cells. These procedures allow investigation of the airway epithelium under more physiologically relevant conditions than those found in undifferentiated monolayer cultures. Together these methods represent a novel platform for improved airway stem cell growth and differentiation that is compatible with high-throughput, high-content translational lung research as well as human airway tissue engineering and clinical cellular therapy.

Keywords: Lung, Stem cells, Epithelial cells, Goblet cells, Cilia, Adult stem cells, Cell culture techniques, Primary cell culture

1 Introduction

Human airways represent a key environmental barrier whose dysregulation in diseases such as asthma, chronic obstructive pulmonary disease (COPD), cystic fibrosis, and cancer is a major cause of worldwide morbidity and mortality [1]. Human airways are composed of a pseudostratified ciliated and mucosecretory epithelium responsible for protecting against inhaled particulate matter, pathogens, and other airborne toxicants. Supporting this epithelium are multipotent basal stem cells characterized primarily by cytokeratin 5 (CK5) and P63 expression [2, 3]. Given the clear physiological importance of maintaining human airway homeostasis, establishing techniques for investigating basal stem cell growth and differentiation is of significant fundamental and translational biomedical relevance [4].

Historically, the majority of airway stem cell research has involved either in vitro studies of undifferentiated, immortalized cell cultures or in vivo animal models that exhibit only limited human translational applicability [1]. To overcome this, researchers have traditionally used human air–liquid interface (ALI) cultures in which a confluent monolayer of human airway basal cells is grown at an air interface in medium containing retinoic acid to encourage ciliated and mucosecretory cell differentiation [5, 6]. Despite the use of primary human cells in ALI models [7], this technique remains extremely time-consuming and is poorly suited to high throughput, high content translational medicine approaches. In addition, evidence suggests that basal cells rapidly lose their capacity for multipotent differentiation and undergo premature senescence after only a small number of passages. This failure to maintain airway stem cells in vitro presents a considerable barrier for their use in translational medicine.

In this chapter, we describe a protocol for improved human airway stem cell expansion that maintains their multipotent differentiation capacity. Originally developed for epidermal keratinocyte stem cells, this technique relies upon the coculture of primary human epithelial cells with mitotically inactivated 3T3-J2 fibroblast feeder cells [8]. Cocultures of epithelial and J2 feeder cells are grown in medium containing the Rho-associated kinase (ROCK) inhibitor Y-27632 [9–11]. Successful application of this procedure to airway epithelial cells provides large numbers of basal cells that exhibit physiologically relevant differentiation, karyotype stability, and maintenance of telomere length [12]. In addition, we also provide two distinct methods for the differentiation and visualization of lung basal cells as 3D airway tracheospheres. These techniques build on previous 3D culture systems in which either ciliated or mucosecretory differentiation was achieved [2, 13, 14]. The methods described here, based on several recently published studies, permit normal ciliated and mucosecretory cell differentiation [12, 15, 16] and are compatible with high-throughput, high-content analyses. Altogether, these methods provide a platform to expand and differentiate large numbers of airway basal epithelial cells from patient endobronchial biopsy samples. These techniques have uses in translational medicine, tissue engineering, and human cellular therapy.

2 Materials

2.1 Human Airway Epithelial Cell Isolation

1. RPMI medium with L-glutamine (Gibco, Thermo Fisher). Storage at 4 °C.
2. Dispase (Corning). Storage at −20 °C.
3. 0.25 % trypsin–EDTA (Gibco, Thermo Fisher). Storage at −20 °C.
4. Neutralization medium: Dulbecco's modified Eagle's medium (DMEM) with 4.5 g/L D-glucose, L-glutamine, and pyruvate

(Gibco, Thermo Fisher) plus 10 % fetal bovine serum (FBS) and 1× penicillin–streptomycin (Gibco, Thermo Fisher). Storage at 4 °C.

5. Transport medium: αMEM (Gibco, Thermo Fisher) plus 1× penicillin–streptomycin (Gibco, Thermo Fisher), 1× gentamicin (Gibco, Thermo Fisher) and 1× amphotericin B (Fisher Scientific). Storage at 4 °C.

2.2 3T3-J2 Feeder Cell Culture

1. Dulbecco's modified Eagle's medium (DMEM) with 4.5 g/L D-glucose, L-glutamine, and pyruvate (Gibco, Thermo Fisher). Storage at 4 °C.

2. Bovine serum (Gibco, Thermo Fisher). Storage at −80 °C.

3. Penicillin–streptomycin (Gibco, Thermo Fisher). Storage at −20 °C.

4. Mitomycin C (Sigma-Aldrich): 0.4 mg/ml in sterile PBS. Storage at −20 °C.

5. 0.05 % trypsin–EDTA (Gibco, Thermo Fisher). Storage at −20 °C, 4 °C short-term.

6. Complete fibroblast culture medium: 500 ml DMEM plus 45 ml bovine serum and 1× penicillin–streptomycin. Storage at 4 °C.

2.3 Human Airway Epithelial Cell Culture

1. Phosphate-buffered saline (PBS; Sigma).

2. Dulbecco's modified Eagle's medium (DMEM) with 4.5 g/L D-glucose, L-glutamine, and pyruvate (Gibco, Thermo Fisher). Storage at 4 °C.

3. Fetal bovine serum (FBS; Gibco, Thermo Fisher; *see* **Note 1**). Storage at −80 °C long-term, −20 °C until use.

4. Penicillin–streptomycin (Gibco, Thermo Fisher). Storage at −20 °C.

5. Ham's F-12 nutrient mixture with L-glutamine (Gibco, Thermo Fisher). Storage at 4 °C.

6. Gentamicin (Gibco, Thermo Fisher): 100×. Storage at 4 °C.

7. Amphotericin B (Fisher Scientific): 250 μg/ml, 1000×. Storage at −20 °C.

8. Hydrocortisone (Sigma-Aldrich). *See* **Note 2** for storage conditions.

9. Recombinant human EGF (Thermo Fisher). *See* **Note 2** for storage conditions.

10. Insulin, 1000× stock (Sigma-Aldrich): 5 mg/ml in distilled water (add glacial acetic acid dropwise to dissolve and sterile filter). Storage at −20 °C.

11. Cholera toxin (Sigma-Aldrich): 1 mg/ml (11.7 μM) in distilled water. Storage at 4 °C.
12. Y-27632 (Cambridge Bioscience): 5 mM in distilled water. Storage at −20 °C.
13. Complete epithelial culture medium: 373 ml serum-containing DMEM (500 ml DMEM plus 50 ml FBS and 1× penicillin–streptomycin), 125 ml Ham's F-12, 0.5 ml gentamicin, 0.5 ml amphotericin B, 0.5 ml hydrocortisone/EGF (*see* **Note 2**), 0.5 ml insulin, 0.5 ml Y-27632, 4.3 μl cholera toxin. Storage at 4 °C for up to 2 weeks.
14. Profreeze Chemically Defined Freezing Medium (2×; Lonza).

2.4 3D Human Tracheospheres (Lumen-in)

1. Matrigel, basement membrane matrix, growth factor reduced (BD Biosciences; *see* **Note 3**). Storage at −20 °C.
2. Ultra-low attachment 96-well plate (Corning).
3. Bronchial epithelial growth medium with supplements (BEGM; Lonza). Medium stored at 4 °C, supplements at −20 °C.
4. Dulbecco's modified Eagle's medium (DMEM) with 4.5 g/L D-glucose, L-glutamine, and pyruvate (Gibco, Thermo Fisher). Storage at 4 °C.
5. Y-27632 (Cambridge Bioscience): 5 mM in distilled water. Storage at −20 °C.
6. All-*trans* retinoic acid (Sigma): 10 mM in 100 % EtOH (50 mg in 16.642 ml). Storage at −80 °C (*see* **Note 4**).
7. Tracheosphere medium: 50 % DMEM (no serum, no antibiotics), 50 % bronchial epithelial basal medium (plus all of the BEGM supplements except amphotericin B, triiodothyronine, and retinoic acid). Storage at 4 °C. Supplemented with 5 μM Y-27632 for cell seeding (but not subsequent feeds). Always supplemented with 100 nM all-*trans* retinoic acid at the time of use.

2.5 Immunofluorescence Staining of Tracheospheres

1. 4 % (w/v) paraformaldehyde in PBS. Heat to 65 °C to dissolve and adjust pH to 7.2.
2. HistoGel Specimen Processing Gel (Thermo Scientific; *see* **Note 5**).
3. Blocking solution (10 % FBS in PBS).
4. DAPI (4′,6-diamidino-2-phenylindole; Molecular Probes). Stock solution: 10 μg/ml.
5. Mounting medium (Immu-mount, Thermo Scientific).

3 Methods

3.1 Preparation of 3T3-J2 Feeder Layers

1. Maintain 3T3-J2 cells at low passage numbers (*see* **Note 6**) under subconfluent conditions (*see* **Note 7**) in 37 °C, 5 % CO_2 incubators with weekly splits of 1:8–1:10.
2. For feeder layer preparation, add mitomycin C (4 μg/ml; final concentration) to flasks of 3T3-J2 cells in fibroblast culture medium for 2 h (*see* **Note 8**).
3. Wash cells with PBS, harvest cells with 0.05 % trypsin–EDTA and replate at a density of at least 20,000 feeder cells/cm^2 in fibroblast culture medium (*see* **Note 9**).
4. Allow 3T3-J2 cells to adhere and spread overnight before adding epithelial cells the following day (*see* **Note 10**).

3.2 Isolation of Human Airway Epithelial Cells

1. Derivation of human cells should be approved by the relevant local ethics committee.
2. Human airway epithelial cells are derived from brushings or biopsies taken during bronchoscopy procedures. Samples are transported to the laboratory on ice as soon as possible.

3.2.1 Cell Seeding from Brushings

1. Brushes are collected in 15 ml falcon tubes in transport medium.
2. Thoroughly vortex tubes to remove cells from the brush.
3. Centrifuge tubes, with brushes inside, at $300 \times g$ for 5 min.
4. Carefully remove transport medium and resuspend the cell pellet in epithelial culture medium (*see* **Note 11**).
5. Seed cell suspension on preprepared T25 feeder layers, incubate at 37 °C (5 % CO_2) and change the medium after 2 days.
6. Epithelial colonies become evident after 2–3 days (*see* **Note 12**).

3.2.2 Cell Seeding from Biopsies

1. For explant cultures, biopsies are seeded directly onto preprepared feeder layers with a minimal covering of epithelial culture medium (Fig. 1a; *see* **Note 13**) or they can be digested to form a cell suspension prior to culture.
2. To digest a biopsy, carefully transfer the biopsy to an eppendorf tube containing 16 U/ml dispase in RPMI. Digest for 20 min at room temperature.
3. Neutralize the digestion by adding an equal volume of neutralization medium.
4. Transfer the tissue to a sterile petri dish and dissect away epithelium (*see* **Note 14**).

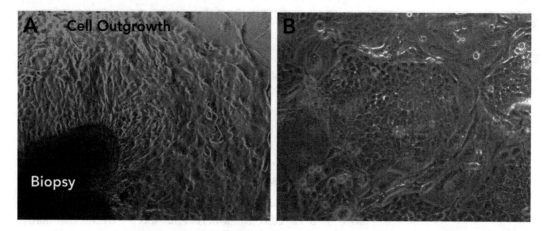

Fig. 1 (**a**) Phase-contrast image of epithelial cell outgrowth from an endobronchial biopsy grown on a 3T3-J2 feeder layer in medium containing the ROCK inhibitor Y-27632 (3T3+Y). (**b**) Phase-contrast image of colonies of subcultured human airway epithelial cells grown in 3T3+Y

5. Transfer to an eppendorf containing 0.1 % trypsin–EDTA (diluted in RPMI) and incubate at 37 °C for 30 min. Remove cells from the incubator and agitate by pipetting every 10 min.
6. Neutralize the trypsin by addition of at least the same volume of neutralization medium and pipette thoroughly.
7. Combine the neutralized digest with the neutralized dispase solution from **step 3**.
8. Centrifuge cells at $300 \times g$ for 5 min.
9. Remove supernatant and resuspend the pellet in epithelial culture medium.
10. Seed cell suspension on preprepared T25 feeder layers, incubate at 37 °C (5 % CO_2) and change the medium after 2 days.
11. Epithelial colonies become evident after 2–3 days.

3.3 Maintenance of Human Airway Basal Epithelial Cells

1. During normal maintenance epithelial cells are fed with fresh complete epithelial medium three times per week (10 ml per T75 or equivalent for other vessel surface areas) and stored in 37 °C, 5 % CO_2 incubators.
2. Passage epithelial cell cultures at 80–90 % confluence (Fig. 1b).
3. Remove epithelial culture medium and wash cells once with PBS.
4. Add 0.05 % trypsin–EDTA (*see* **Note 15**) for 2 min at room temperature, gently tap flasks and confirm the removal of feeder cells, which are substantially more trypsin-sensitive than epithelial cells, under a microscope.

5. Remove the trypsin and wash with PBS to remove remaining feeder cells.

6. Add 0.05 % trypsin–EDTA and incubate at 37 °C until epithelial cells detach from the flask (typically this takes 5 min; see **Note 16**).

7. Neutralize the trypsin solution with the same volume of epithelial culture medium, centrifuge at $300 \times g$ for 5 min and resuspend the pellet in epithelial culture medium.

8. For continued passage, perform a 1:5 split and seed cells onto a preprepared feeder layer in complete epithelial culture medium. Incubate in a 37 °C, 5 % CO_2 incubator.

9. To cryopreserve, freeze 0.5×10^6 cells/vial in 200 µl total volume using Profreeze freezing medium according to the manufacturer's instructions.

3.4 3D Tracheosphere Culture

1. Coat the base of a non-adherent 96-well plate with 25 % prechilled Matrigel in tracheosphere medium (30 µl per well) and transfer to an incubator (37 °C, 5 % CO_2) for at least 20 min (see **Note 17**).

2. Differentially trypsinize cultured airway basal cells from feeder cells as described above (**step 3** in Section 3.3). For one 96-well plate, resuspend 270,000 basal cells in 7 ml tracheosphere medium containing 5 % prechilled Matrigel and 5 µM Y-27632. Add 65 µl cell suspension per well (2500 cells per well, see **Note 18**). Return the plate to an incubator (37 °C, 5 % CO_2; see **Note 19**).

3. Feed cells by adding 70 µl tracheosphere medium (without Y-27632) on days 3, 8, and 14 of culture (Fig. 2a).

4. On day 18 of culture tracheospheres are well differentiated with basal cells, goblet cells and ciliated cells present (Fig. 1; see **Note 20**).

3.5 Immunofluorescence Staining of Tracheospheres

1. To collect tracheospheres, pipette the well contents into chilled PBS on ice using 200 µl pipette tips with the ends cut off. Centrifuge at $200 \times g$ for 3 min. Fix tracheospheres by resuspension in 4 % PFA for 30 min.

2. Using the same centrifuge settings, wash once with PBS and resuspend the tracheosphere pellet in pre-warmed HistoGel specimen processing gel. Pipette the HistoGel onto Parafilm as small droplets appropriate for embedding and allow it to gel at room temperature for 10 min.

3. Process, embed and stain the tracheospheres according to standard histology protocols (Fig. 2b–d).

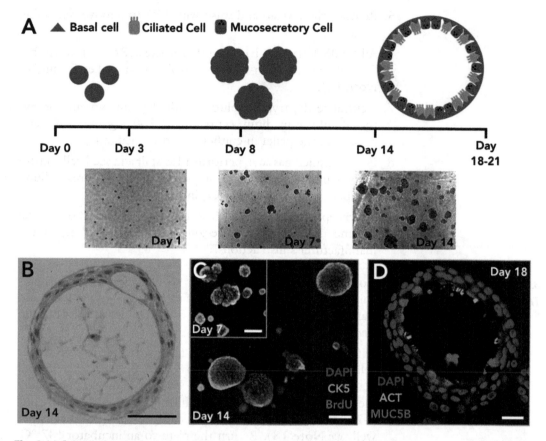

Fig. 2 (**a**) Schematic representation of airway tracheosphere formation in 3D Matrigel culture with brightfield images showing the time course of tracheosphere growth. (**b**) Hematoxylin and eosin (H&E) staining of a tracheosphere. Scale bar = 50 μm. (**c**) Whole-mount immunofluorescence staining of tracheospheres at day 7 and day 14 shows that BrdU uptake is high in the early stages of culture, as tracheospheres form, but reduces over time. Scale bar = 100 μm. (**d**) Immunofluorescence staining of tracheospheres shows the differentiation of cultured primary human airway basal cells into both ciliated (ACT; *green*) and mucosecretory (MUC5B; *red*) cells in tracheospheres. Scale bar = 20 μm

4 Notes

1. Batch testing of FBS for the ability to support rapid expansion and serial passage of primary human epithelial cells is essential.

2. Hydrocortisone/EGF stock is prepared by dissolving hydrocortisone at 0.5 mg/ml in 100 % ethanol. 1 ml hydrocortisone is added to 19 ml DMEM containing 2.5 μg recombinant human EGF. Hydrocortisone/EGF stock aliquots are stored at −20 °C.

3. A bottle of Matrigel must be thawed on ice for >6 h. Once thawed, Matrigel should be aliquoted into prechilled, sterile eppendorf tubes and stored at −20 °C. Freeze-thaw Matrigel a

maximum of three times. Matrigel should be kept on ice throughout the tracheosphere seeding process.

4. Retinoic acid is extremely light sensitive. Care should be taken to minimize light exposure during preparation and storage, for example by foil wrapping falcon tubes. Aliquoted 10 mM stocks can be stored for up to 1 month at −80 °C. We prepare a 1:1000 (10 μM) stock and dilute this 1:100 to prepare tracheosphere medium (final conc. 100 nM).

5. A 10 ml vial of HistoGel should be preheated to 65 °C in a water bath for 30 min. HistoGel can be aliquoted into 1.5 ml eppendorf tubes to reduce this time to 10 min and sets quickly at room temperature so should be kept at 65 °C until use.

6. We thaw fresh 3T3-J2 cells after passage 12 as the cells are susceptible to undergoing spontaneous transformation, proliferating more rapidly and adopting a cuboidal morphology compared with the usual spread appearance of fibroblasts.

7. 3T3-J2 cultures should never be allowed to become over confluent as their morphology and sensitivity to irradiation/mitomycin C can be affected, leading to poor quality feeder layers.

8. Alternatively, 3T3-J2s can be mitotically inactivated by 40 Gy (4000 rd) irradiation.

9. After mitotic inactivation, 3T3-J2 cells are fragile and care should be taken when pipetting the cells as excessive stress can cause feeder layers to degenerate more rapidly.

10. Co-seeding of 3T3-J2 and epithelial cells in complete epithelial culture medium produces poor quality feeder layers. Mitotically inactivated fibroblasts do not spread, as when seeded in fibroblast culture medium, and feeder layers degenerate more quickly. Thus, it is important to be well organized in the planning of experiments using cocultured cells.

11. Care should be taken to wash the brush with epithelial culture medium to free as many cells as possible. We use at least two rounds of washing with fresh medium.

12. Contamination of cultures with human fibroblasts has not been a problem in our lab. It is thought that the combination of the feeder layers and culture medium additives suppress the growth of fibroblasts.

13. We do not change the medium on explant biopsies for the first 7 days of culture in order to avoid the risk of detaching the biopsy by moving the flask. Epithelial outgrowths are normally visible in the second week and can be serially passaged as described in **step 3** in Section 3.3.

14. In practice, it is difficult to see the epithelium at this stage—after proper dispase digest the biopsy will appear "fluffy"—but

we do find that pulling apart the biopsy at this stage improves the efficiency of the trypsin step.

15. TrypLE Select (Gibco, Thermo Fisher), a recombinant trypsin that avoids the use of porcine trypsin in human cultures, can be used in place of 0.05 % trypsin–EDTA for passaging epithelial cells with no negative impact on the quality of cultures.

16. Epithelial cells should be treated with care during passaging. We remove cells following trypsinization by gently tapping flasks rather than by washing. Care should also be taken to avoid excessive pipetting of epithelial cells during resuspension.

17. To maximize the well volume available for addition of culture medium, the Matrigel layer at the well base is minimal. Non-adherent culture plates are used to prevent cells in close proximity to the culture plastic from losing 3D structure.

18. Video microscopy studies indicate that tracheospheres formed from cells at this density are not clonal but multiple basal cells in the initial culture can migrate and interact to form tracheospheres.

19. Tracheospheres are very sensitive to culture conditions. We use dedicated incubators, with extra humidification, for airway differentiation assays in order to reduce the number of times incubator doors are opened and to better maintain constant humidity, CO_2, etc.

20. The abundance of cilia in tracheospheres increases over time. Cilia are observed at day 14, but longer culture periods produce better results. This must be balanced with the ability to continue to add culture medium to wells. End-point assays can be performed between day 14 and day 21.

Acknowledgments

We thank Prof. Richard Schlegel (Georgetown University, USA), Xuefeng Liu (Georgetown University, USA), and Dr. Henry Danahay (University of Sussex, UK) for sharing their laboratories protocols during the development of those described here.

This work was supported by a BBSRC-CASE studentship with industrial support from Unilever (R.E.H.), a Wellcome Trust Clinical Research Fellowship (C.R.B.), an ERC Starting Grant (A.G.), a Wellcome Trust Senior Fellowship (S.M.J.) and was undertaken at UCLH/UCL who receive funding from the Department of Health's NIHR Biomedical Research Centre's funding scheme and the UCL Experimental Cancer Medicine Centre (S.M.J.).

References

1. Hogan BL, Barkauskas CE, Chapman HA, Epstein JA, Jain R, Hsia CC, Niklason L, Calle E, Le A, Randell SH, Rock J, Snitow M, Krummel M, Stripp BR, Vu T, White ES, Whitsett JA, Morrisey EE (2014) Repair and regeneration of the respiratory system: complexity, plasticity, and mechanisms of lung stem cell function. Cell Stem Cell 15:123–138
2. Rock JR, Onaitis MW, Rawlins EL, Lu Y, Clark CP, Xue Y, Randell SH, Hogan BL (2009) Basal cells as stem cells of the mouse trachea and human airway epithelium. Proc Natl Acad Sci U S A 106:12771–12775
3. Teixeira VH, Nadarajan P, Graham TA, Pipinikas CP, Brown JM, Falzon M, Nye E, Poulsom R, Lawrence D, Wright NA, McDonald S, Giangreco A, Simons BD, Janes SM (2013) Stochastic homeostasis in human airway epithelium is achieved by neutral competition of basal cell progenitors. Elife 2, e00966
4. Hynds RE, Giangreco A (2013) Concise review: the relevance of human stem cell-derived organoid models for epithelial translational medicine. Stem Cells 31:417–422
5. de Jong PM, van Sterkenburg MA, Hesseling SC, Kempenaar JA, Mulder AA, Mommaas AM, Dijkman JH, Ponec M (1994) Ciliogenesis in human bronchial epithelial cells cultured at the air-liquid interface. Am J Respir Cell Mol Biol 10:271–277
6. Fulcher ML, Gabriel S, Burns KA, Yankaskas JR, Randell SH (2005) Well-differentiated human airway epithelial cell cultures. Methods Mol Med 107:183–206
7. Mathis C, Poussin C, Weisensee D, Gebel S, Hengstermann A, Sewer A, Belcastro V, Xiang Y, Ansari S, Wagner S, Hoeng J, Peitsch MC (2013) Human bronchial epithelial cells exposed in vitro to cigarette smoke at the air-liquid interface resemble bronchial epithelium from human smokers. Am J Physiol Lung Cell Mol Physiol 304:L489–L503
8. Rheinwald JG, Green H (1975) Serial cultivation of strains of human epidermal keratinocytes: the formation of keratinizing colonies from single cells. Cell 6:331–343
9. Chapman S, Liu X, Meyers C, Schlegel R, McBride AA (2010) Human keratinocytes are efficiently immortalized by a Rho kinase inhibitor. J Clin Invest 120:2619–2626
10. Suprynowicz FA, Upadhyay G, Krawczyk E, Kramer SC, Hebert JD, Liu X, Yuan H, Cheluvaraju C, Clapp PW, Boucher RC Jr, Kamonjoh CM, Randell SH, Schlegel R (2012) Conditionally reprogrammed cells represent a stem-like state of adult epithelial cells. Proc Natl Acad Sci U S A 109:20035–20040
11. Liu X, Ory V, Chapman S, Yuan H, Albanese C, Kallakury B, Timofeeva OA, Nealon C, Dakic A, Simic V, Haddad BR, Rhim JS, Dritschilo A, Riegel A, McBride A, Schlegel R (2012) ROCK inhibitor and feeder cells induce the conditional reprogramming of epithelial cells. Am J Pathol 180:599–607
12. Butler CR, Hynds RE, Gowers KH, Lee Ddo H, Brown JM, Crowley C, Teixeira VH, Smith CM, Urbani L, Hamilton NJ, Thakrar RM, Booth HL, Birchall MA, De Coppi P, Giangreco A, O'Callaghan C, Janes SM (2016) Rapid Expansion of Human Epithelial Stem Cells Suitable for Airway Tissue Engineering. Am J Respir Crit Care Med 194(2):156–168. doi:10.1164/rccm.201507-1414OC, PubMed PMID: 26840431
13. Wu X, Peters-Hall JR, Bose S, Pena MT, Rose MC (2011) Human bronchial epithelial cells differentiate to 3D glandular acini on basement membrane matrix. Am J Respir Cell Mol Biol 44:914–921
14. Hegab AE, Ha VL, Darmawan DO, Gilbert JL, Ooi AT, Attiga YS, Bisht B, Nickerson DW, Gomperts BN (2012) Isolation and in vitro characterization of basal and submucosal gland duct stem/progenitor cells from human proximal airways. Stem Cells Transl Med 1:719–724
15. Danahay H, Pessotti AD, Coote J, Montgomery BE, Xia D, Wilson A, Yang H, Wang Z, Bevan L, Thomas C, Petit S, London A, LeMotte P, Doelemeyer A, Velez-Reyes GL, Bernasconi P, Fryer CJ, Edwards M, Capodieci P, Chen A, Hild M, Jaffe AB (2015) Notch2 is required for inflammatory cytokine-driven goblet cell metaplasia in the lung. Cell Rep 10:239–252
16. Gao X, Bali AS, Randell SH, Hogan BL (2015) GRHL2 coordinates regeneration of a polarized mucociliary epithelium from basal stem cells. J Cell Biol 211:669–682

Human Pluripotent Stem Cells (iPSC) Generation, Culture, and Differentiation to Lung Progenitor Cells

Mahboobe Ghaedi and Laura E. Niklason

Abstract

Induced pluripotent stem (iPS) cells are the product of adult somatic cell reprogramming to an embryonic-like state by inducing a "forced" expression of specific genes. They are similar to natural pluripotent stem cells, such as embryonic stem (ES) cells, in many aspects, such as the expression of certain stem cell genes and potency and differentiability. Human iPS cells are invaluable resource for basic research, cell therapy, drug discovery, and human organ tissue engineering. iPS cells can be derived from the patient to be treated and thus are genetically identical cells that may avoid immune rejection. The following protocols offer a general guideline for the induction of iPSCs from fibroblasts, and for culture and expansion to produce lung precursor cells.

Keywords: Human induced pluripotent stem cell, Reprogramming, Expansion, Cell culture, Differentiation, Progenitors

1 Introduction

Induced pluripotent stem cells (iPSCs) are typically derived by introducing a specific set of pluripotency-associated genes, or "reprogramming factors," into an adult cell type. These cells show qualities very similar to human embryonic stem cells. The original set of reprogramming factors (also called Yamanaka factors) are the genes Oct4 (Pou5f1), Sox2, cMyc, and Klf4. There are multiple methods to generate iPSCs, including retrovirus or lentivirus-mediated gene transduction and chemical induction. The retroviral vectors require integration into host chromosomes to express reprogramming genes, but DNA-based vectors and plasmid vectors do not generally integrate to the cell genome [6, 11]. To generate the iPSCs, each of the pluripotency factors can be also replaced by related transcription factors, miRNAs or small molecules [5, 14, 16]. iPSC generation is a slow and inefficient process. It takes 1–2 weeks for mouse cells and 3–4 weeks for human cells and the efficiencies are as low as 0.01–0.1 %. Recently, many advances have been made in improving the efficiency and the time it takes to derive iPSCs. After introduction of reprogramming factors, cells begin to form colonies very similar to human embryonic stem cells [1]. These iPSC colonies can be isolated based on their morphology, expression of pluripotent genes and surface markers and can be expanded in an appropriate culture system to keep pluripotency over several passages. The following protocols provide a general guideline for the induction of iPSCs from fibroblasts using an inducible lentiviral system.

Traditional human embryonic stem cells (hES cells) and induced pluripotent stem cells (hiPS cells) culture methods require the use of mouse or human fibroblast feeder layers [11, 16]. The feeder cell preparation requires significant time and effort, is labor-intensive and hard to scale. More recently, researchers have developed feeder free systems for both hES and hiPS cell culture. These systems include Matrigel, or other extracellular matrix (ECM) proteins such as vitronectin, to maintain hESCs and hiPSCs and the reduction or complete removal of serum from human stem cell culture [8]. To move toward feeder-free culture systems, researchers have also designed MTeSR™ and Essential 8™ medium. MTeSR™1 is the most widely published feeder-free cell culture medium for ES and iPS cells, with established protocols for applications ranging from derivation to differentiation. It has been used to successfully maintain hundreds of ES cell and iPS cell lines and has supported the main pluripotency genes expression in ES and iPS cells. Essential 8™ Medium is another xeno-free and feeder-free medium specially formulated for the growth and expansion of human pluripotent stem cells. Essential 8™ Medium has been extensively tested in multiple iPSC lines. In addition, both MTeSR™ and Essential 8™ Medium have been used to scale up production of iPSCs and have been shown to support iPSC growth for >50 passages without any signs of karyotypic abnormalities, along with maintaining the ability of iPSCs to differentiate into all three germ line lineages [9].

The lack of an abundant source of human lung epithelial cells is a major limitation for studying the lung disease phenotypes, drug screening and clinical application of these cells in respiratory disease. Recent advances in the stem cells field suggest that the use of induced pluripotent stem cells (iPSC) may be the most effective strategy to develop functional lung epithelial cells [3, 4, 7].

Embryonic lung arises from definitive endoderm (DE). Following developmental paradigms, directed differentiation of iPS cells to lung progenitors should proceed by generation of definitive endoderm, followed by patterning into anterior foregut endoderm (AFE). The AFE will then differentiate to lung progenitor cells and finally to lung alveolar and airway epithelial cells that cover the respiratory airways and alveoli. The following protocols offer a guideline for a stepwise differentiation of iPSC to human lung epithelial progenitors via definitive endoderm and anterior foregut endoderm (AFE) [4, 7, 12, 13, 15].

2 Materials

2.1 Feeder-Dependent iPSC Culture Protocol

Prepare all equipment, reagents, and solutions listed below:

2.1.1 Equipment and Supplies

1. Biosafety cabinet.
2. Centrifuge.
3. Microscope.
4. 37 °C–5 % CO_2 incubator.
5. 37 °C water bath.
6. Cryogenic handling gloves and eye protection.
7. 6-well plates (Costar, 3516).
8. 5 mL sterile serological pipettes (Corning, 4487) and 10 mL sterile serological (Corning, 4488) pipettes or equivalent.
9. 2 mL aspirating pipette (Falcon, 357558).
10. 15 mL conical tube (Corning, 430791).
11. 50 mL conical tube (Corning, 430829).
12. 70 % ethanol (Dean lab Inc, 2701)

2.1.2 Reagents

1. DMEM/F-12 medium (Life Technologies, Gibco, 11330-032).
2. Knockout Serum Replacer (KOSR) (Invitrogen, 10828-028).
3. L-glutamine, non-animal, cell culture tested (Life Technologies, Gibco, 25030081).
4. MEM nonessential amino acid solution (Life Technologies, Gibco, 11140-050).
5. Basic fibroblast growth factor (b-FGF) (Life Technologies, Gibco, PHG0021) or equivalent.
6. β-mercaptoethanol (Life Technologies, Gibco, 21985-023).
7. High glucose DMEM Medium (Life Technologies, Gibco, 11965092).
8. Penicillin–streptomycin (Life Technologies, Gibco, 1510-122).
9. Fetal bovine serum (Hyclone, SH30084.03).
10. PBS without $CaCl_2$ or $MgCl_2$ (Invitrogen, 14190-144).
11. PBS with $CaCl_2$ and $MgCl_2$ (Invitrogen, 14040-141).
12. Bovine serum albumin (Sigma, A2153).
13. 2 % gelatin solution (Sigma G1393).
14. Mouse embryonic feeder (Global stem cells GSC-6201).

15. Rock Inhibitor (Y-27632 dihydrochloride; Ascent Scientist, Asc-129).
16. Sterile water (Sigma, W4502).
17. Collagenase Type IV (Invitrogen, 17104-019).
18. Defined FBS (Hyclone, SH30070.01).
19. Dimethyl sulfoxide (DMSO) (Sigma-Aldrich, D2650).
20. Dispase (StemCell Technologies, 07923).

2.1.3 Solution

Mouse Embryonic Feeder (MEF) Medium (250 mL)

To make 250 mL feeder culture medium combine following components, filter-sterilize, store at 4 °C for up to 1 month.

- 217.5 mL high glucose DMEM medium.
- 25 mL fetal bovine serum.
- 2.5 mL 200 mM L-glutamine.
- 2.5 mL MEM nonessential amino acids.
- 2.5 mL penicillin–streptomycin.

Stem Cell Culture Medium (250 mL)

To make 250 mL stem cell culture medium combine following components, filter-sterilize, store at 4 °C for up to 14 days.

- 195 mL DMEM/F-12 medium.
- 50 mL Knockout Serum Replacer.
- 2.5 mL 200 mM L-glutamine.
- 2.5 mL MEM nonessential amino acids.
- 0.5 mL 2 μg/mL Basic FGF solution (*See* below).
- 450 μL BME.

2 μg/mL Basic FGF Solution

To make 2 μg/mL Basic FGF Solution, for stem cell culture medium, dissolve 10 μg Basic FGF in 5 mL 0.1 % BSA in PBS with $CaCl_2$ and $MgCl_2$. Aliquot 0.5 mL/tube and store at −20 °C for up to 6 months. Each aliquot is enough to make 250 mL of stem cell culture medium. Thaw aliquot just prior to making Stem Cell Medium. Do not refreeze aliquots.

0.1 % Gelatin Solution

To prepare 0.1 % gelatin for coating the plate, dilute the 2 % gelatin solution with PBS with $CaCl_2$ and $MgCl_2$ to make 0.1 % gelatin solution. To make 200 mL 0.1 % gelatin solution, add 2 mL gelatin to 200 mL PBS with $CaCl_2$ and $MgCl_2$ and autoclave it and store it at 4 °C up to 6 months.

Rock Inhibitor Solution

To make 10 mM Rock Inhibitor stock solution, dilute 1 mg Rock Inhibitor (FW 320.26) into 295 μL sterile water (*see* **Note 1**). Aliquot 20–50 μL/tube and store at −80 °C for up to 1 year. Aliquots can be stored up to 2 months at 4 °C. Rock Inhibitor working stock solution will be used at 1 μL to 1 mL final medium volume.

1 mg/mL Collagenase Solution	To make 1 mg/mL collagenase solution, dissolve the 25 mg collagenase Type IV powder in 25 mL DMEM/F-12, filter it and aliquot it to 5 mL and store at −20 °C. Collagenase solution can be stored at 4 °C for up to 14 days.
2× Cryopreservation Medium	To make 10 mL, 2× cryopreservation medium, add 4 mL defined FBS to 4 mL stem cell culture medium, filter-sterilize and then add 2 mL sterile DMSO. Keep on ice until use.

2.2 Feeder-Free Human iPSC Culture Protocol (Matrigel/MTeSR Medium)

The mTeSR™1 medium from StemCell Technologies and Matrigel™ Matrix from BD Bioscience, have been shown to be a successful combination for feeder-free maintenance of different human ES and iPS cell lines for up to 20 passages. BD Matrigel hESC-qualified Matrix is compatible with the MTeSR™1 medium from StemCell Technologies, in order to provide the reproducibility for hESC and hiPSC culture [2, 8, 9, 10].

Prepare all equipment, reagents, and solutions listed below:

2.2.1 Equipment and Supplies

1. Biosafety cabinet.
2. Centrifuge.
3. Microscope.
4. 37 °C / 5 % CO_2 incubator.
5. 37 °C water bath.
6. Cryogenic handling gloves and eye protection.
7. 6-well plates (Costar, 3516).
8. 5 mL sterile serological pipettes (Corning, 4487) and 10 mL sterile serological (Corning, 4488) pipettes or equivalent.
9. 2 mL aspirating pipette (Falcon, 357558).
10. 15 mL conical tube (Corning, 430791).
11. 50 mL conical tube (Corning, 430829).
12. 70 % ethanol (Dean lab Inc, 2701).

2.2.2 Reagents

1. hESC qualified Matrigel™ (BD Biosciences®, 354277).
2. DMEM/F-12 Medium (Life Technologies, Gibco, 11330-032).
3. MTeSR™1 medium (StemCell Technologies, 05850).
4. Dispase (StemCell Technologies 07923).
5. Defined FBS (Hyclone, SH30070.01).
6. Dimethyl sulfoxide (DMSO) 10 mL ampoules (Sigma-Aldrich, D2438).

2.2.3 Solution

Preparing MTeSR™ Medium

Thaw the Supplement (50×) at 2–8 °C overnight. Do not thaw at 37 °C. To prepare 500 mL of complete MTeSR™ Medium, add the 100 mL supplement to the 400 mL basal medium. Aliquot it in 50 mL and store it in −20. Thaw each aliquot at 4 °C before use.

Aliquot Matrigel™

1. Keep the Matrigel™ frozen in −80 °C until you are ready to aliquot it. Matrigel™ is frozen at −20 °C to −80 °C, liquid at 4 °C, and gels rapidly at room temperature.
2. Calculate the volume of Matrigel™ needed per plate based on supplier instruction. Depending on volume of cell culture performed in the laboratory, different sized aliquots can be prepared. For provided catalog number above, 100 μL is enough for coating one 6-well plate. Label the concentration clearly on each tube. Each Matrigel™ aliquot is intended for one use.
3. Thaw the Matrigel™ overnight on ice in 4 °C refrigerator.
4. 2–3 h before aliquot the Matrigel™, place unopened box of the appropriate sized pipette tips, 1.5 mL tubes in the −20 °C or −80 °C freezer to keep them cold.
5. Aliquot the Matrigel™ while is on an ice using cold pipette tips in a cold 1.5 mL eppendorf tubes in sterile biosafety cabinet. Place each aliquot on ice and transfer them in −80 °C soon after finishing aliquot (*see* **Note 2**).

2× Cryopreservation Medium

To make 10 mL of 2× Cryopreservation Medium, add 4 mL defined FBS to 4 mL MTeSR™ Medium, filter-sterilize and then add 2 mL Sterile DMSO. Keep it on ice until use.

2.3 Feeder-Free Human iPSC Culture Protocol (Vitronectin/Essential 8™ Medium)

Prepare all equipment, reagents and solutions listed below:

2.3.1 Equipment and Supplies

1. Biosafety cabinet.
2. Centrifuge.
3. Microscope.
4. 37 °C/5 % CO_2 incubator.
5. 37 °C water bath.
6. Cryogenic handling gloves and eye protection.
7. 6-well plates (Costar, 3516).
8. 5 mL sterile serological pipettes (Corning, 4487) and 10 mL sterile serological (Corning, 4488) pipettes or equivalent.
9. 2 mL aspirating pipette (Falcon, 357558).
10. 15 mL conical tube (Corning, 430791).
11. 50 mL conical tube (Corning, 430829).
12. 70 % Ethanol (Dean lab Inc, 2701).

2.3.2 Reagents

1. Essential 8™ Medium, consisting of Essential 8™ Basal Medium and Essential 8™ Supplement (Life Technologies, A1517001).
2. Vitronectin, truncated recombinant human (Life Technologies, A14700).
3. PBS without calcium and magnesium (Invitrogen 14190-144).
4. UltraPure™ 0.5 M EDTA, pH 8.0 (Life Technologies 15575-020).
5. Dimethyl sulfoxide (DMSO) (Sigma-Aldrich, D2650).

2.3.3 Solutions

Preparing Essential 8™ Medium

Thaw the Essential 8™ Supplement (50×) at 2–8 °C overnight. Do not thaw at 37 °C. To prepare 500 mL of complete Essential 8™ Medium, add the 10 mL Essential 8™ supplement to the 490 mL basal medium and store at 2–8 °C for up to 2 weeks (*see* **Note 3**).

0.5 mM EDTA

To prepare 50 mL of 0.5 mM EDTA in PBS, add 50 μL of 0.5 M EDTA to the 50 mL of sterile PBS. Filter-sterilize the solution and store it at room temperature for up to 6 months.

Aliquot Vitronectin

Thaw the vial of vitronectin on ice. Calculate the volume of vitronectin needed per plate based on supplier instructions. Depending on volume of cell culture performed in the laboratory, different sized aliquots may be preferred. For provided cat number above, 60 μL is enough for coating one 6-well plate. Make sure the concentration is clearly labeled on the tube. Freeze the aliquots at −80 °C or use immediately. Each aliquot is intended for one use.

Cryopreservation Medium

To make 10 mL of Cryopreservation Medium, combine 9 mL Complete Essential 8™ Medium with 1 mL, sterile DMSO in a sterile 15-mL tube. Keep it on ice until use.

2.4 Differentiation of Human iPSC to Lung Progenitors

Prepare all equipment, reagents and solutions listed below:

2.4.1 Equipment and Supplies

1. Biosafety cabinet.
2. Centrifuge.
3. Microscope.
4. 37 °C/5 % CO_2 incubator.
5. 37 °C water bath.
6. Cryogenic handling gloves and eye protection.
7. 6-well plates (Costar, 3516).
8. 5 mL sterile serological pipettes (Corning, 4487) and 10 mL sterile serological (Corning, 4488) pipettes or equivalent.
9. 2 mL aspirating pipette (Falcon, 357558).

10. 15 mL conical tube (Corning, 430791).
11. 50 mL conical tube (Corning, 430829).
12. 70 % Ethanol (Dean lab Inc, 2701).

2.4.2 Reagents

1. RPMI 1640 (Life Technologies, Gibco 12633-012).
2. Recombinant human Activin A (Life Technologies, Gibco, PHG9014).
3. Penicillin–streptomycin (Life Technologies Gibco, 1510-122).
4. 100 mM L-glutamine, non-animal, cell culture tested (Life Technologies, Gibco, 25030081).
5. MEM nonessential amino acid solution (Life Technologies, Gibco, 11140-050).
6. 50× B27 supplement (Life Technologies Gibco, 17504-044).
7. Fetal bovine serum (Hyclone, SH30084.03).
8. Sodium pyruvate (Life Technologies Gibco 11360-070).
9. IMDM (Life Technologies, Gibco, 12440-053).
10. Recombinant human Noggin (Life Technologies, Gibco, PHC1506).
11. SB431542 (Tocris BioScience, 1614).
12. Basic fibroblast growth factor (Life Technologies, Gibco, PHC9394).
13. Bone morphogenetic protein-4 (BMP4) (Life Technologies, Gibco, PHC9531).
14. Wnt3a (R&D system, 5036 WN).
15. Retinoic acid (Sigma, R-2625).
16. Keratinocyte growth factor (KGF) (Life Technologies, Gibco, PHG0094).
17. Epidermal growth factor (EGF) (Life Technologies, Gibco, PHG0311).
18. Fibroblast growth factor-10 (FGF10) (Life Technologies, Gibco, PHG0204).

2.4.3 Reagents

Differentiation of iPCs to Definitive Endoderm

Prepare all solutions listed below:

RPMI Basal Medium

To make 250 mL of RPMI as a basal medium for definitive endoderm differentiation combine the following components, filter-sterilize, store at 4 °C for up to 2 weeks.

- 245 mL RPMI 1640 Medium.
- 2.5 mL 100X penicillin–streptomycin.
- 2.5 mL 100 mM L-glutamine.

100 µg/mL Activin A Solution	To make 100 µg/mL Activin A Solution, for definitive endoderm medium, dissolve 5 µg Activin A in 5 mL 0.1 % BSA in PBS with $CaCl_2$ and $MgCl_2$. Aliquot 1 mL/tube and store at $-20\ °C$ for up to 6 months. Each aliquot is enough to make 10 mL of definitive endoderm medium. Thaw aliquot just prior to making the medium. Do not refreeze aliquots.
B27 Supplement	Thaw the B27 supplement at 4 °C overnight. Aliquot 0.5 mL/tube and store at $-20\ °C$ for up to 6 months. Each aliquot is enough to make 25 mL of definitive endoderm medium. Thaw aliquot just prior to making the medium.
Definitive Endoderm Medium (10 mL)	To make 10 mL of definitive endoderm differentiation medium for the first 48 h of differentiation, combine the following components, filter-sterilize, store at 4 °C for up to 4 days.

- 10 mL RPMI 1640 basal medium.
- 1 mL of 100 µg/mL Activin A solution.

After 48 h, switch the medium to the following medium: to make 10 mL of the medium to be used from day3 to day 6, combine the following components:

- 10 mL RPMI 1640 basal medium.
- 1 mL of 100 µg/mL active solution.
- 0.2 mL of 50X B27 supplement.
- 50 µL 100 mM sodium butyrate.

Differentiation of Definitive Endoderm to Anterior Foregut Endoderm	To make 250 mL of IMDM/10 % FBS as a basal medium for anterior foregut endoderm differentiation, combine the following components, filter-sterilize, store at 4 °C for up to 4 days.
IMDM Basal Medium	

- 217.5 mL IMDM Medium.
- 25 mL FBS.
- 2.5 mL 100× penicillin–streptomycin.
- 2.5 mL 100 mM L-glutamine.
- 2.5 mL MEM nonessential amino acids.

200 µg/mL Noggin Solution	To make 200 µg/mL Noggin Solution, for anterior foregut endoderm differentiation, reconstitute the 20 µg Noggin in 10 mL 0.1 % BSA in PBS with $CaCl_2$ and $MgCl_2$. Aliquot 1 mL/tube and store at $-20\ °C$ for up to 6 months. Each aliquot is enough to make 10 mL of anterior foregut endoderm medium. Thaw aliquot just prior to making the medium. Do not refreeze aliquots.
10 mM SB431542 Solution	To make 10 mM SB431542 Solution, for anterior foregut endoderm medium, dissolve 1 mg of SB431542 (MW: 420.42) in 2.379 mL DMSO. Aliquot 100 mL/tube and store at $-20\ °C$ for up to 6 months.

Human Extracellular Matrix Solution	Human ECM protein comes in 1 mL volumes. Calculate the required amount of ECM protein solution for coating enough required wells. To make the human ECM solution, dilute it 1:100 in DMEM-F12 and use it freshly. Aliquot the human ECM in 100 μL/well and store it in −20 °C for up to 6 months. Thaw aliquot just prior to making the medium. Do not refreeze and thaw the aliquots.
Anterior Foregut Endoderm Medium (10 mL)	To make 10 mL of anterior foregut endoderm differentiation medium, combine the following components, filter-sterilize, store at 4 °C for up to 4 days.

- 10 mL IMDM basal medium.
- 1 mL of 200 μg/mL NOGGIN solution.
- 10 μL of 10 mM SB431542.

Differentiation of Anterior Foregut Endoderm to Lung Progenitors *10 μg/mL BMP4 Solution*	To make 10 μg/mL BMP4 Solution, reconstitute 10 μg BMP4 in 1 mL 0.1 % BSA in PBS with $CaCl_2$ and $MgCl_2$. Aliquot 50 μL/tube and store at −80 °C for up to 6 months. Each aliquot is enough to make 50 mL of differentiation medium. Thaw aliquot just prior to making the medium.
10 μg/mL KGF Solution	To make 10 μg/mL KGF (FGF7) Solution, reconstitute 10 μg KGF in 1 mL 0.1 % BSA in PBS with $CaCl_2$ and $MgCl_2$. Aliquot 50 μL/tube and store at −80 °C for up to 6 months. Each aliquot is enough to make 50 mL of differentiation medium. Thaw aliquot just prior to making the medium.
10 μg/mL bFGF Solution	To make 10 μg/mL bFGF Solution, dissolve 10 μg bFGF in 1 mL 0.1 % BSA in PBS with $CaCl_2$ and $MgCl_2$. Aliquot 50 μL/tube and store at −80 °C for up to 6 months. Each aliquot is enough to make 50 mL of differentiation medium. Thaw aliquot just prior to making the medium.
100 μg/mL Wnt3a Solution	To make 100 μg/mL Wnt3a Solution, dissolve 100 μg Wnt3a in 1 mL 0.1 % BSA in PBS with $CaCl_2$ and $MgCl_2$. Aliquot 100 μL/tube and store at −80 °C for up to 6 months. Each aliquot is enough to make 10 mL of differentiation medium. Thaw aliquot just prior to making the medium.
0.5 mM Retinoic Acid Solution	To make 25 mM retinoic acid solution dissolve 100 mg (MW = 300.44) of retinoic acid powder in 12 mL DSMO. Aliquot 500 μL/tube and store at −80 °C for up to 6 months. To make 0.5 mM retinoic acid solution, add 100 μL of 25 mM retinoic acid solution to 500 μL DMSO.

Lung Progenitor Differentiation Medium (10 mL)

To make 10 mL of lung progenitor differentiation medium, combine following components and store at 4 °C for up to 4 days.

- 10 mL IMDM basal medium.
- 10 μL of 10 μg/mL BMP4 solution.
- 10 μL of 10 μg/mL bFGF solution.
- 100 μL of 10 μg/mL Wnt3a solution.
- 10 μL of 10 μg/mL KGF solution.

10 μL of 0.5 mM retinoic acid solution.

2.5 iPSC Generation Using Lentiviral Vectors Protocol

Prepare all Equipment, Reagents and Solutions Listed Below:

2.5.1 Equipment and Supplies

1. Biosafety cabinet.
2. Centrifuge.
3. Microscope.
4. 37 °C/5 % CO_2 incubator.
5. 37 °C water bath.
6. Cryogenic handling gloves and eye protection.
7. 5 mL sterile serological pipettes (Corning, 4488).
8. 10 mL sterile serological pipettes (Corning, 4487).
9. 2 mL aspirating pipettes (Falcon, 357558).
10. 6-well tissue culture plates (Costar, 3516).
11. 24-well tissue culture plates (Costar, 3526).
12. 48-well tissue culture plates (Costar, 3548).
13. 96-well tissue culture plates (Costar, 3596).
14. 10 cm culture plate (Corning, 430167).
15. 15 mL conical tubes (Corning, 430791).
16. 50 mL conical tubes (Corning, 430829).
17. 70 % Ethanol (Dean lan Inc, 2701).

2.5.2 Required Reagents

1. FUW-tetO-lox-hOCT4 plasmid (Add gene Plasmid 20728).
2. FUW-tetO-lox-SOX2 plasmid (Add gene Plasmid 20729).
3. FUW-tetO-lox-hKLF4 (Add gene Plasmid 20727).
4. FUW-tetO-lox-hNanog (Add gene Plasmid 20727).
5. FUW-tetO-lox-hcMyc (Add gene Plasmid 20324).
6. FUW-tetO-lox-m2rt TA (Add gene Plasmid 20342).
7. pMD2.G (Plasmid 12259).
8. psPAX2 (Plasmid 12260).

9. DMEM-F12 Medium (Life Technologies, Gibco, 11330-032).
10. High glucose DMEM Medium (Life Technologies, Gibco, 11965092).
11. Penicillin–streptomycin (Life Technologies, Gibco, 1510-122).
12. Fetal bovine serum (Hyclone, SH30084.03).
13. PBS without $CaCl_2$ or $MgCl_2$ (Invitrogen, 14190-144).
14. Knockout Serum Replacer (KOSR) (Invitrogen, 10828-028).
15. L-glutamine, non-animal, cell culture tested (Life Technologies, Gibco, 25030081).
16. MEM nonessential amino acid solution (Life Technologies, Gibco, 11140-050).
17. Basic fibroblast growth factor (b-FGF) (Life Technologies, Gibco, PHG0021) or equivalent.
18. β-mercaptoethanol (Life Technologies, Gibco, 21985-023).
19. PBS with $CaCl_2$ and $MgCl_2$ (Invitrogen, 14040-141).
20. Bovine serum albumin (Sigma, A2153).
21. 2 % gelatin solution (Sigma G1393).
22. Mouse embryonic feeder (Global stem cells GSC-6201).
23. Rock Inhibitor (Y-27632 dihydrochloride; Ascent Scientist, Asc-129).
24. Sterile water (Sigma, W4502).
25. Collagenase Type IV (Invitrogen, 17104-019).
26. Defined FBS (Hyclone, SH30084.01).
27. Dimethyl sulfoxide (DMSO) (Sigma-Aldrich, D2650).
28. Dispase (StemCell Technologies 07923).
29. Polyberene (Sigma 107689, H9268).
30. Human HEK293 cell line (ATCC CRL1573™).
31. Doxycycline (Dox) (Sigma: D9891-1G).
32. Valproic acid (PVA) (Sigma, P4543).
33. Opti-MEM® Reduced Serum Medium (Life Technologies 31985070).
34. Lipofectamine® 2000 Transfection Reagent (Life Technologies 11668027).

2.5.3 Solution

DMEM Medium (To Culture HEK293T, Human Fibroblast and Mouse Embryonic Feeder (MEF) (250 mL)

To make 250 mL feeder culture medium combine following components, filter-sterilize, store at 4 °C for up to 1 month.

- 217.5 mL high glucose DMEM Medium.
- 25 mL fetal bovine serum.
- 2.5 mL 200 mM L-glutamine.

- 2.5 mL MEM nonessential amino acids.
- 2.5 mL penicillin–streptomycin.

Cryopreservation Medium (HEK293T, Human Fibroblast)

To make 10 mL of cryopreservation medium, combine 9 mL fibroblast culture medium with 1 mL, sterile DMSO in a sterile 15-mL tube. Keep it on ice until use.

Stem Cell Culture Medium (250 mL)

To make 250 mL stem cell culture medium, combine following components, filter-sterilize, store at 4 °C for up to 14 days.

- 195 mL DMEM/F-12 medium.
- 50 mL Knockout Serum Replacer.
- 2.5 mL 100 mM L-glutamine.
- 2.5 mL MEM nonessential amino acids.
- 0.5 mL 2 µg/mL Basic FGF solution (*See* below).
- 450 µL BME.

2 µg/mL Basic FGF Solution

To make 2 µg/mL Basic FGF Solution, for stem cell culture medium, dissolve 10 µg Basic FGF in 5 mL 0.1 % BSA in PBS with $CaCl_2$ and $MgCl_2$. Aliquot 0.5 mL/tube and store at −20 °C for up to 6 months. Each aliquot is enough to make 250 mL of stem cell culture medium. Thaw aliquot just prior to making stem cell culture medium. Do not refreeze aliquots.

0.1 % Gelatin Solution

To prepare 0.1 % gelatin for coating the plates, dilute the 2 % gelatin solution with PBS with $CaCl_2$ and $MgCl_2$ to make 0.1 % gelatin solution. To make 200 mL 0.1 % gelatin solution, add 2 mL gelatin to 200 mL PBS with $CaCl_2$ and $MgCl_2$ and autoclave it and store it at 4 °C

1 µg/mL Doxycycline (Dox) Solution

Reconstitute 10 mg of powder in 10 mL PBS and filter with 0.2 µm filter, aliquot it and store at −20 °C.

1 M Valproic acid (VPA)

Reconstitute 166 mg of VPA in 1 mL sterile H_2O to make 1 M solution. Add 1 µL to 1 mL medium to get 1000× dilution. Sometimes VPA is toxic, and sodium butyrate can be used instead of VPA.

Dox Induction Medium

To make 10 mL Dox induction medium combine the following components.

- 10 mL hESC Culture Medium.
- 10 µL Dox Solution (2 mg/mL).

Thaw Dox Solution on ice and add to pre-warmed hESC medium.

Polyberene Solution

Polybrene is a polycation that increases binding between the pseudoviral capsid and the cellular membrane. Prepare a 6 mg/mL Polybrene stock solution in deionized, sterile water. Filter-sterilize it and aliquot the stock solution at 100 μL/tube and store at −20 °C for up to 1 year. The working stock can be stored at 4 °C for up to 2 weeks. Do not freeze/thaw the stock solution more than three times as this may result in loss of activity.

3 Methods

3.1 Feeder-Dependent iPSC Culture Protocol

3.1.1 Prepare Mouse Embryonic Feeder (MEF) Plates

1. Sterilize the biosafety cabinet for 20 min with UV light.
2. Turn on the blower and spray down the whole surface with ethanol and allow it to evaporate for 20 min prior to initiating cell culture.
3. Coat two 6-well plate with 0.1 % gelatin solution at least 2 h prior to thawing the MEF.
4. Remove a frozen vial of MEF (2×10^6 cells) from the liquid nitrogen tank and thaw by immersing the vial in a 37 °C water bath without submerging the cap. Swirl the vial gently (*see* **Note 4**).
5. Remove the MEF vial from the water bath when only a small ice crystal remains.
6. Spray the vial with a 70 % ethanol to sterile the outside of the tube and transfer it into the sterile biosafety cabinet.
7. Transfer the MEF cells gently into a sterile 15 mL conical tube using a 1 mL sterile serological pipette.
8. Add 4 mL of warmed DMEM-10%FBS medium dropwise to the MEF cells in the 15 mL conical tube. To reduce osmotic shock to the cells, move the tube back and forth gently to mix the cells, while adding the medium.
9. Centrifuge the cells at $233 \times g$ for 5 min.
10. Aspirate and discard the supernatant with a sterile aspirating pipette.
11. Resuspend the cell pellet in 24 mL of MEF medium; 2 mL for every well that will receive cells (2 M MEF cells are enough for 12 wells of a 6-well plate. It is based on the thaw recommendation from Global stem cells to get 2×10^5 cells per one well of 6-well plate).
12. Aspirate gelatin from each well and wash with PBS prior to transferring MEF into the gelatin coated plates.
13. Gently pipette cells up and down few times and add 2 mL of medium containing MEF cells to each well of 6-well plate.
14. Transfer the plates to 37 °C, 5 % CO_2 incubator (*see* **Note 5**).

3.1.2 Thawing the iPSC onto the Feeder (MEF)

1. 2–3 h prior to thawing the iPSC, aspirate MEF medium from each well, rinse it with 1 mL DMEM-F12 and add 1 mL of stem cell culture medium to each well of 6-well plates.
2. Transfer the MEF plate back to the 37 °C incubator.
3. Remove the frozen vial of iPSC from the liquid nitrogen storage tank.
4. Quickly remove the label or copy the information written on the tube in your notebook (*see* **Note 6**).
5. Immerse the vial in a 37 °C water bath and swirl the vial gently without submerging the cap. Remove the vial from the water bath when no ice crystals remain.
6. Spray the vial with a 70 % ethanol to sterilize the outside of the tube and transfer it into the sterile biosafety cabinet.
7. Transfer the cells gently into a sterile 15 mL conical tube using a 1 mL sterile pipette.
8. Slowly, add 4 mL of warmed stem cell medium dropwise to cells in the 15 mL conical tube. Gently move the tube back and forth to mix the cells while adding the medium.
9. Centrifuge the cells at $149 \times g$ for 5 min.
10. Aspirate and discard the supernatant with a 2 mL sterile aspirating pipette.
11. Resuspend the cell pellet in 2.5 mL stem cell culture medium for every well that will receive cells (*see* **Note 7**).
12. To increase the iPS cell viability and attachment, add 1 mM Rock Inhibitor to stem cell medium only at first 24 h.
13. Gently pipette cells up and down in the tube a few times.
14. Transfer iPSC cells onto the feeder layer and place the iPSC plate into the incubator and gently shake the plate back and forth and side to side to evenly distribute the cells. Avoid circular motions to prevent pooling iPSCs colonies in the center of the well.
15. The next day, remove the spent medium using a sterile aspirating pipette and gently add 3 mL of stem cell culture medium to the well.
16. Place the plate into a 37 °C incubator.
17. Feed the iPS cells daily until ready to passage or freeze.

3.1.3 Feeding iPSC Cultured on MEF

1. Warm enough stem cell culture medium to feed the iPSC. Generally, 2–3 mL for each well that will be fed.
2. Aspirate the spent medium with a 2 mL sterile aspirating pipette.
3. Add 2 mL of warmed stem cell culture medium to each well for first 2 days. Starting from day 3, add 3 mL medium to each well (*see* **Note 8**).

4. Return the 6-well plate to the 37 °C incubator.

5. Feed the iPSCs daily until ready to passage or freeze (*see* **Note 9**).

3.1.4 Passaging iPSC Cultured on MEF

In general, iPS cells should be split when the MEF feeder layer is 10 days old or iPSC colonies are becoming too dense or too large or increased differentiation occurs. There are two methods for passaging the iPSC. An enzymatic method which is a standard passaging method, recommended when more than 10–20 colonies are in each well. And, a nonenzymatic or manual passaging method is recommended for sparse iPSC colonies or normally fewer than 10–20 colonies in each well. This method is also recommended when there is significant differentiation present and the culture must be maintained.

1. Prepare enough MEF plate as described in Section 3.1.1, the day prior to passaging the iPS cells.

2. Two hours prior to passaging the iPS cells, aspirate the MEF medium from the wells, and rinse with 1 mL/well warmed sterile DMEM/F-12.

3. Add 1 mL stem cell medium to each well and return the MEF plates to 37 °C incubator.

4. Remove differentiated colonies before passaging the iPS cells if necessary. Transfer the plate into a hood equipped with a stereomicroscope and remove all areas of differentiation with a sterile modified pastor pipette or sterile micropipette tip.

5. Return the plate in the biosafety cabinet. Aspirate the spent medium from the wells to be passaged with an aspirating pipette to remove floating picked differentiated colonies in medium.

6. Rinse it once with 1 mL warmed DMEM-F12.

7. Add 1 mL room temperature collagenase solution to each well to be passaged.

8. Incubate for 5–7 min at 37 °C incubator (*see* **Note 10**).

9. Check the iPSC colonies under a microscope to confirm appropriate incubation time. When the edges of the colonies just slightly folded back, aspirate the collagenase without disturbing the attached iPS cell colonies.

10. Gently add 1 mL of warmed DMEM/F-12 to each well with a 5 mL pipette and then aspirate off the medium.

11. Rinse each well with 1 mL of warmed DMEM/F-12 one more time (*see* **Note 11**).

12. Add 1 mL of stem cell culture medium to each well.

13. Gently scrape the surface of the plate using a sterile 2 mL glass pipette, while holding the pipette perpendicular to the plate and simultaneously dispensing medium.

14. Repeat it at different direction to cut colonies to small pieces as scrape the cells off the surface of the plate.
15. Pipette the medium slowly up and down to wash the cells off the surface.
16. Transfer iPSC pieces to 15 mL sterile conical tube after all wells are scraped and the cells are removed from the surface of the well.
17. Pipette cells up and down gently a few times in the conical tube to further break-up iPSCs colonies if needed.
18. Add 1–2 mL stem cell culture medium to the first well to wash and collect residual cells. Then take up the medium and transfer it into each subsequent well to collect all residual cells.
19. Determine how much additional medium is required based on the split ratio and the number of wells that will be used (*see* **Note 12**).
20. Add Rock Inhibitor at 1 µM concentration to the stem cell culture medium during the first 24 h after passaging the iPSC. This increases the viability and attachment of iPS cells.
21. Add 2 mL of cell suspension to each well of the new plate. In general, there should be a total of 3 mL of stem cell culture medium and cells in each of the new wells (2 mL of cell suspension + 1 mL of pre-plated stem cell culture medium on MEF cells).
22. Return the plate to the 37 °C incubator after plating the cells. Move the plate back-and-forth and side-to-side for a few times to further disperse cells across the surface of the wells (*see* **Note 13**).
23. Incubate cells overnight to allow colonies to attach.
24. Feed the iPSC as previously described until ready to passage or freeze (*see* **Note 14**).

3.1.5 Freezing iPSC Cultured on MEF

1. Label cryovials with the cell line, passage number, the freeze date, and your initials. Use an alcohol proof pen or labels that resist liquid nitrogen and ethanol.
2. Spray down the whole surface with ethanol and allow it to evaporate for 20 min prior to initiating cryopreservation.
3. Prepare the isopropanol freezing container "Mr. Frosty" and keep it in 4 °C, about 30 min before start (*see* **Note 15**).
4. Prepare the required amount of cryopreservation medium and keep it on ice until ready to use. 1 mL freezing media will be needed for every vial. Always make a little extra to account for pipet error.
5. Remove differentiated colonies before freezing the iPS cells if necessary.

6. Aspirate stem cell medium from each well to remove floating picked differentiated colonies in medium.

7. Add 1 mL of room temperature collagenase solution to each well of each 6-well plate.

8. Incubate cells for 5–7 min at 37 °C incubator.

9. Check the plate under the microscope every 3 min and look for the edge of the colonies to be slightly folded back.

10. Aspirate collagenase solution from each well, rinse it with DMEM-F12 two times. Take care not to remove any floating colonies.

11. Add 1 mL of stem cell culture medium to each well. Take up 1 mL of medium from each well using 2 mL serological glass pipette and scrape the colonies of the plate while slowly expelling the medium to wash the cells off the surface.

12. Repeat several times and at different direction in each well to remove the cells from the well surface (see **Note 16**).

13. Collect the cells in a sterile 15 mL conical tube. Wash each plate with 2 mL of stem cell culture medium, transferring the medium from well to well and add the medium to the 15 mL conical tube.

14. Centrifuge at 149 × g for 5 min.

15. Aspirate the supernatant and resuspend each cell pellet in enough stem cell culture medium. Stem cells from each well go to 0.5 mL stem cell culture medium.

16. Freeze iPSC at 1 well/cryovial. Add 0.5 mL cryopreservation medium to each 0.5 mL of iPSC cells. For example, for three vials, add 1.5 mL of cryopreservation medium.

17. Add the cryopreservation medium very slowly and dropwise. Gently pipette up and down two times to mix. Do not break up the colonies.

18. Distribute 1 mL of cell suspension to each of the prepared cryovials, tighten caps and place cryovials into an isopropanol containing freezing container.

19. Place the freezing containers in the −80 °C freezer overnight and transfer cell vials to liquid nitrogen storage the following day (see **Note 17**).

3.2 Feeder-Free Human iPSC Culture Protocol (Matrigel/ MTeSR Medium)

3.2.1 Matrigel™ Coated Plates

1. Thaw as many as aliquots of Matrigel™ may be needed depending on volume of cell culture on ice.

2. For coating one 6-well plate, transfer 5 mL cold, sterile DMEM/F-12 medium to the 15 mL conical tube in sterile biosafety cabinet.

3. Add 1 mL of cold DMEM/F-12 medium to one of the 100 μL Matrigel™ aliquot and transfer it to 15 mL conical tube using 1 mL sterile pipette and pipette up and down to dissolve it.

4. Add 1 mL of Matrigel™ solution to each well of a 6-well plate. This will be enough for one full 6-well plate.
5. Incubate the Matrigel™ coated plate 1 h at room temperature or 37 °C incubator before use or storage (*see* **Note 18**).

3.2.2 Thawing the iPSC on Matrigel

1. Remove the frozen vial of iPSC from the liquid nitrogen storage tank.
2. Quickly remove the label or copy the information written on the tube in your notebook (*see* **Note 19**).
3. Immerse the vial and swirl the vial gently in a 37 °C water bath, without submerging the cap. Remove the vial from the water bath when no more ice crystal remains.
4. Spray the vial with a 70 % ethanol to sterilize the outside of the tube and transfer the vial in the sterile biosafety cabinet.
5. Transfer the cells gently into a sterile 15 mL conical tube using a 1 mL sterile pipette.
6. Add 4 mL of warmed MTeSR™ medium dropwise to cells in the 15 mL conical tube. To reduce osmotic shock to the cells, gently move the tube back and forth to mix the cells while adding the medium.
7. Centrifuge the cells at 149 rpm for 5 min.
8. Aspirate the supernatant with a sterile aspirating pipette.
9. Resuspend the cell pellet in 2 mL of MTeSR™ medium for every well that will receive cells (*see* **Note 20**).
10. To increase the iPS cell attachment to the Matrigel™, add 1 mM Rock Inhibitor to MTeSR™ medium only for the first day.
11. Gently pipette cells up and down in the tube a few times.
12. Aspirate the Matrigel™ from each well gently without damaging the Matrigel™ layer.
13. Transfer iPSC cells onto the Matrigel™ layer and place the iPSC plate into the incubator and gently move the plate back and forth and side to side to evenly distribute the cells.
14. The next day, remove the spent medium using a sterile aspirating pipette.
15. Gently add 2.5 mL of MTeSR™ Medium to the each well of a 6-well plate.
16. Place plate back into a 37 °C incubator.
17. Feed iPS cells daily until ready to passage or freeze.

3.2.3 Feeding iPSC Cultured on Matrigel

1. Warm enough MTeSR™ medium to feed 2–2.5 mL for each well that will be fed.
2. Aspirate the spent medium with a sterilized aspirating pipette.

3. Add 2 mL of warmed MTeSR™ medium to each well (*see* **Note 21**).

4. Return the 6-well plate to the 37 °C incubator.

5. Repeat procedure daily until ready to passage or freeze (*see* **Note 22**).

3.2.4 Passaging iPSC Cultured on Matrigel

Split iPS cells when iPSC colonies are becoming too dense or too large or increased differentiation occurs. iPS cells will grow at a different rates and the split ratio will need to be adjusted every single time you passage the cells.

1. Prepare Matrigel™ plate as described in Section 3.2.1, 1 h or a day prior to passaging the iPS cells.

2. Remove all areas of differentiation in each well before passaging the iPS cells if necessary. Transfer the plate into the hood equipped with a stereomicroscope and remove differentiated colonies with a sterile modified pipette or sterile micropipette tip.

3. Transfer the plate in the biosafety cabinet and aspirate the spent medium from the wells to remove floating picked differentiated colonies in medium.

4. Rinse it once with 1 mL warmed DMEM-F12.

5. Add 1 mL room temperature dispase solution to each well to be passaged.

6. Incubate for 3–5 min at 37 °C incubator.

7. Check the colonies under a microscope to confirm appropriate incubation time. When the colonies appear to slightly fold back, aspirate the dispase solution without disturbing the attached iPS cell colonies.

8. Gently add 1 mL of warmed DMEM/F-12 to each well and then aspirate off the medium (*see* **Note 23**).

9. Repeat rinsing each well with 1 mL of warmed DMEM/F-12 one more time.

10. Add 2 mL of MTeSR™ Medium to each well.

11. Using a sterile cell scraper, gently scrape the iPS cells colonies from the plate. Repeat in different directions if necessary.

12. Pipette the medium slowly up and down using 5 mL sterile pipette to wash the cells off the surface.

13. Transfer iPSC colonies to 15 mL sterile conical tube after all wells are scraped and the cells are removed from the surface of the well.

14. Pipette cells up and down gently a few times in the conical tube to further break-up iPS cell colonies if needed.

15. Take up 1–2 mL MTeSR™ in a 5 mL pipette and add it to the first well to wash and collect residual cells. Take up the medium and transfer it into each subsequent well to collect cells.

16. Determine how much additional medium is required. This is dependent on the split ratio and the number of wells used. There should be a total of 2 mL of MTeSR™ medium and cells in each of the new wells (*see* **Note 24**).

17. Add 1 mM Rock Inhibitor to the MTeSR™ medium during first 24 h after passaging the iPSC to increase viability and attachment of cells to the Matrigel™.

18. Add 2 mL of cell suspension to each well of the new plate.

19. Return the plate to the incubator after plating the cells. Move the plate in several quick, short, back-and-forth and side-to-side motions to further disperse cells across the surface of the wells.

20. Incubate cells overnight to allow colonies to attach (*see* **Note 25**).

3.2.5 Freezing iPSC Cultured on Matrigel™

1. Label enough cryovials with the cell line, passage number, the freeze date, and your initials with alcohol proof pen or labels that resist liquid nitrogen and ethanol.

2. Place the isopropanol freezing container and keep it in 4 °C, about 30 min before start.

3. Prepare the required amount of cryopreservation medium and keep it on ice until ready to use. Prepare 1 mL freezing medium for every vial plus a little extra for pipet error.

4. Remove differentiated colonies before passaging the iPS cells if necessary. Aspirate MTeSR™ medium from each well to remove floating picked differentiated colonies in medium.

5. Add 1 mL of room temperature dispase solution to each well of each 6-well plate.

6. Incubate cells for 3–5 min at 37 °C incubator.

7. Check the plate under the microscope every 2 min and look for the edge of the colonies to slightly be folded back.

8. Aspirate dispase solution from each well, rinse it with DMEM-F12 two times, taking care not to remove any floating colonies.

9. Add 1 mL of MTeSR™ medium to each well.

10. Using a sterile cell scraper, gently scrape the iPS cell from the plate. Repeat in different directions if necessary.

11. Pipette the medium slowly up and down to wash the cells off the surface.

12. Transfer iPSC pieces to 15 mL sterile conical tube after all wells are scraped and the cells are removed from the surface of the well.
13. Pipette cells up and down gently a few times in the conical tube to further break-up cell colonies if needed. Wash and collect residual cells from each well with 1–2 mL MTeSR™ medium and add them to 15 mL conical tube (*see* **Note 26**).
14. Centrifuge at 149 × *g* for 5 min.
15. Aspirate the supernatant and resuspend each cell pellet in enough MTeSR™ medium (stem cells from each well go to 0.5 mL MTeSR™ medium).
16. Freeze iPSC at 1 well/cryovial. Add 0.5 mL cryopreservation medium to each 0.5 mL of iPSC cells very slowly and dropwise. Pipette up and down two times to mix.
17. Distribute 1 mL of cell suspension to each of the prepared cryovials, tighten caps and place cryovials into an isopropanol containing freezing container.
18. Place the freezing containers in the −80 °C freezer overnight and transfer cell vials to liquid nitrogen storage the following day.

3.3 Feeder-Free Human iPSC Culture Protocol (Vitronectin/Essential 8™ Medium)

3.3.1 Vitronectin Coated Plates

1. Thaw as many as aliquots of vitronectin may be needed depending on volume of cell culture on ice (each 60 μL is enough for one 6-well plate to get the concentration of 5 μg/mL).
2. For coating the wells of one 6-well plate, transfer 6 mL cold, sterile PBS to the 15 mL conical tube in sterile biosafety cabinet.
3. Add 60 μL of thawed vitronectin into a 15-mL conical tube and pipette up and down to resuspend it.
4. Add 1 mL of vitronectin solution to each well of a 6-well plate. Allow to set 1 h at room temperature or 37 °C incubator before use or storage (*see* **Note 27**).

3.3.2 Passaging iPSC Cultured in Essential 8™ Medium

Split cells when iPSC colonies are becoming too dense or too large or if increased differentiation occurs. iPS cells will grow at a different rates and the split ratio will need to be adjusted every single time when you passage the cells.

1. Prepare vitronectin coated plate as described in Section 3.3.1 1 h or a day prior to passaging the iPS cells.
2. One hour before starting, warm the vitronectin-coated plate to room temperature in the biosafety cabinet.
3. Aspirate residual vitronectin solution from each well. It is not necessary to rinse off the well after removal of vitronectin. Cells can be passaged directly onto the vitronectin-coated plate.

4. Add 1 mL pre-warmed Essential 8™ Medium to each well of a coated 6-well plate and leave it in the biosafety cabinet.

5. Pre-warm the required volume of Essential 8™ Medium at room temperature until it is no longer cool to the touch (*see* **Note 28**).

6. Remove differentiated colonies before passaging the iPS cells if necessary in the hood equipped with a stereomicroscope. Remove all areas of differentiation with a sterile modified pipette or sterile micropipette tip.

7. Transfer the plate in the biosafety cabinet and aspirate the spent medium from the wells to remove floating picked differentiated colonies in medium.

8. Rinse it once with 1 mL warmed PBS without Ca^{+2} and Mg^{+2}.

9. Add 1 mL room temperature 0.5 mM EDTA solution to each well to be passaged.

10. Incubate the plate at room temperature for 5–8 min or 37 °C for 4–5 min (*see* **Note 29**).

11. Check the colonies under a microscope to confirm appropriate incubation time. When the cells start to separate and round up, and the colonies appear to have holes in them, they are ready to be removed from the plate (*see* **Note 30**).

12. Aspirate the EDTA solution gently from each well using 2 mL aspirating pipette.

13. Add 1 mL of warmed complete Essential 8™ Medium to each well. The initial effect of the EDTA will be neutralized quickly by adding the medium.

14. Wash the cells from the surface of the well by pipetting the colonies up using a 5 mL sterile pipette. Do not scrape the cells from the dish (*see* **Note 31**).

15. Collect cells in a 15 mL sterile conical tube after all the cells are removed from the surface of the well.

16. Pipette cells up and down gently a few times in the conical tube to further break-up iPS cell colonies if needed.

17. Determine how much additional medium is required. This is dependent on the split ratio and the number of wells used. There should be a total of 2 mL of Essential 8™ medium and cells in each of the new wells after the cell suspension has been added to each well (*see* **Note 32**).

18. Add 2 mL of cell suspension to each well of the new plate.

19. Return the plate to the incubator after plating the cells. Move the plate back-and-forth and side-to-side a few times to further disperse cells across the surface of the wells.

20. Incubate cells overnight to allow colonies to attach (*see* **Note 33**).
21. Feed iPSC cells beginning the second day after splitting. Then feed them daily until ready to passage or freeze.

3.3.3 Freezing iPSC Cultured in Essential 8™ Medium

1. Label cryovials with the cell line, passage number, the freeze date, and your initials. Use an alcohol proof pen or labels that resist liquid nitrogen and ethanol.
2. Place the isopropanol freezing container and keep it in 4 °C, about 30 min before start.
3. Prepare the required amount of cryopreservation medium and keep it on ice until ready to use. Prepare 1 mL cryopreservation medium for every vial plus a little extra for pipet error.
4. Pre-warm the required volume of Essential 8™ Medium at room temperature.
5. Remove differentiated colonies before passaging the iPS cells if necessary in the hood equipped with a stereomicroscope with a sterile modified pipette or sterile micropipette tip.
6. Transfer the plate in the biosafety cabinet and aspirate the spent medium from the wells to remove floating picked differentiated colonies in medium.
7. Rinse once with 1 mL warmed PBS without Ca^{+2} and Mg^{+2}.
8. Add 1 mL room temperature 0.5 mM EDTA solution to each well to be passaged.
9. Incubate the plate at room temperature for 5–8 min or 37 °C for 4–5 min.
10. Check the colonies under a microscope to confirm appropriate incubation time. When the cells start to separate and round up, and the colonies appear to have holes in them, they are ready to be removed from the plate.
11. Aspirate the EDTA solution gently from each well.
12. Add 1 mL of warmed complete Essential 8™ Medium to each well. Wash the cells from the surface of the wells by pipetting the colonies up using a 5 mL pipette. Do not scrape the cells from the dish.
13. Collect cells in a 15 mL sterile conical tube after all the cells are removed from the surface of the well.
14. Pipette cells up and down gently a few times in the conical tube to further break up iPS cell colonies if needed (*see* **Note 34**).
15. Centrifuge at $149 \times g$ for 5 min.
16. Aspirate the supernatant and resuspend each cell pellet in enough cryopreservation medium.

17. Pipette up and down two times to mix. Freeze iPSC at 1 well/cryovial (stem cells from each well go into 1 mL cryopreservation medium).
18. Distribute 1 mL of cell suspension to each of the prepared cryovials, tighten caps and place cryovials into an isopropanol containing freezing container.
19. Place the freezing containers in the −80 °C freezer overnight and transfer cell vials to liquid nitrogen storage the following day.

3.4 Differentiation of Human iPSC to Lung Progenitors

3.4.1 Differentiation of iPSCs to Definitive Endoderm

1. Aspirate the human stem cell medium or MTeSR medium from each well and add 1.5 mL of the endoderm differentiation medium directly to the each well of iPSC in a 6-well plate while iPSC colonies are still on MEF or Matrigel.
2. Change the medium the next day.
3. After 48 h, switch the medium to the same definitive endoderm medium, supplemented with $1 \times$ B27 supplement and 0.5 mM sodium butyrate.
4. Culture the cells in this medium for another 3–4 days with changing the medium daily.
5. Stain one well of differentiated cells to DE cells for CXCR4/c-Kit or SOX17 to evaluate the efficiency of induction (*see* **Note 35**).

3.4.2 Differentiation of Definitive Endoderm to Anterior Foregut Endoderm

1. Thaw as many as aliquots of human ECM gel may be needed, depending on volume of differentiation, on ice.
2. For coating one 6-well plate, transfer 9 mL cold, sterile DMEM/F-12 medium to a 15 mL conical tube in sterile biosafety cabinet.
3. Add 1 mL of cold DMEM/F-12 medium to one of the 100 μL human ECM gel aliquot and transfer it to 15 mL conical tube using 1 mL sterile pipette and pipette up and down to dissolve it.
4. Add 1 mL of human ECM gel solution to each well of a 6-well plate.
5. Incubate it in 37 °C incubator 24 h before use (*see* **Note 36**).
6. Aspirate the DE medium from each well using 2 mL sterile aspirating pipette.
7. Wash each well with warmed DMEM-F12 to remove floating dead cells in medium.
8. Trypsinize the DE cells with diluted 0.25 % trypsin (1:3) for 1 min and check cells every 30 s to confirm appropriate incubation time.
9. When the cells start to be round and half are detached from the plate, add 1 mL warmed FBS to neutralize trypsin quickly.

10. Collect the cells and spin them down at 233 × *g* for 3 min.
11. Determine how much AFE medium is required based on the split ratio (The split ratio is 1:2) and the number of wells that will be used. In general, there should be a total of 2 mL of AFE medium in each of the new wells.
12. Aspirate excess ECM from each well gently without damaging the ECM on the plates.
13. Resuspend the cell pellets in the required amount of AFE medium and transfer to human ECM coated plate. Add 2 mL of cell suspension to each well of the ECM coated plate.
14. Add Rock Inhibitor at 1 µM concentration to the AFE medium during first 24 h after transferring DE cells onto ECM coated plate to increase the viability and attachment of DE cells.
15. Return the plate to the 37 °C incubator after plating the cells. Move the plate back-and-forth and side-to-side a few times to further disperse cells across the surface of the wells.
16. Change the AFE medium after 24 h.
17. Culture the cells in the AFE medium 48 h.

3.4.3 Differentiation of DE to AFE Cells While They Are Still on MEF or Matrigel

1. Aspirate the DE medium from each well using sterile aspirating pipette.
2. Wash the cells with 1 mL warmed RPMI to remove floating dead cells in the medium.
3. Add 2 mL of the anterior foregut endoderm differentiation medium directly to the each well of iPSC in a 6-well plate, while iPSC-derived DE are still on MEF or Matrigel.
4. Change the medium the next day.
5. Culture the cells in this medium for another 24 h (*see* **Note 37**).

3.4.4 Differentiation of Anterior Foregut Endoderm to Lung Progenitors

1. Beginning on day 9 of differentiation, switch the medium to the lung progenitor differentiation medium. Aspirate AFE medium from each well gently using 2 mL sterile aspiration pipette.
2. Add 2 mL of the lung progenitor differentiation medium to each well.
3. Return the plate to the incubator.
4. Change the medium every other day until day 15 of differentiation.
5. To check the differentiation efficiency to lung progenitor cells, stain cells for the co-expression of FOXA2 and NKX2.1 at day 15.

3.5 iPSC Generation Using Lentiviral Vectors Protocol (Reprogramming Fibroblasts by an Inducible Lentiviral System)

3.5.1 Skin Biopsy and Isolating Fibroblasts

1. To isolate the fibroblasts from the skin biopsy, cut the skin biopsy into several small pieces with razor blade.
2. Put each small piece in one well of the 6-well plate and cover it with coverslip. The coverslip helps to keep skin from moving around.
3. Add 2 mL DMEM medium to each well of the 6-well plate.
4. After 3 days, the skin cut will attach and fibroblasts from skin start to sprout out. The coverslip can be taken out from each well and let the fibroblast cells grow.
5. Change the medium every 2 days.
6. When each well of a 6-well plate becomes confluent, expand the isolated fibroblasts to a 10 cm cell culture dish.
7. Aspirate the DMEM medium from the culture dish using sterile 2 mL aspirating pipette.
8. Rinse each well with 1 mL PBS without $CaCl_2$ or $MgCl_2$ using 5 mL serological pipette.
9. Add 1 mL pre-warmed 0.25 % trypsin solution to the each well of a 6-well plate.
10. Incubate the cells with trypsin for 3 min at 37 °C incubator.
11. Check the plate under the microscope every 2 min to find the appropriate incubation time.
12. When the majority of the cells look round and start detaching from the surface, add 2 mL of DMEM medium to each well to neutralize the trypsin enzyme.
13. Pipette the medium slowly up and down to wash the cells off the surface.
14. Transfer cells to 15 mL sterile conical tube after all the cells are removed from the surface of the flask.
15. Centrifuge at $149 \times g$ for 5 min.
16. Resuspend cells in 1 mL of DMEM medium.
17. Count the cells number using hemacytometer.
18. Plate 5×10^5 fibroblast cells in each 10 cm plate in 10 mL DMEM medium.
19. Return the plates to the 37 °C incubator and let the cells attach to the plate.
20. Feed the fibroblast cells every 2 days until ready to passage or freeze.

3.5.2 Freezing Human Fibroblasts

1. When each 10 cm plate becomes confluent, freeze the fibroblast at passage 2 or 3. Fibroblasts can be reprogrammed more efficiently at low passage numbers.
2. Label the cryovials with the cell line, passage number, the freeze date and your initials.

3. Aspirate the DMEM medium from the flask using sterile 2 mL aspirating pipette.

4. Rinse each well with 1 mL PBS without $CaCl_2$ or $MgCl_2$ using 5 mL serological pipette.

5. Add 2 mL pre-warmed 0.25 % trypsin solution to the flask.

6. Incubate the cells with trypsin for 3 min at 37 °C incubator.

7. Check the plate under the microscope every 2 min. When the majority of the cells look round and start detaching from the surface, add 5 mL of DMEM medium to the flask to neutralize the trypsin enzyme.

8. Pipette the medium slowly up and down to wash the cells off the surface.

9. Transfer cells to 15 mL sterile conical tube after all the cells are removed from the surface of the flask.

10. Centrifuge at 149 × g for 5 min.

11. Resuspend cells in 1 mL of DMEM medium.

12. Count the cells number using hemacytometer.

13. Resuspend fibroblast cells in enough freezing medium. Freeze fibroblast cells at 5×10^4/cryovial.

14. Tighten caps and place cryovials into isopropanol containing freezing container.

15. Place the freezing containers in the −80 °C freezer overnight and transfer cell vials to liquid nitrogen storage the following day.

3.5.3 Thawing the HEK293T

1. Turn on the blower and spray down the whole surface of the hood with ethanol and allow it to evaporate for 20 min prior to initiating cell culture.

2. Remove a frozen vial of HEK293T (2×10^6 cells) from the liquid nitrogen tank and thaw by immersing the vial in a 37 °C water bath without submerging the cap. Swirl the vial gently.

3. Remove the HEK293T vial from the water bath when only a small ice crystal remains.

4. Spray the vial with a 70 % ethanol to sterilize the outside of the tube and transfer it into the sterile biosafety cabinet.

5. Transfer the HEK293 cells gently into a sterile 15 mL conical tube using a 1 mL serological pipette.

6. Slowly, add 4 mL of warmed DMEM medium dropwise to cells in the 15 mL conical tube. Gently move the tube back and forth to mix the cells, while adding the medium. This reduces osmotic shock to the cells.

7. Centrifuge the cells at 233 × g for 5 min.
8. Aspirate and discard the supernatant with a sterile aspirating pipette.
9. Resuspend the cell pellet in 10 mL DMEM medium; gently pipette cells up and down few times and add 2.5×10^6 cells/10 mL to the T-75 cell culture flask.
10. Transfer the flask to 37 °C incubator (*see* **Note 38**).

3.5.4 Prepare HEK-293 Plates

To start generating the virus, split the HEK293T flask into a 10 cm dish when it is 70 % confluent.

1. Aspirate the DMEM medium from the flask using sterile 2 mL aspirating pipette.
2. Add 5 mL PBS without $CaCl_2$ or $MgCl_2$ to the flask using 5 mL serological pipette.
3. Aspirate the PBS using sterile 2 mL aspiration pipette.
4. Add 2 mL pre-warmed 0.25 % trypsin solution to the flask.
5. Incubate the cells with trypsin for 3 min in a 37 °C incubator.
6. Check the plate under the microscope every 2 min to find the appropriate incubation time.
7. When majority of the cells look round and start detaching from the surface, add 5 mL of DMEM medium to the flask to neutralize the trypsin enzyme.
8. Pipette the medium slowly up and down to wash the cells off the surface.
9. Transfer cells to 15 mL sterile conical tube after all the cells are removed from the surface of the flask.
10. Centrifuge at 149 × g for 5 min.
11. Resuspend cells in 1 mL of DMEM medium.
12. Count the cells number using hemacytometer.
13. Transfer the 5×10^6 HEK293T cells into 10 cm culture dish so at the day of transfection (next day) ideally you have around 6×10^6 cells/10 cm dish (around 70 % confluence) (*see* **Note 39**).
14. Return the plates to the 37 °C incubator and let the cells attach to the plate.

3.5.5 Transfection of HEK-293

1. Two hours before transfection, replace the medium with 10 mL of fresh preheated at 37 °C DMEM medium without antibiotics.

2. For each 10 cm dish, prepare the following transfection mix in a 1.5 mL eppendorf tube and incubate at room temperature for 5 min:
   - 10 μg vector plasmid (with gene of interest, e.g., Oct4, Sox2)
   - 2.5 μg envelop plasmid (pMD2G codes for the broad range VSV-G envelope)
   - 7.5 μg packaging plasmid psPAX2 (It codes for packaging protein and is suitable for most studies)
   - 1.5 mL Opti MEM without FBS.

3. To prepare the Lipofectamine 2000 transfection solution, mix 36 μL of Lipofectamine 2000 with 1.5 mL Opti MEM without FBS in a 1.5 mL eppendorf tube and incubate for 5 min at room temperature.

4. Add the plasmid mix to the Lipofectamine solution dropwise, and then mix them gently by pipetting up and down.

5. Incubate the transfection mix for 20 min at room temperature.

6. Add dropwise 3 mL/dish of the transfection mix, and mix gently by rotating the 10 cm dish.

7. Transfer the plates to 37 °C incubator.

8. Remove medium around 14–16 h post-transfection and put 10 mL/dish of fresh warmed DMEM medium.

9. Collect supernatant for the first time 48 h after transfection.

10. Supernatant can be harvested two times, every 24 h. Collect supernatant for the second time 72 h after transfection. Keep supernatant at 4 °C during the collecting period.

11. Pool the collected supernatants for each individual virus and centrifuge 5 min at $524 \times g$ to remove cell debris.

12. Filter the supernatant for each virus separately through 0.4 μm filter.

13. The cleared supernatants can be kept at 4 °C for 4–5 days. Supernatants can be used directly, stored at −80 °C in aliquots, or concentrated if needed.

3.5.6 Virus Concentration

To make more concentrated virus, each virus can be concentrated by ultracentrifuge.

1. Transfer the virus solution to the sterile ultracentrifuge.

2. Ultracentrifuge it at $47.000 \times g$ for 2 h at 16 °C in a swinging rotor (or alternatively at 19.500 rpm in a Beckman SW32 Ti rotor).

3. After centrifuge is finished, transfer the tube to the biosafety cabinet.

4. Discard supernatant and resuspend pellet of each virus in 100–200 µL of sterile cold PBS (try to make a 100 or 1000-fold concentration).

5. The concentrated virus can be used directly or aliquot and store at-80 °C for future use.

3.5.7 Virus Titration

Titer the generated virus using a p24 ELISA kit, to confirm the success of virus packaging reaction and to avoid wasting time with your experiments. To perform consistent experiments, calculate the sufficient MOI for fibroblast transduction.

1. Allow all reagents to reach room temperature (18–25 °C).

2. Select a sufficient number of 8-well strips to accommodate all standards, test samples, controls, and culture medium blanks in duplicate.

3. Label wells according to sample identity using the letter/number on the plastic frame.

4. Dispense 200 µL of each standard, sample, and blank into corresponding labeled duplicate wells and follow the manufacturer's instruction of Lenti-X™ p24 Rapid Titer.

5. After reading the absorbance values, calculate the virus particle from each plate based on P24 protein concentration. The following values and calculations can be used to determine approximate titers, and are based on the observation that each lentiviral particle (LP) contains approximately 2000 molecules of p24:

   - 1 LP contains 8×10^{-5} pg of p24 (derived from $(2000) \times (24 \times 103$ Da$)/(6 \times 1023)$.
   - ng p24 is equivalent to $\sim 1.25 \times 10^7$ LPs.
   - For a typical lentiviral vector, there is 1 IFU for every 100–1000 LPs.
   - Therefore, a supernatant titer of 10^7 IFU/mL $\approx$ 109–1010 LP/mL or 80–800 ng p24/mL.

3.5.8 Prepare Fibroblast Cells for Reprogramming

The following protocol is based on human fibroblasts and will be applicable to most other cell types. If using skin fibroblast cells, follow the plating protocol listed below. If using another cell type, a different plating density may be required. Follow the instructions provided at the beginning of this section to prepare MEF medium, hESC medium, bFGF solution, etc. before starting the reprogramming process.

1. Plate 2×10^5 fibroblast cells at passage 2 or 3 on one well of a gelatin coated 6-well plate (~70 % confluent) or 1×10^5 cells in 3.5 cm dish in DMEM medium, one day before viral infection, including one well for GFP control. Incubate overnight at 37 °C to reach up to 80 % confluency.

2. The following day, dilute enough amount of each lentivirus to make the recommended MOI below for each lentivirus to 1 mL culture medium without FBS and add polybrene at a concentration of 2 μg/mL.
    - FUW-tetOLoxP lentiviruses of hOct4: MOI = 15.
    - FUW-tetOLoxP lentiviruses of hSox4: MOI = 15.
    - FUW-tetOLoxP lentiviruses of hKlf4: MOI = 15.
    - FUW-tetOLoxP lentiviruses of hNanog: MOI = 15.
    - FUW-tetOLoxP lentiviruses of hc-Myc: MOI = 6.
    - M2rtTA: MOI = 30.
3. Aspirate fibroblast medium from 6-well plate and add mix virus (1 mL) to each well of cells to be reprogrammed in the 6-well.
4. Move the cell culture dish gently side-to-side and back-and-forth to ensure that the medium is evenly distributed.
5. Incubate for 4 h at 37 °C.
6. Add 1 mL of fibroblast medium supplemented with 20 % FBS.
7. Return the plate to a CO_2 incubator and incubate overnight.
8. The following day, aspirate virus-containing medium and add 2 mL of fresh fibroblast medium to each reprogrammed well.
9. Change medium again after 48 h.

3.5.9 Determine Infection Efficiency

48 hours post infection; the efficiency can be assessed by immunostaining for 2–3 transcription factors.

1. Plate the extra human fibroblasts into a 24 well plate.
2. Transduce the cells with the same MOI and concentration of virus that was used to reprogrammed human fibroblasts, follow the instructions described in Section 3.5.8.
3. About 48 h after infection, perform immunocytochemistry staining for 1 transcription factor per well (*see* **Note 40**).

3.5.10 Re-plate Infected Cells

24 hours to 72 hours post viral treatment, re-plate the fibroblast cells onto a MEF layer.

1. Prepare the MEF in a 10 cm dish, the day prior to splitting the infected cells. Seed 1.2×10^7 MEF cells in MEF medium in each 10 cm dish.
2. Wash the well of infected fibroblasts with 2 mL of PBS using 5 mL sterile pipette.
3. Aspirate the PBS and add 1 mL of warmed 0.05 % trypsin enzyme.
4. Incubate the plate for 2 min at 37 °C.

5. When the majority of the cells look round and start detaching from the surface, add 2 mL of medium to each well to neutralize the trypsin enzyme.

6. Pipet the medium across the surface of the well until the cells appear completely detached.

7. Transfer cells to 15 mL sterile conical tube after all the cells are removed from the surface of the wells.

8. Centrifuge the cells for 5 min at $233 \times g$.

9. Remove the supernatant.

10. Resuspend cells 1 mL of fibroblast medium and count the total number of cells in solution using a hemacytometer.

11. Add the appropriate amount of fibroblast medium to the 15 mL conical tube to bring the cell suspension to 2.5×10^5 cells/mL.

12. Mix the cell solution gently in order to create a uniform suspension of single cells.

13. Seed the 2×10^5 infected skin fibroblasts onto MEF in DMEM medium.

14. Return the plates to the 37 °C incubator and let the cells attach to the plate.

15. About 24 h after plating of infected human fibroblast on MEF, aspirate the fibroblast medium from the plate.

16. Add 10 mL of Dox induction medium supplemented with 2 % FBS and 1 μL Dox and 1 mM VPA to induce the expression of pluripotency genes.

17. Feed the cells with 10 mL of Dox induction medium every 24 h for 6 days (Day 2 to day 7 after re-plating infected cells onto MEF) (*see* **Note 41**).

18. After 6 days, switch the medium to hESC medium supplemented with only 1 μg/mL of Dox for 14 days (Day 8 to day 25 after re-plating infected cells onto MEF).

19. Wait for 2–4 weeks for iPSC colonies to form (*see* **Note 42**).

3.5.11 Picking and Expansion of iPSC Colonies

1. Once iPSC colonies form, prepare the MEF cells in a 24-well plate at a concentration of 9×10^4 per well (80 % confluent).

2. Three hours before picking the colony, aspirate the MEF medium and rinse it with pre-warmed DMEM-F12 medium.

3. Add 200 μL of stem cell culture medium to each well and transfer the plate to the 37 °C incubator.

4. Pick colonies manually and transfer one colony into the one new MEF in 24-well plate (*see* **Note 43**).

5. When MEF cells become too old (about 2 weeks) or a lot of iPSC colonies have developed in the 24-well plate, prepare new

a MEF feeder layer in 6-well plates as described before to expand the iPSC colonies.

6. Using the same procedures described in the beginning of the chapter, iPSCs can be further expanded to meet your lab needs for iPSC analysis or down-stream applications.

4 Notes

1. If FW of material is not 320.26, dilute appropriately to achieve a 10 mM solution.
2. Work quickly; if the Matrigel™ is allowed to warm at all, it will become gel and will not be appropriate for plating. Matrigel™ cannot be thawed and refrozen.
3. Warm complete medium required for that day at room temperature until it is no longer cool to the touch. Do not warm the medium at 37 °C.
4. Vials stored in liquid nitrogen may accidentally explode when warmed. Wear ultra-low temperature cryo gloves and also wear eye protection.
5. The feeder should be used within a week.
6. Wear ultra-low temperature cryo gloves and eye protection when taking the cells from nitrogen tank.
7. Number of wells receiving cells is based on the thaw recommendation found in the certificate of analysis if you bought it from the company. If you thaw the frozen vial prepared previously in laboratory follow the lab's instruction. For example: When the thaw recommendation is to thaw 1 vial into 1 well, resuspend the pellet in 3 mL of stem cell culture medium.
8. To reduce the contamination potential, do not reinsert a used pipette into sterile medium for any reason. If feeding more than one plate, use a different pipette for each plate to reduce risk of contamination.
9. Observe the pluripotent stem cells using a microscope. If they require passaging, follow the passaging protocol below.
10. At least one well of cells should be left and used as a backup to protect against problems with the split.
11. Make sure that the cells remain adhered to the plate. To avoid the iPSC colonies peeling off, do not dispense the medium in a continuous stream.
12. iPS cells will grow at a different rate, and the split ratio will need to be adjusted every single time the iPSC cells are passaged. The split ratio is variable and generally is between 1:2 and 1:4. Always, as a general rule, observe the iPSC colonies

from the last split ratio and adjust the ratio according to the appearance of the iPSC colonies. If the cells look healthy and colonies have enough space, split the iPSC using the same ratio, if they are dense and crowding, increase the ratio, and if the cells are sparse, decrease the ratio. iPS cells will need to be split every 4–7 days based on the morphology of the colonies.

13. After splitting the iPSC, while cells are attaching, open and close the incubator carefully. This will prevent disturbing the even distribution of cells to the surface of the well.

14. iPSC can be passaged with the same method using dispase enzyme solution (1 mg/mL). To use dispase at step 7, add 1 mL room temperature dispase instead of adding 1 mL room temperature collagenase solution to each well to be passaged. Incubate for 3–5 min at 37 °C and continue with the following steps after step 7.

15. The isopropanol must be replaced every five uses.

16. Try not to break iPSC colonies up into small clumps. If iPSCs are frozen in large aggregates, they will recover from the thaw more efficiently.

17. Once cells are in contact with DMSO, they should be aliquoted quickly and initiate freezing within 2–3 min.

18. Wrap the extra plates in Parafilm and store in refrigerator at 2–8 °C and use the plates within 7 days after preparation. If any portion of the well dries out, do not use the well.

19. Vials stored in liquid nitrogen may accidentally burst when warmed due to influx of liquid nitrogen into the vial (rare). Wear ultra-low temperature cryo gloves and eye protection.

20. number of wells receiving cells is based on the thaw recommendation found in the certificate of analysis if you bought the iPSC line from company. If you thaw the frozen vial of iPSC prepared previously in the laboratory, follow the lab instructions. For example: When the thaw recommendation is to thaw 1 vial into 1 well, resuspend the pellet in 2.5 mL of MTeSR™ medium.

21. To reduce the contamination potential, do not reinsert a used pipette into sterile medium. If feeding more than one plate, use a different pipette for each plate.

22. Observe the pluripotent stem cells using a microscope. Follow the passaging protocol below when they require passaging.

23. Make sure that the cells remain adhered to the plate. Do not dispense the medium in a continuous stream in one spot to avoid of detaching iPSC from the wells.

24. The split ratio is variable, and is generally between 1:2 and 1:4. Always as a general rule, observe the last split ratio and adjust

the ratio according to the appearance of the iPSC colonies in the wells. If the colonies on Matrigel™ have enough space, split using the same ratio, if they are dense and crowding, increase the ratio, and if the cells are sparse, decrease the ratio. Cells will need to be split every 5–7 days based on the iPSC cell colonies' appearance.

25. To prevent disturbing the even distribution of cells to the surface of the well, try to limit opening and closing the incubator for few first hours after passaging the iPSC, while cells are attaching. Feed the iPSC daily until ready to passage or freeze.

26. Try not to break the iPSC to small clumps. Cells will recover from the thaw more efficiently if frozen in large aggregates.

27. Wrap the extra plates in Parafilm and store in refrigerator at 2–8 °C. The plates should be used within a week after preparation. Do not allow the wells to dry. If any portion of the well dries out, do not use the well. Prior to use, pre-warm the plate to room temperature for at least 1 h.

28. Warm complete medium required for that day at room temperature until it is no longer cool to the touch. Do not warm the medium at 37 °C.

29. With certain cell lines, this may take longer than 5 min.

30. At least one well of cells should be left and used as a backup to protect against problems with the split.

31. Work quickly to remove cells after adding Essential 8™ Medium to the well. Do not passage more than one to three wells at a time.

32. The split ratio is variable, though generally between 1:2 and 1:4. A general rule is to observe the last split ratio and adjust the ratio according to the appearance of the iPSC colonies. If the cells look healthy and colonies have enough space, split using the same ratio, if they are dense and crowding, increase the ratio, and if the cells are sparse, decrease the ratio.

33. While cells are attaching, try to limit opening and closing the incubator doors, and if you need to access the incubator, open and close the doors carefully. This will prevent disturbing the even distribution of cells to the surface of the well.

34. Try not to break clumps into little collections of cells. Cells will recover from the thaw more efficiently if frozen in large aggregates.

35. Proper maintenance of human iPSC in culture is critical for efficiency of endoderm induction. If the efficiency gradually decreases over several differentiations for a specific cell line, check the maintenance methods and reagents.

36. Prepare a 6-well plate with human ECM protein 24–48 h before transferring DE cells onto the ECM coated plate. Wrap the extra plates in Parafilm and store in a refrigerator at 2–8 °C and use the plates within 7–10 days after preparation. If any portion of the well dries out, do not use the well.

37. To assess the appropriate anterior foregut endoderm induction, stain a couple of wells without switching to lung progenitor differentiation medium for SOX2 and FOXA2.

38. It is important to use low passage HEK293T cells for the production of viruses. To make sure the cells are always in the fastest growth phase, never let the cells grow to 100 % confluence.

39. Prepare eight plates of HEK293T, each plate for making one type of virus.

40. For each transcription factor to be tested, incubate a Dox-induced (1 µL Dox for 48 h) and non-induced well with both the primary and secondary antibodies. The remaining Dox-induced well should be used as a negative control by incubating with the secondary antibody only.

41. Add 10 mL of human iPSC culture medium to one plate, which will serve as a negative control.

42. to ensure completion of the reprogramming process, it is necessary to remove Dox from the induction medium once colony with good morphology are observed. We recommend removing Dox from the medium at day 25. Removal of Dox ensures that the iPS cell colonies picked and passaged around day 30 are reliant on endogenous expression of pluripotency genes and are not the result of sustained induction of ectopic transcription factor expression by Dox.

43. Some transduced fibroblasts in a 10 cm dish may reprogram later. To get the maximum reprogramming efficiency, trypsinize the rest of the cells into a 10 cm plate. Transfer the cells from each of the 10 cm plates into a new MEF cell plate and change medium every 48 h with hESC culture medium. Wait for another 1–2 weeks for iPSC colonies to develop.

References

1. Aasen T, Raya A, Barrero MJ, Garreta E, Consiglio A, Gonzalez F, Vassena R, Bilić J, Pekarik V, Tiscornia G et al (2008) Efficient and rapid generation of induced pluripotent stem cells from human keratinocytes. Nat Biotechnol 26 (11):1276–1284
2. BD Biosciences. (2013) Assay methods protocol: human embryonic stem cell culture. https://www.daigger.com:1-7
3. Ghaedi M, Calle EA, Mendez JJ, Gard AL, Balestrini J, Booth A, Bove PF, Gui L, White ES, Niklason LE (2013) Human iPS cell-derived alveolar epithelium repopulates lung extracellular matrix. J Clin Invest 123 (11):4950–4962. doi: 10.1172/JCI68793
4. Huang SX, Islam MN, O'Neill J, Hu Z, Yang YG, Chen YW, Mumau M, Green MD, Vunjak-Novakovic G, Bhattacharya J et al (2014)

Efficient generation of lung and airway epithelial cells from human pluripotent stem cells. Nat Biotechnol 32(1):84–91. doi: 10.1038/nbt.2754

5. Huangfu D, Maehr R, Guo W, Eijkelenboom A, Snitow M, Chen AE, Melton DA (2008) Induction of pluripotent stem cells by defined factors is greatly improved by small-molecule compounds. Nat Biotechnol 26(7):795–797

6. Kim D, Kim CH, Moon JI, Chung YG, Chang MY, Han BS, Ko S, Yang E, Cha KY, Lanza R et al (2009) Generation of human induced pluripotent stem cells by direct delivery of reprogramming proteins. Cell Stem Cell 4(6):472–476

7. Longmire TA, Ikonomou L, Hawkins F, Christodoulou C, Cao Y, Jean JC, Kwok LW, Mou H, Rajagopal J, Shen SS et al (2012) Efficient derivation of purified lung and thyroid progenitors from embryonic stem cells. Cell Stem Cell 10(4):398–411

8. Ludwig TE, Levenstein ME, Jones JM, Berggren WT, Mitchen ER, Frane JL, Crandall LJ, Daigh CA, Conard KR, Piekarczyk MS et al (2006) Derivation of human embryonic stem cells in defined conditions. Nat Biotechnol 24(2):185–187

9. Stein GS, Borowski M, Luong MX, Shi M-J, Smith KP, Vazquez P (2010) Human embryonic stem cell culture on BD Matrigel™ with mTeSR®1 medium. Human Stem Cell Technol Biol. doi:10.1002/9780470889909.ch11

10. Stem Cell Technologies. (2012) Technical manual: maintenance of human pluripotent stem cells in mTeSR™1 and TeSR™2. http://www.stemcell.com:1-24

11. Takahashi K, Tanabe K, Ohnuki M, Narita M, Ichisaka T, Tomoda K, Yamanaka S (2007) Induction of pluripotent stem cells from adult human fibroblasts by defined factors. Cell 131(5):861–872

12. Van Haute L, De Block G, Liebaers I et al. (2009) Generation of lung epithelial-like tissue from human embryonic stem cells. Respir Res 10:105

13. Wang D, Haviland DL, Burns AR, Zsigmond E, Wetsel RA (2007) A pure population of lung alveolar epithelial type II cells derived from human embryonic stem cells. Proc Natl Acad Sci U S A 104(11):4449–4454

14. Woltjen K, Michael IP, Mohseni P, Desai R, Mileikovsky M, Hämäläinen R, Cowling R, Wang W, Liu P, Gertsenstein M et al (2009) piggyBac transposition reprograms fibroblasts to induced pluripotent stem cells. Nature 458(7239):766–770

15. Wong AP, Bear CE, Chin S, Pasceri P, Thompson TO, Huan LJ, Ratjen F, Ellis J, Rossant J (2012) Directed differentiation of human pluripotent stem cells into mature airway epithelia expressing functional CFTRTR protein. Nat Biotechnol 30(9):876–882

16. Yu J, Hu K, Smuga-Otto K, Tian S, Stewart R, Slukvin II, Thomson JA (2009) Human induced pluripotent stem cells free of vector and transgene sequences. Science 324(5928):797–801

Organoid Culture of Lingual Epithelial Cells in a Three-Dimensional Matrix

Hiroko Hisha and Hiroo Ueno

Abstract

A novel lingual epithelial organoid culture system using a three-dimensional (3D) matrix and growth factors has recently been established. In the culture system, organoids with multilayered squamous keratinized epithelium and typical morphological features of filiform papillae are generated from single lingual epithelial cells at a high efficiency. The culture system is created in order to observe the differentiation and maturation process of each lingual epithelial stem cell and to observe abnormal organoid formation from malignant cells obtained from carcinogen-treated mice. Thus, our culture system will contribute to the advancement of research into the regulatory mechanism of lingual epithelium and the underlying mechanisms of carcinogenesis.

Keywords: Tongue, Epithelial cells, Organoid, 3D culture, Matrigel

1 Introduction

Tongue epithelium is one of epithelial tissues showing rapid turnover rates; lingual epithelial cells (LECs) are continuously replaced every 6–7 days [1]. Recent studies using powerful in vivo lineage tracing methods have revealed that lingual epithelial stem cells (LESCs) are located in the suprabasal or basal layer of lingual epithelium. They showed that LESCs expressing NTPDase2 [2], Tcf3 [3], or Bmi1 (B cell-specific Moloney murine leukemia virus integration site 1) [4] have the ability to maintain the lingual epithelium for a long time: 150, 180, and 336 days, respectively.

Fetal tongues have been cultured since the 1990s [5, 6]. The fetal organ culture system closely mimics physiological conditions; however, it cannot be applied to adult tongues. Recently, two methods for the culture of adult LECs have been discovered: the culture of adult LECs on an extracellular matrix [7] and on a collagen gel containing feeder cells (organotypic raft culture) [8]. The former can generate an epithelial cell monolayer but not a multilayered epithelium. The latter can generate keratinized multilayered epithelium, but prior preparation of the feeder layer is required and experimental results are influenced by the quality of the feeder layers. In both systems, the separation of specific LECs is required and the seeding efficiency is very low (0.78 %) [8]. In light of this, a novel, simple, and easy culture technique that can generate a stratified epithelial cell layer at a higher efficiency, even from unseparated LECs, was required.

We have recently established a new lingual organoid culture system in which the 3D growth of single unseparated LEC is induced at a seeding efficiency of 6.9 % in Matrigel feed in a culture medium containing growth factors—epidermal growth factor (EGF), noggin, and R-spondin1 (Fig. 1a, b) [9]. Three different types of organoids are generated: round-shaped organoids with concentric cell arrangements, and rugged- and round-shaped organoids with a reticulated cell arrangement, at a generation ratio of 5:4:1 (Fig. 1c, d). The round-shaped organoids with concentric cell arrangements have multilayered keratinized epithelium and a stratum corneum (Fig. 1d). This morphology is characteristic of

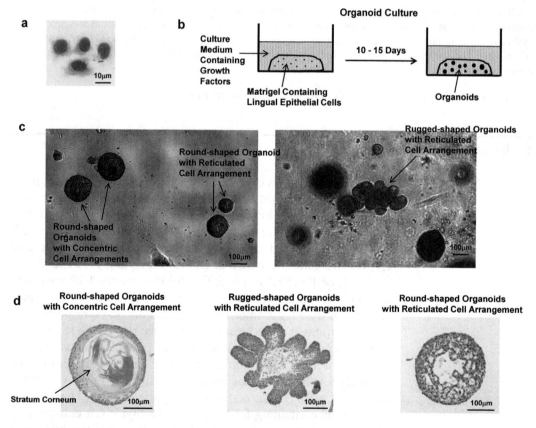

Fig. 1 (a) Representative morphological features of separated lingual epithelial cells (LECs) (Fraction 1). Cells were morphologically similar to epithelial cells; they were larger with round or elliptical-shaped nuclei and a wide cytoplasm. Hematoxylin-eosin (HE) staining. (b) Image of organoid culture. (c) Overview of the organoids generated on day 9 of culture. Three different types of organoids were observed. Phase-contrast image. (d) Morphological features of the three different types of organoids. Organoids were separated from Matrigel using dispase on day 14 of culture and their paraffin sections were stained with HE reagents

filiform papillae, which are representative of papillae in mouse tongues. Therefore, it is possible to investigate the differentiation and maturation ability of each LESC with this system. We have shown that single Bmi1-positive cells, which are labeled with red, orange, or blue fluorescent color after tamoxifen has been injected into *Bmi1-CreER;Rosa26*-rainbow (Bmi1-rainbow) mice to induce Cre-mediated recombination, can generate organoids with a multilayered keratinized epithelium (Fig. 2a, b) [9]. Moreover, it has been shown that immature 3-day cultured organoids harvested from the culture could be grafted in recipient mouse tongues and maturate in the tongue [9], indicating its possible application in regenerative medicine. We also observed that organoids generated from the LECs of carcinogen-treated mice showed an abnormal morphology (Fig. 2c) [9] and therefore, our system could also contribute to the study of lingual carcinogenesis.

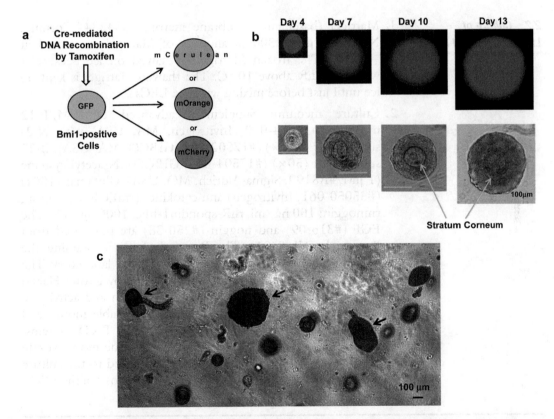

Fig. 2 (a) *Cre*-mediated fluorescent color change in Bmi1-rainbow mice. In Bmi1-rainbow mice, fluorescent colors of Bmi1-positive cells changed from green to a different color (*red, orange, or blue*) due to *Cre*-mediated excision of floxed cassettes induced by tamoxifen. Their descendant cells retain the changed color. (b) Representative time course of organoid growth in *blue-colored, round-shaped* organoids with a concentric cell arrangement. LECs from Bmi1-rainbow mice injected with tamoxifen 2 days before culturing using the organoid culture system. Fluorescent image and phase-contrast image. (c) Overview of organoids generated on day 15 of culture. LECs from carcinogen-treated mice were cultured using the organoid culture system. Very large or slender organoids (indicated by *arrows*) were detected. Phase-contrast image

2 Materials

2.1 Separation of Lingual Epithelial Cells

1. Cold PBS: ice-cold Ca-Mg free PBS containing 500 μm dithiothreitol (DTT).
2. Chelating buffer: 27 mM trisodium citrate, 5 mM Na_2HPO_4, 94 mM NaCl, 8 mM KH_2PO_4, 1.5 mM KCl, 500 μm DTT, 55 mM D-sorbitol and 44 mM sucrose, pH 7.3 [10]. About 800 ml of distilled water is added into a graduated cylinder. In total, 7.94 g trisodium citrate, 1.79 g Na_2HPO_4, 5.61 g NaCl, 1.09 g KH_2PO_4, 112 mg KCl, 10.0 g D-sorbitol, and 15.1 g sucrose are weighed and transferred into the cylinder. A 1 L solution is prepared with distilled water, filtered through a 0.22 μm filter to sterilize it, and stored at 4 °C. Before use, 1 ml of sterilized 500 mM DTT stock solution (stored at −20 °C) is added to the 1 L solution.

2.2 Culture of Lingual Epithelial Cells

1. Matrigel (basement membrane matrix, #354234, Corning, NY): 200 μl or 300 μl aliquots of Matrigel are stored at −20 °C. The frozen Matrigel is thawed on ice because it would solidify above 10 °C. The thawed Matrigel is kept on ice until just before mixing with the LECs.
2. Culture medium: Supplement advanced D-MEM/F-12 medium (#12634-010, Invitrogen, MA, USA) with N-2-supplement (100×) (#17502-048, GIBCO, MA, USA), B-27 supplement (50×) (#17504-044, GIBCO), N-acetyl cysteine (1 μM, #A8199, Sigma-Aldrich, MO, USA), Glutamax (100x) (#35050-061, Invitrogen) and cytokines (rmEGF: 50 ng/ml, rmnoggin: 100 ng/ml, rhR-spondin1-hFc: 1000 ng/ml). The EGF (#315-09) and noggin (#250-38) are purchased from Peprotech (NJ, USA). The R-spondin1-hFc containing the C-terminal of human IgG is produced in our laboratory. The cDNA of rhR-spondin1 was kindly donated by Kyowa Hakko Kirin (Tokyo, Japan). Similar rates of growth and activity of LECs are observed when commercially available mouse and human R-spondin1 (#347-RS, #4645-RS, R&D systems, MN, USA) are used. Y-27632 dihydrochloride monohydrate (10 μM, #Y0503, Sigma-Aldrich) is also added to the culture medium for the first 2–3 days to prevent apoptosis of the LECs.

3 Methods

3.1 Separation of Lingual Epithelial Cells

1. The mouse tongue is obtained, washed in cold PBS, and cut into about 2-mm size fragments in cold PBS. The tongue fragments are transferred into a 35-mm dish containing 2 ml of 50 units/ml dispase (pH 7.4, #354235, BD Biosciences, CA, USA) and are incubated in a 5 % CO_2 incubator for

60 min at 37 °C. Thereafter, the fragments are transferred to a 35-mm dish containing about 2 ml of cold PBS. This process is repeated again to remove the dispase completely.

2. The washed fragments are incubated in 10 ml chelating buffer in a 10-ml or 20-ml beaker at 4 °C, stirring constantly for 10 min. The cells released from the fragments into the chelating buffer are collected in a 50-ml centrifuge tube by passing the mixture through a cell strainer (40-μm mesh size, #REF352340, Corning, NC, USA). Centrifugation at 400 × g for 5 min at 4 °C is used to remove the supernatant. The collected cells are described as Fraction 1 (Fig. 1a).

3. The tissue fragments are transferred to 20 ml of fresh cold chelating buffer in a 50-ml centrifuge tube and shaken vigorously by hand (20 inversions). Cells released from the lingual fragments into the chelating buffer are collected as Fraction 2 (*see* **Note 1**).

3.2 Preparation of Matrigel Containing LECs and Their Culture

1. For 3 wells of 24-well plate, thawed Matrigel (200 μl) is added to a pellet with $2–4 \times 10^4$ LECs and suspended using a 200 μl tip, while taking care not to make bubbles.

2. Matrigel (50 μl) containing LECs are plated into a pre-warmed 24-well plate ($0.5–1 \times 10^4$ cells/50 μl of Matrigel/well). The cell-containing Matrigel solidifies within 3–4 min at room temperature.

3. The solidified Matrigel is covered with 0.75 ml culture medium. Distilled water is added to the surrounding wells to prevent drying. The 24-well plate is incubated at 37 °C in a 5 % CO_2 incubator.

4. All culture medium in the wells is removed every 2 or 4 days, and 0.5 ml of fresh culture medium is added to the wells (*see* **Notes 2–5**).

3.3 Separation of Organoids from Matrigel

1. All culture medium in the wells is removed and 0.5–1 ml of pre-warmed dispase (50 units/ml) is added to the Matrigel containing organoids in the wells. Incubation is conducted for about 10 min to depolymerize the Matrigel.

2. Organoids from the depolymerized Matrigel are released by vigorous pipetting using a 1000 μl tip (5–6 times). The organoids and gel fragments are transferred to a 15-ml tube and centrifuged at 110 × g for 1 min at 4 °C, after which the supernatant and gels are removed.

3. For histological analysis, the organoids are fixed with 4 % paraformaldehyde and frozen. Once frozen, these are cut into sections and stained with various antibodies. Paraffin-embedded sections could also be prepared after fixation.

4 Notes

1. Further shaking of the lingual fragments is not necessary because only a very small number of LECs are obtained by further shaking. The cell number for Fraction 1 is about 1.7 times higher than that of Fraction 2 (not significant), and $(2.8 \pm 1.0) \times 10^4$ cells could be obtained from one tongue [9]. There are no obvious morphological differences between cell populations in Fractions 1 and 2 [9].

2. Frequency of changing the medium depends on the number and size of the organoids per gel. During the early stages of culture, when the organoids are small, it is sufficient to change the medium every 3 or 4 days. Thereafter, more changes are needed. Organoid growth gradually slows down after 11 days of culture. Active proliferation of stromal cells with fibroblast-like morphology is frequently observed within the Matrigel and on the well surface [9].

3. Typically, the formation of the stratum corneum is not evident in organoids until day 3–4 of culture, but thereafter, the formation of a concentrically structured stratum corneum is observed (Fig. 2b).

4. We recently showed that Bmi1-positive cells could generate lingual epithelium of filiform papillae [4] by using tamoxifen-induced lineage tracing methods using *Bmi1-CreER;Rosa26-*rainbow (Bmi1-rainbow) mice [11, 12]. When tamoxifen is injected into Bmi1-rainbow mice, a fluorescent color change (from green to red, orange, or blue) is induced in Bmi1-positive cells only (Fig. 2a). LECs collected from the mice 2 days later are cultured in the organoid culture system. Figure 2b shows the time course for the formation of a representative blue-colored organoid, indicating that Bmi1-positive cells can generate organoids in the culture system [9]. This result has confirmed our pervious study showing that Bmi1 is a marker of LESCs [4].

5. When LECs obtained from 4-nitroquinoline 1-oxide (4-NQO) [13] treated mice, which had carcinomas in situ in some parts of the lingual epithelium, were cultured, abnormal shaped organoids (very large or slender organoids) were observed on day 15 of culture [9] (Fig. 2c).

Acknowledgement

This work was financially supported by the Funding Program for Next Generation World-Leading Researchers, The Mochida Memorial Foundation, The Naito Memorial Foundation, The

Cell Science Research Foundation, The Uehara Memorial Foundation, The Mitsubishi Foundation, and The Yasuda Memorial Foundation to H.U., and a grant from a Grant-in-aid for Scientific Research (C) 23590953 to H.H.

References

1. Hume WJ (1986) Kinetics of cell replacement in the stratum granulosum of mouse tongue epithelium. Cell Tissue Kinet 19:195–203
2. Li F, Cao J, Zhou M (2012) NTPDase 2$^+$ cells generate lingual epithelia and papillae. Front Genet 3:255. doi:10.3389/fgene.2012.00255
3. Howard JM, Nuguid JM, Ngole D, Nguyen H (2014) Tcf3 expression marks both stem and progenitor cells in multiple epithelia. Development 141:3143–3152
4. Tanaka T, Komai Y, Tokuyama Y, Yanai H, Ohe S, Okazaki K, Ueno H (2013) Identification of stem cells that maintain and regenerate lingual keratinized epithelial cells. Nat Cell Biol 15:511–518
5. Mbiene JP, Maccallum DK, Mistretta CM (1997) Organ cultures of embryonic rat tongue support tongue and gustatory papilla morphogenesis in vitro without intact sensory ganglia. J Comp Neurol 377:324–340
6. Zhou Y, Liu HX, Mistretta CM (2006) Bone morphogenic proteins and noggin: inhibiting and inducing fungiform taste papilla development. Dev Biol 297:198–213
7. Ookura T, Kawamoto K, Tsuzaki H, Mikami Y, Ito Y, Oh SH, Hino A (2002) Fibroblast and epidermal growth factors modulate proliferation and neural cell adhesion molecule expression in epithelial cells derived from the adult mouse tongue. In Vitro Cell Dev Biol Anim 38:365–372
8. Luo X, Okubo T, Randell S, Hogan BLM (2009) Culture of endodermal stem/progenitor cells of the mouse tongue. In Vitro Cell Dev Biol Anim 45:44–54
9. Hisha H, Tanaka T, Kanno S, Tokuyama Y, Komai Y, Ohe S, Yanai H, Omachi T, Ueno H (2013) Establishment of a novel lingual organoid culture system: generation of organoids having mature keratinized epithelium from adult epithelial stem cells. Sci Rep 3:3224. doi:10.1038/srep03224
10. Flint N, Cove FL, Evans GS (1991) A low-temperature method for the isolation of small-intestinal epithelium along the crypt-villus axis. Biochem J 280:331–334
11. Red-Horse K, Ueno H, Weissman IL, Krasnow M (2010) Coronary arteries form by developmental reprogramming of venous cells. Nature 464:549–553
12. Rinkevich Y, Lindau P, Ueno H, Longaker MT, Weissman IL (2011) Germ-layer and lineage-restricted stem/progenitors regenerate the mouse digit tip. Nature 476:409–413
13. Tang X-H, Knudsen B, Bemis D, Tickoo S, Gudas LJ (2004) Oral cavity and esophageal carcinogenesis modeled in carcinogen-treated mice. Clin Cancer Res 10:301–313

Generation of Functional Kidney Organoids In Vivo Starting from a Single-Cell Suspension

Valentina Benedetti*, Valerio Brizi*, and Christodoulos Xinaris

Abstract

Novel methods in developmental biology and stem cell research have made it possible to generate complex kidney tissues in vitro that resemble whole organs and are termed organoids. In this chapter we describe a technique using suspensions of fully dissociated mouse kidney cells to yield organoids that can become vascularized in vivo and mature and display physiological functions. This system can be used to produce fine-grained human–mouse chimeric organoids in which the renal differentiation potential of human cells can be assessed. It can also be an excellent method for growing chimeric organoids in vivo using human stem cells, which can differentiate into specialized kidney cells and exert nephron-specific functions. We provide detailed methods, a brief discussion of critical points, and describe some successfully implemented examples of the system.

Keywords: Kidney organoids, Stem cells, Implantation, Kidney engineering, Glomerulogenesis, Dissociation-reaggregation assay, VEGF, Cell suspensions, Kidney development

1 Introduction

Kidney tissue generated from single cells is a powerful tool for investigating human kidney development, modeling disease, developing new drugs as well as for evaluating novel regenerative medicine strategies.

One classic, pioneering study, reported that after dissociation into a single cell suspension and then reaggregation, mouse metanephric mesenchyme (MM) can be induced by spinal cord cells to generate three-dimensional (3D) renal tissue in vitro that contains rudimental nephron-like structures [1]. Based on this study, a new method has been established wherein whole embryonic kidneys are dissociated and then reaggregated and cultured in vitro in the presence of a Rho-associated protein kinase (ROCK) inhibitor to prevent apoptosis. This protocol leads to the formation of immature nephrons and multiple collecting ducts without using any exogenous tissue [2]. However, in vitro culture systems neither

---

* Author contributed equally with all other contributors.

support the long-term survival of renal tissues nor, more importantly, the development of vascularized glomeruli, both indispensable for achieving filtering function.

To overcome this limitation, we devised a new system that starts with mouse embryonic kidney cell suspensions to generate, in vitro, mouse renal tissues—called renal organoids—that can grow and mature in vivo after implantation under the kidney capsule of rat hosts [3]. The resulting tissue displays a high degree of maturation, including vascularized glomeruli containing fully differentiated podocytes [3, 4]. Furthermore, the organoid exerts kidney-specific functions, such as permselective blood filtration, tubular reabsorption of filtered macromolecules, and the production of erythropoietin. The following two crucial steps are key to our technology's success: the construction of large cell aggregate (LCA)-derived organoids that enable organoid survival and growth in vivo; and treating the organoids and rat hosts with vascular endothelial growth factor (VEGF), a molecule essential for both glomerulogenesis and nephrogenesis [5]. These technical maneuvers allow implanted organoids to integrate into the host tissue, and grow and develop functional nephrons with filtering glomeruli [3, 4].

The ability of mouse embryonic kidney cells to self-organize and generate kidney tissue after dissociation and reaggregation has been exploited to incorporate and evaluate the nephrogenic potential of different human cell types in vitro, such as amniotic fluid stem cells (AFSCs) [6], adult proximal tubule cells reprogrammed into nephron progenitors [7], bone marrow stromal cells reprogrammed into proximal tubule-like cells [8] and human pluripotent stem cell-derived renal progenitor cells [9–12]. Very recently, by mixing mouse embryonic kidney cells with AFSCs modified to temporarily express glial cell-derived neurotrophic factor (GDNF)—a molecule secreted by MM—we generated functional human–mouse chimeric organoids in vivo. The human cells preferentially localized into vascularized glomeruli in which they differentiated into highly specialized and functional podocytes [4].

Our technology can be used as a platform for testing the nephrogenic potential of human stem or renal progenitor cells, and for growing chimeric organoids in vivo where human stem cells can differentiate into specialized and functional kidney cells. Here, we provide detailed instructions of our methods for (1) the construction and culture of mouse and human–mouse chimeric organoids in vitro, (2) immunofluorescence analysis of organoids in vitro, (3) organoid implantation, and (4) histological analysis of organoids in vivo.

2 Materials

2.1 Construction and Culture of Mouse and Chimeric Organoids In Vitro

1. Glass Pasteur pipettes.
2. P1000 and P200 micropipettes.
3. 3.5 cm and 6 cm sterile petri dishes.

4. 1.5 ml Eppendorf tubes and 3–5 ml plastic tubes.
5. Embryonic kidney isolating medium: Eagle's Minimum Essential Medium (MEM; Sigma-Aldrich, St. Louis, MO, USA).
6. Culture medium: Advanced DMEM (Gibco, Invitrogen Corporation, Grand Island, NY) supplemented with 2 % Embryonic Stem cells Fetal Bovine Serum (ES-FBS, Gibco), 1 % L-glutamine (Invitrogen Corporation, Carlsbad, CA), and 1 % penicillin/streptomycin (Invitrogen).
7. Trypsin–EDTA solution 0.1 % in phosphate-buffered saline (PBS) 1× without Ca^{2+} (Biochrom AG, Berlin, Germany).
8. Rho-associated protein kinase (ROCK) inhibitor. We use glycyl-H1152 dihydrochloride (Tocris) at a final concentration of 1.25 µM in culture medium.
9. Tracker to detect human cells. We use green fluorescent chloromethyl derivative of fluorescein diacetate probe CellTracker (Molecular Probes Inc., Eugene, OR, USA).
10. 40 µm cell strainer (BD Falcon, Oxford, UK).
11. 5 µm pore polycarbonate filter (Merck Millipore Ltd., Ireland).
12. Stainless steel culture grids. These are used as a support of the culture filter at the air–medium interface. A description of how to produce grids was reported previously [13]. Briefly, pieces of stainless steel mesh are cut in the shape of small squares with sides 1.5–2 cm long, and the corners are bent down to function as grid "legs" of approximately 2–3 mm.
13. Trypan blue (Sigma-Aldrich).
14. Automatic cell counter or Burker counting chamber.
15. Microcentrifuge.
16. We use the ZOOM 2000 Model Z45 E and M205 FA (Leica) stereomicroscopes.

2.2 Immunofluorescence Analysis of Organoids In Vitro

1. Paraformaldehyde 8 % aqueous solution (Electron Microscopy Sciences, Hatfield, PA, USA) made up to 4 % in PBS 2×.
2. Methanol stored at −20 °C.
3. Phosphate-buffered saline (PBS).
4. Primary antibodies: chicken anti-laminin (Sigma-Aldrich), mouse anti-calbindin D28k (Abcam, Cambridge, MA, USA), rabbit anti-paired box 2 (Pax-2) (Zymed Laboratories, San Francisco, CA, USA), goat anti-megalin (Santa Cruz Biotechnology, Santa Cruz, CA, USA), 5B8 anti-NCAM (1:2; developed by Jessel T.M., Dodd J. and Brenner-Morton S. from the Developmental Studies Hybridoma Bank, University of Iowa, Iowa City, IA), mouse anti-synaptopodin (Progen Biotechnik GmbH, Heidelberg, Germany), mouse anti-E-cadherin (BD Biosciences, Franklin Lakes, NJ, USA), FITC-conjugated human nuclear antigen (HNA) (Merck Millipore Ltd.).

5. Secondary antibodies (Jackson ImmunoResearch Laboratories, West Grove, PA).

6. FITC-conjugated *Bandeiraea simplicifolia* Isolectin B4 (BSLB4) (Vector Laboratories, Burlingame, CA).

7. Dako Fluorescence Mounting Medium (DAKO Corporation, Denmark).

8. Optional: green fluorescent chloromethyl derivative of fluorescein diacetate (CMFDA) probe CellTracker (Molecular Probes Inc., Eugene, OR).

9. We use the inverted confocal laser scanning microscope LS 510 Meta Zeiss (Carl Zeiss, Jena, Germany) to reconstruct three-dimensional (3D) images of organoid tissues. We also use the Apotome fluorescence microscope Axio Vision Imager 2Z Zeiss.

2.3 Implantation of Mouse and Chimeric Organoids

Animal studies must be approved by the Institutional Animal Care and Use Committees and conducted according to the guidelines, in compliance with national and international law and policies. Given the complexity and invasiveness of the surgery required for these experiments (uninephrectomy and implantation under the kidney capsule) they should be performed on anesthetized animals, only by experienced personnel.

Below we provide a list of the materials necessary for treating and handling the organoids generated for in vivo implantation.

1. P200 micropipettes and glass Pasteur pipettes.

2. 96-well plate.

3. Recombinant rat vascular endothelial growth factor (VEGF) protein (Gibco, Invitrogen) reconstituted according to manufacturer's instructions to 0.1 mg/ml.

4. Culture medium.

5. Catheter. This homemade device is composed of a sterile polypropylene 1-ml syringe with a 5–6 cm long rubber cannula instead of the needle and a 3–4 cm long piece of the tapered end of a glass Pasteur pipette.

2.4 Renal Histology of Organoids In Vivo

1. Periodate-lysine paraformaldehyde (PLP).

2. Optimal Cutting Temperature compound (OCT).

3. Cryostat.

4. Hematoxylin and eosin (Bio-Optica, Milan, Italy).

5. Dako Faramount Aqueous Mounting Medium (DAKO Corporation).

6. Light microscope. We use Olympus BH2-RFCA (Olympus America Inc., Melville, NY, USA).

2.5 Animals	1. CD1 mice (Charles River Italia SpA, Calco, Italy).
	2. Male 6–8-week-old athymic nude rats (Harlan Laboratories Inc., Indianapolis, IN, USA).

3 Methods

3.1 Construction and Culture of Mouse Organoids In Vitro

The dissection and isolation of embryonic day (E) 11.5 or 12.5 CD1 mouse kidneys have been described in detail previously [13]. The technology we describe here involves meticulous micromanipulation of tissues in open-air and in media buffered against 5 % CO_2. Therefore it is crucial to observe best-practice sterile techniques while preparing and handling instruments, solutions, and equipment to avoid culture contamination.

1. Isolate fresh E11.5 CD1 mouse kidneys in isolating medium.

2. Prepare petri dishes, culture filters, and grids for later use. Cut small squares of 5 µm polycarbonate filter about 8 mm per side using sterile scissors and tweezers. Place the metal grid in a 6 cm petri dish and add 7–8 ml (or until the medium level reaches grid height and the filter on the grid is wet) of culture medium containing 1.25 µM Glycyl-H1152 dihydrochloride. Place the filter on top of the grid and the whole dish in the 37 °C, 5 % CO_2 incubator.

3. Using the glass Pasteur pipette, transfer the embryonic kidneys to a 3.5 cm petri dish containing 2.5 ml trypsin–EDTA 0.1 % and incubate E11.5 kidneys for 3 min and E12.5 kidneys for 4 min at 37 °C, 5 % CO_2.

4. Using a P1000 micropipette, transfer trypsin-treated embryonic kidneys to a 3.5 cm petri dish containing 3 ml of isolating medium supplemented with 10 % ES-FBS to quench the trypsin–EDTA action.

5. Using a P1000 micropipette, transfer the embryonic kidneys into a 1.5 ml Eppendorf tube containing 300 µl of isolating medium supplemented with 10 % ES-FBS. Dissociate the kidneys into single-cell suspensions by pipetting them up and down through a P200 micropipette tip.

6. Filter the single-cell suspension through the 40 µm cell strainer. Wash the Eppendorf tube with 150 µl of isolating medium and filter the washing medium. Then, wash the filter with 150 µl of isolating medium (again) and finally transfer the cell suspension into a new 1.5 ml Eppendorf tube.

7. To determine cell viability, mix 20 µl cell suspension with 20 µl Trypan Blue and visually examine cells using a microscope to determine whether they internalize or exclude the dye. If dissociation and filtering steps are performed rapidly and carefully, more than 92 % of cells should be viable.

8. Count the cells using an automatic cell counter or a Burker counting chamber. Place aliquots of 1.2×10^5 or 4×10^5 freshly dissociated renal cells for in vitro and in vivo studies, respectively, in new 1.5 ml Eppendorf tubes. Add fresh medium to obtain a final volume of 500–600 μl per tube and mix the suspension gently using a P1000 micropipette.

9. Centrifuge the cell suspensions at $900 \times g$ for 4 min to form a pellet.

10. During centrifugation, place the previously prepared petri dishes (see **step 2**) on the stereoscope. Check there are no air bubbles under the filter, because they would disturb tissue development by precluding nutritional supply to the cells. Repeat this step for each petri dish.

11. To detach the pellet from the tube, aspirate a little medium using a P200 micropipette within the centrifuged Eppendorf tube and then expel it very carefully over the top portion of the pellet (see **Note 1**).

12. Using a glass Pasteur pipette, immediately collect the floating pellet and gently place it on top of the filter. Repeat this step for each pellet.

13. Place the petri dish containing the pellet in the incubator in a humidified atmosphere with 5 % CO_2 at 37 °C. After only a few minutes, ureteric bud (UB) cells already start to reaggregate into multiple UB structures.

14. After 24 h change the medium by replacing it with fresh warm culture medium without ROCK inhibitor. At this stage, the pellet grows into tissue—defined mouse renal organoid—containing UB tubules and metanephric mesenchyme (MM) derivatives.

15. Change medium every 2 days. We cultured mouse organoids for up to 21 days (Fig. 1).

3.2 Construction and Culture of Human–Mouse Chimeric Organoids In Vitro

For chimeric organoid construction, single-cell suspensions of E11.5 or E12.5 mouse kidney cells are mixed with human cells, aggregated and cultured as above. The human cell types that were tested are amniotic fluid stem cells (AFSCs) [4]; HK2 renal proximal tubular epithelial cells, and bone marrow-mesenchymal stem cell-derived renal proximal tubular-like epithelial cells [8]; embryonic stem cells (ESCs) and induced pluripotent stem cells (iPSCs) (see **Note 2**).

1. Trypsinize human cells into homogenous single-cell suspensions and harvest them. Make sure there are no residual cell clusters. Cell viability can be evaluated by Trypan Blue exclusion test. Count human cells using an automatic cell counter or a Burker counting chamber.

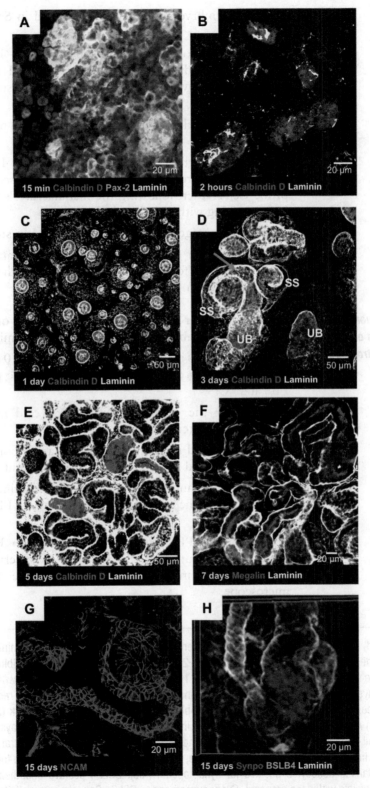

Fig. 1 In vitro development of mouse renal organoids. (**a**) At 15 min, ureteric bud (UB) epithelia expressing calbindin D28k (*red*) were reformed. A number of cells were positive for the UB- and nephron marker Pax-

2. Dissociate embryonic kidneys as described in Section 3.1 in **steps 3–8**.

3. Mix 1.2×10^4 or 4×10^4 human cells with 1.2×10^5 or 4×10^5 freshly dissociated mouse renal cells (1:10, human–mouse), respectively. Mix the chimeric cell suspension by pipetting.

4. Centrifuge the cell suspensions at $900 \times g$ for 5 min to form a pellet.

5. Human stem cells can also be labeled with 4 μM green-fluorescent chloromethyl derivative of fluorescein diacetate probe CellTracker, following the manufacturer's instructions, before mixing with mouse cells.

6. Proceed with stages described in Section 3.1 in **steps 10–14**.

7. Change medium every 2 days. Chimeric organoids can be cultured in vitro for several days. We cultured them for up to 5 days (Fig. 2a–f).

3.3 Immunofluorescence Analysis of Organoids In Vitro

1. Transfer the filter with the mouse or chimeric organoid to a closable plastic tube with PBS and wash for 10 min.

2. Fix the organoids in 4 % paraformaldehyde for 10 min at room temperature. Once fixed, the organoids can be preserved in PBS at 4 °C.

3. Permeabilize the organoids with 100 % cold methanol for 10 min at room temperature.

4. Wash in PBS and replace the PBS with a solution of primary antibodies diluted in PBS and incubate overnight at 4 °C.

5. Wash in PBS and incubate with the specific secondary antibodies (and the lectin if necessary) diluted in PBS overnight at 4 °C.

6. Wash again in PBS, then mount with Dako Fluorescence Mounting Medium and observe with an inverted confocal laser scanning microscope or an Apotome fluorescence microscope.

Fig. 1 (Continued) 2 (*green*). Cells positive for the general basement membrane marker laminin (*white*) were randomly distributed within the reforming tissue. (**b**) At 2 h, reformed UB cells expressed calbindin D28k (*red*) and some of these were laminin-positive (*white*). (**c**) At 1 day, laminin positive membranes (*white*) surrounded UBs. (**d**) At 3 days, S-shaped bodies (SS) were connected to calbindin D28k-positive (*red*) UBs. *Arrow*: connection between UB and SS. (**e**) At 5 days, developing tubuli expressed calbindin D28k (*red*) in UB and distal domains. (**f**) At 7 days, megalin (*red*) was found in the proximal portions. (**g**) At 15 days, an elongated nephron with well-defined tubular portions and glomerular pole was visualized, by neural cell adhesion molecule (NCAM) immunostaining. (**h**) At the same time, the 3D image showed more mature tubuli connected to glomeruli containing podocytes positive for synaptopodin (*red*). BSLB4-positive endothelial progenitors (*green*) were also visible within the organoid. *Synpo* synaptopodin, *BSLB4 Bandeiraea simplicifolia* Isolectin B4

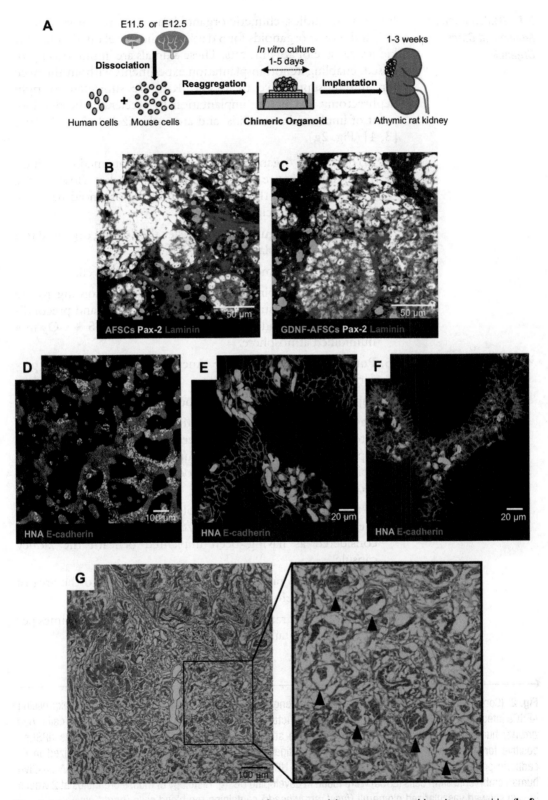

Fig. 2 (a) In vitro construction and implantation of mouse and human–mouse chimeric organoids. (b–f) Integration of human cells into the chimeric organoids after 2 days in vitro. (b) AFSCs (*green, arrows*) were

3.4 Implantation of Mouse and Chimeric Organoids

For in vivo studies, chimeric organoids are cultured in vitro for 1 day, and mouse organoids for 5 days, and then implanted under the kidney capsule of athymic rats. These animals are immunocompromised, enabling xenotransplantation experiments without the need for immunosuppressive treatment. Rats are subjected to right nephrectomy just before implantation to accelerate the development of implanted organoids, and are euthanized 1–3 weeks later [3, 4] (Fig. 2g).

1. To detach the organoid from the filter, aspirate medium using a P200 micropipette and then carefully expel it close to the organoid. Repeat this procedure until the organoid has been detached completely.
2. Using a glass Pasteur pipette, transfer the floating organoid to a well of a 96-well plate (*see* **Note 3**).
3. Remove as much medium as possible from the well.
4. Add 2 μg recombinant rat VEGF prepared according to the manufacturer's instructions, to the well to soak and precondition the tissue, and incubate for 4 h at 37 °C, 5 % CO_2 in a humidified atmosphere.
5. Put the 96-well plate on ice. Anesthetize athymic rat and perform unilateral nephrectomy.
6. Aspirate the organoid in the catheter.
7. Carefully insert the end of the catheter containing the organoid into the hole that has been generated previously in the kidney capsule, and gently expel the solution drop by drop until organoid comes out. The organoid looks whitish and roundish (*see* **Note 4**).
8. Extract the catheter from the hole, taking care not to damage the kidney capsule, and cauterize. The organoid can now be considered as having been implanted beneath the kidney capsule.
9. Inject recipient rat locally with VEGF (1 μg) into the area of implantation before performing the suture.
10. Inject recipient rat intravenously with VEGF (1 μg, 3 times per week) into the tail vein until euthanasia.

Fig. 2 (Continued) mainly concentrated in areas among Pax-2-positive renal structures. (**c**) GDNF-expressing AFSCs integrated into Pax-2-positive developing structures more efficiently compared with control cells. *Red arrows*: human cells integrated into Pax-2-positive structures. (**d**) Induced pluripotent stem cells (iPSCs), positive for the human marker human nuclear antigen (HNA, *green*) were almost entirely localized in E-cadherin-positive UB. (**e**) Human–mouse chimeric UB at higher magnification. (**f**) Similarly, HNA-positive human embryonic stem cells (ESCs) were found in developing UB. (**g**) Histology of mouse organoids at 2 weeks in vivo showed vascularized glomeruli (*inset, arrowheads*) containing red blood cells (*bright pink*)

3.5 Renal Histology of Organoids Grown In Vivo

1. Euthanize the host rat by CO_2 inhalation 1–3 weeks after implantation and remove the rat kidney.
2. Fix the implanted organoids in PLP and embed in OCT compound.
3. Stain 3-µm cryosections with hematoxylin for 15 min at room temperature
4. After washing in running water, stain the slices with eosin for 5 min at room temperature [3, 4].
5. Wash in running water, mount with Dako Faramount Aqueous Mounting Medium and observe by light microscopy.

4 Notes

1. Detaching the pellet from the Eppendorf is a very delicate step because of the extreme fragility of the pellet. To avoid pellet fragmentation be careful to expel the aspirated medium gently but firmly. Note that repeating centrifugations may reduce cell viability.
2. Previous studies reported that when mouse embryonic kidneys cells were reaggregated with undifferentiated human pluripotent stem cells (PSCs) (i.e., hESCs or hiPSCs) and cultured for 3–4 days, human cells did not integrate into renal structures and severely disrupted 3D renal tissue development [10, 11]. In contrast to these data, in our experimental conditions we observed a massive integration of both undifferentiated hESCs and hiPSCs almost exclusively into the developing UB epithelia (Fig. 2d–f). This is an important point to take into consideration in studies aimed at validating the integration potential of in vitro hPSC-derived renal progenitors by using undifferentiated hPSCs as negative controls in reaggregation assays, especially if the in vitro differentiation protocol is designed to generate UB progenitor cells.
3. To aspirate the LCA for transplantation without damaging it, we suggest using a glass Pasteur pipette broken manually in order to obtain a larger hole at the end.
4. During LCA implantation under the kidney capsule, we recommend pushing the syringe plunger very slowly and making sure that the implant is positioned away from the intervention point. This will minimize the risk of it sliding out from the site of implantation.

References

1. Auerbach R, Grobstein C (1958) Inductive interaction of embryonic tissues after dissociation and reaggregation. Exp Cell Res 15(2):384–397
2. Unbekandt M, Davies JA (2010) Dissociation of embryonic kidneys followed by reaggregation allows the formation of renal tissues. Kidney Int 77(5):407–416
3. Xinaris C, Benedetti V, Rizzo P, Abbate M, Corna D, Azzollini N, Conti S, Unbekandt M, Davies JA, Morigi M, Benigni A, Remuzzi G (2012) In vivo maturation of functional renal organoids formed from embryonic cell suspensions. J Am Soc Nephrol 23(11):1857–1868
4. Xinaris C, Benedetti V, Novelli R, Abbate M, Rizzo P, Conti S, Tomasoni S, Corna D, Pozzobon M, Cavallotti D, Yokoo T, Morigi M, Benigni A, Remuzzi G (2015) Functional human podocytes generated in organoids from amniotic fluid stem cells. J Am Soc Nephrol 27(5):1400–1411
5. Kitamoto Y, Tokunaga H, Tomita K (1997) Vascular endothelial growth factor is an essential molecule for mouse kidney development: glomerulogenesis and nephrogenesis. J Clin Invest 99(10):2351–2357
6. Siegel N, Rosner M, Unbekandt M, Fuchs C, Slabina N, Dolznig H, Davies JA, Lubec G, Hengstschlager M (2010) Contribution of human amniotic fluid stem cells to renal tissue formation depends on mTOR. Hum Mol Genet 19(17):3320–3331
7. Hendry CE, Vanslambrouck JM, Ineson J, Suhaimi N, Takasato M, Rae F, Little MH (2013) Direct transcriptional reprogramming of adult cells to embryonic nephron progenitors. J Am Soc Nephrol 24(9):1424–1434
8. Papadimou E, Morigi M, Iatropoulos P, Xinaris C, Tomasoni S, Benedetti V, Longaretti L, Rota C, Todeschini M, Rizzo P, Introna M, Grazia de Simoni M, Remuzzi G, Goligorsky MS, Benigni A (2015) Direct reprogramming of human bone marrow stromal cells into functional renal cells using cell-free extracts. Stem Cell Reports 4(4):685–698
9. Mae S, Shono A, Shiota F, Yasuno T, Kajiwara M, Gotoda-Nishimura N, Arai S, Sato-Otubo A, Toyoda T, Takahashi K, Nakayama N, Cowan CA, Aoi T, Ogawa S, McMahon AP, Yamanaka S, Osafune K (2013) Monitoring and robust induction of nephrogenic intermediate mesoderm from human pluripotent stem cells. Nat Commun 4:1367
10. Xia Y, Nivet E, Sancho-Martinez I, Gallegos T, Suzuki K, Okamura D, Wu MZ, Dubova I, Esteban CR, Montserrat N, Campistol JM, Izpisua Belmonte JC (2013) Directed differentiation of human pluripotent cells to ureteric bud kidney progenitor-like cells. Nat Cell Biol 15(12):1507–1515
11. Takasato M, Er PX, Becroft M, Vanslambrouck JM, Stanley EG, Elefanty AG, Little MH (2014) Directing human embryonic stem cell differentiation towards a renal lineage generates a self-organizing kidney. Nat Cell Biol 16(1):118–126
12. Taguchi A, Kaku Y, Ohmori T, Sharmin S, Ogawa M, Sasaki H, Nishinakamura R (2014) Redefining the in vivo origin of metanephric nephron progenitors enables generation of complex kidney structures from pluripotent stem cells. Cell Stem Cell 14(1):53–67
13. Davies JA (2010) The embryonic kidney: isolation, organ culture, immunostaining and RNA interference. Methods Mol Biol 633:57–69

Efficient Culture of Intestinal Organoids with Blebbistatin

Zhen Qi and Ye-Guang Chen

Abstract

The intestinal epithelium is one of the most rapidly self-renewing tissues throughout life in mammals. A small population of stem cells at the base of crypt in the epithelium can continually self-renew and give rise to differentiated epithelial cells. The self-renewal and differentiation of intestinal stem cells are under a tight control during homeostasis, and disruption of this balancing regulation leads to intestinal degeneration or tumorigenesis. Accordingly, exploration of the mechanism underlying the regulation of stem cells is essential for the understanding and treatment of intestinal disorders. As traditional methods using mice models are costly and time-consuming, the recently established ex vivo intestinal organoids model provides an ideal tool to investigate the mechanisms regulating the self-renewal and differentiation of intestinal stem cells. The intestinal organoids recapitulate major characteristics in both structure and function of intestinal epithelium in vivo. Here, we describe a new protocol to generate the intestinal organoids from both crypts and single stem cells with a higher efficiency using the small molecule blebbistatin and provide an approach to assess the self-renewal and differentiation of stem cells in intestinal organoids.

Keywords Intestinal organoids, Lgr5+ stem cells, Blebbistatin, Culture, Self-renewal, Differentiation

1 Introduction

The intestinal epithelium is a monolayer of epithelial cells in the inner lining of the intestine and has important physiological functions in food digestion, nutrient absorption, and defense against bacteria [1]. In the intestine, the epithelium is composed of numerous crypt-villus units. Villi are fingerlike protrusions containing various kinds of differentiated cells, and crypts are epithelial invaginations that mainly consist of intestinal stem cells and progenitor cells. The intestinal epithelium is one of the most rapidly self-renewing tissues in mammals, with a turnover rate of 4–5 days [2]. This prominent ability of self-renewal is imposed by persistent loss of epithelial cells during the digestion and absorption each day [3]. Lgr5+ intestinal stem cells are located at the base of crypts and can indefinitely self-renew and generate epithelial cells throughout the life [4]. Lgr5+ stem cells firstly give rise to transient amplifying (TA) cells, and the TA cells gradually differentiate into secretory lineages (Paneth cells, goblet cells, and enteroendocrine cells) or absorptive lineages (enterocytes).

The intestinal organoids are epithelial structures cultured in a 3D system that recapitulate key features of in vivo intestinal epithelium [5]. In this culture system, the intestinal crypts or Lgr5+ single stem cells can grow into the organoids in Matrigel supplemented with a cocktail of growth factors, including epidermal growth factor (EGF), Noggin, and R-spondin. The intestinal organoids also contain crypt and villus-like structures, and Lgr5+ stem cells are located at the base of crypt structure. The Lgr5+ stem cells in the organoids can also continually self-renew and give rise to TA cells, which can further differentiate into mature epithelial cells including goblet cells, Paneth cells, enteroendocrine cells, and enterocytes [6].

Blebbistatin is a reversible inhibitor of the nonmuscle myosin II (NMII) heavy chain and can specifically inhibit membrane blebbing in some types of cells [7]. Blebbistatin is reported to significantly improve cell survival and cloning efficiency of both human embryonic stem cells and induced-pluripotent stem cells [8–10]. We have recently shown that blebbistatin can efficiently improve the survival rate of both intestinal crypts and single stem cells and promote the growth of intestinal organoids [11].

In this protocol, we describe step by step how to isolate intestinal crypts from murine intestine and grow the intestinal organoids with a high efficiency in the presence of blebbistatin. We further provide a protocol to assess the status of self-renewal and differentiation of intestinal organoids upon small molecule treatment.

2 Materials

2.1 Intestinal Crypt Isolation

1. 1× phosphate-buffered saline (PBS), pH 7.4, without Ca^{2+} and Mg^{2+}.
2. 2 mM EDTA, stored at 4 °C.
3. 50-mL conical polypropylene tube.
4. 70 µm cell strainer.

2.2 Lgr5+ Stem Cell Sorting

1. *Lgr5-GFP-IRES-CreER* mouse (The Jackson Laboratory).
2. TrypLE (Life Technologies).
3. FACS tube.
4. 40 µm cell strainer.

2.3 Intestinal Organoid/Stem Cell Culture

1. Advanced DMEM/F12 (Life Technologies).
2. Glutamax (Life Technologies).
3. N2 supplement (Life Technologies).
4. B27 supplement (Life Technologies).
5. Penicillin-streptomycin (Life Technologies).
6. *N*-Acetylcysteine (100× stock, 100 mM, Sigma-Aldrich)

7. Matrigel, growth factor reduced, phenol red-free (Coring) (see **Note 1**).
8. Recombinant human EGF (2000× stock, 100 μg/mL, Life Technologies).
9. Recombinant mouse Noggin (500× stock, 50 μg/mL, R&D Systems).
10. Recombinant mouse R-spondin (200× stock, 100 μg/mL, R&D Systems).
11. Recombinant mouse Wnt3a (400× stock, 40 μg/mL, R&D Systems).
12. Jagged-1 (1000× stock, 1 mM in sterile water, Ana Spec).
13. Blebbistatin (2000× stock, 20 mM in DMSO, Calbiochem).
14. Y-27632 (1000× stock, 20 mM in sterile water, Sigma-Aldrich).
15. Basal crypt culture medium: advanced DMEM/F12 supplemented with N2 (1×), B27 (1×), N-Acetylcysteine (1 mM), and penicillin-streptomycin (1×).
16. Complete crypt culture medium: basal crypt culture medium supplemented with EGF (50 ng/mL), Noggin (100 ng/mL), and R-spondin (500 ng/mL) (see **Note 2**).
17. Complete stem cell culture medium: complete crypt culture medium supplemented with Wnt3a (100 ng/mL) and Y-27632 (10 μM).

2.4 Imaging and Immunofluorescence

1. 1.5 mL Snap-Cap microcentrifuge tubes.
2. 4% paraformaldehyde (PFA, Sigma-Aldrich) in PBS.
3. 1% Triton-X100 (Sigma-Aldrich) in 1× PBS.
4. 3% BSA with 0.1% Triton-X 100 in 1× PBS.
5. Primary antibodies: anti-Ki67 (Abcam), anti-Muc2 (Santa Cruz), anti-chromogranin A (Santa Cruz), and anti-lysozyme (Dako).
6. 7-AAD (eBioscience).
7. Propidium iodide (PI) (eBioscience).
8. 4′,6-diamidino-2-phenylindole (DAPI) (Sigma-Aldrich).
9. In situ Cell Death Detection Kit (Roche).
10. Glass slides and coverslips.
11. Mounting solution (ZSGB-BIO).

3 Methods

3.1 Intestinal Organoid Culture

1. Cut the intestine longitudinally and wash the lumen of the intestine three times with cold PBS.
2. Add 10 mL cold PBS in a 10 cm Petri dish and put the intestine into the dish. Keep the villi upward.
3. Carefully scraped away the villi with a dissecting scalpel. Check whether most of villi are removed under the microscope (see **Note 3**).
4. Cut the intestine into small pieces (5 mm).
5. Incubate the pieces in 2 mM EDTA in PBS for 30 min at 4 °C.
6. Transfer the intestine pieces into 10 mL cold PBS in a Petri dish and vigorously suspend them using 10 mL pipette.
7. The suspension is passed through a 70 μm cell strainer and centrifuged at $300 \times g$ for 3 min.
8. The pellet is rich of purified intestinal crypts.
9. Resuspend the crypt pellet with Matrigel (200 crypts/20 μL).
10. Drop 20 μL Matrigel suspension into each well of 48-well plate and put the plate into a CO_2 incubator (5%, 37 °C) for 10–20 min to allow the solidification of Matrigel.
11. Add 250 μL complete crypt culture medium supplemented with blebbistatin (10 μM) per well. Note that blebbistatin significantly improves the survival and growth of intestinal organoids (see Fig. 1).

3.2 Lgr5+ Stem Cell Colony Formation

1. Resuspend the crypts isolated from *Lgr5-GFP-IRES-CreER* mouse with 3 mL prewarmed TrypLE medium.
2. Incubate the suspension at 37 °C for 20 min (see **Note 4**).
3. Vigorously pipet the sample up and down for several times until most of the crypts are dissociated into single cells.
4. The single cell suspension is passed through a 40 μm cell strainer and centrifuged at $400 \times g$ for 3 min, 4 °C.
5. Resuspend the pellet with 1 mL basal crypt culture medium supplemented with blebbistatin and transfer to FACS tube. Add PI or 7AAD into the medium.
6. Sort PI/7AAD negative and GFP-high (account for ~5% of PI/7AAD negative cells) single cells into 1 mL basal crypt culture medium supplemented with blebbistatin.
7. Centrifuge the cell suspension at $400 \times g$ (4 °C, 3 min) to collect single cells.

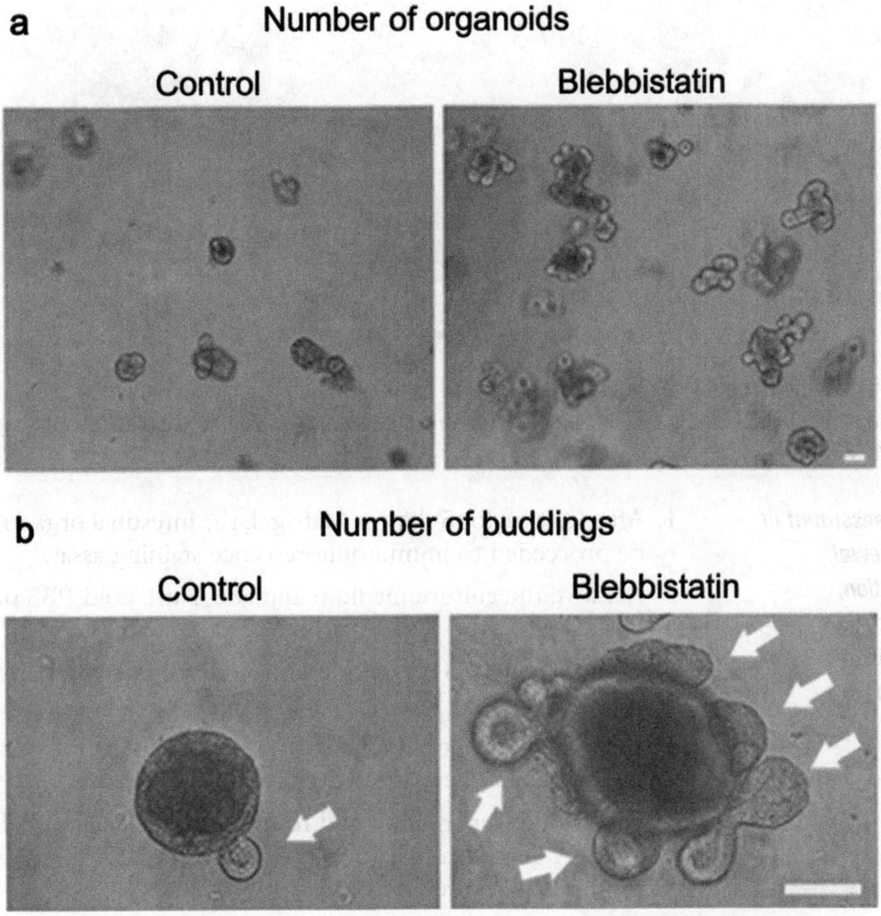

Fig. 1 Assessment of survival, growth, and budding of intestinal organoids. (**a**) Macroscopic images of intestinal organoids cultured for 3 days in different groups. (**b**) Macroscopic images of representative organoids cultured for 3 days in different groups. *Arrows* mark the buddings of organoids. Scale bars, 50 μm. Reproduced from Zhao et al. [11]

8. Resuspend the pellet with Matrigel supplemented with Jagged-1 (1 μM) at the concentration of 10,000 cells/20 μL (see **Note 5**).

9. Add 20 μL Matrigel in each well of 48-well plate and transfer the plate to a CO_2 incubator (5%, 37 °C) for 10–20 min.

10. Add 250 μL complete stem cell culture medium with blebbistatin (10 μM) into each well (see Fig. 2). Blebbistatin can efficiently improve the survival of single stem cells.

This protocol describes how to evaluate the self-renewal, proliferation, differentiation, and apoptosis in organoids using whole mount immunostaining.

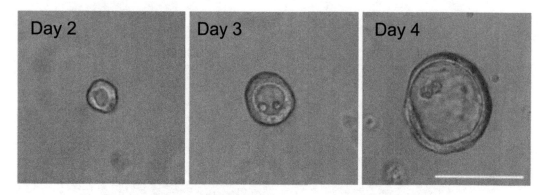

Fig. 2 Time course of Lgr5+ single stem cells growth in the 3D culture system. The Lgr5+ stem cells are sorted and embedded in Matrigel. Colonies grown from single stem cells can be counted and photographed. Scale bars, 50 μm

3.3 Assessment of Self-Renewal, Proliferation, Differentiation, and Apoptosis in Organoids

1. After culture for 5 days in Matrigel, the intestinal organoids can be proceeded to immunofluorescence staining assay.

2. Remove the culture medium and add 1 mL cold PBS per well of 48-well plate.

3. Gently pipet PBS up and down to release intestinal organoids from Matrigel.

4. Transfer the suspension to a 1.5 mL tube and let the organoids sink to the bottom (see **Note 6**).

5. Discard the supernatant and resuspend the pellet with 100 μL 4% PFA gently. Incubate the organoids in PFA for 20 min at room temperature (see **Note 7**).

6. Wash the organoids twice with 500 μL cold PBS, and let the organoids sink to the bottom each time.

7. The organoids are permeabilized by incubation in 1% Triton-X100 for 20 min on ice.

8. Discard the supernatant and resuspend the organoids in blocking buffer (3% BSA in PBS with 0.1% Triton-X100) on ice for 1 h.

9. Discard the supernatant and incubate the organoids with 100 μL primary antibody (anti-Ki67, Muc2, or chromogranin A) diluted in blocking buffer (1:100–300) at 4 °C overnight. To perform TUNEL assay, incubate the organoids with TUNEL reaction mixture at 37 °C for 1 h according to the manufacturer's instructions in the in situ Cell Death Detection Kit.

10. Remove the primary antibody and wash the organoids with cold PBS twice. Always allow the organoids to sink to the bottom instead of centrifugation.

11. Incubated the organoids with secondary antibody diluted in blocking buffer (usually 1:300) at room temperature for 1 h. DAPI (4′,6-diamidino-2-phenylindole) is included in the staining solution at the concentration of 1 μg/mL.
12. Discard the secondary antibody and wash the organoids with cold PBS twice.
13. Resuspend the organoids with 50 μL mounting solution and drop the suspension on glass slide.
14. Put on a clean coverslip and seal the edges with nail polish.
15. Analyze the organoids with confocal microscope (see Fig. 3).

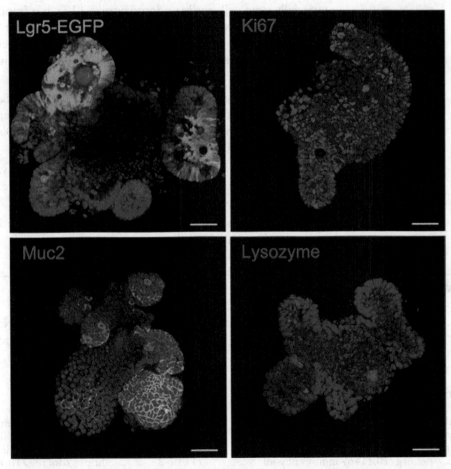

Fig. 3 Immunofluorescence staining of intestinal organoids. Lgr5-EGFP, Lgr5+ stem cells; Ki67, proliferating cells; Muc2, goblet cells; lysozyme, Paneth cells. Nuclei were counterstained with DAPI. Scale bars, 50 μm

4 Notes

1. We recommend to use the Matrigel from Coring (BD Bioscience) instead of other ventures. The Matrigel should be stored at −20/−80 °C and thawed on ice. Make sure to place the Matrigel on ice when it is used.

2. Basal crypt culture medium can be stored at 4 °C for up to 1 month, but the complete crypt culture medium and complete stem cell culture medium are always freshly prepared.

3. Be careful when scraping villi as a proportion of crypts may be lost during this procedure. Check the intestine under light microscope to make sure that villi are removed as much as possible while most crypts are present.

4. As the dissociation process renders the intestinal epithelial cells especially stem cells easily undergo anoikis [12], it is important to avoid incubating the intestinal crypts in TrypLE for more than 20 min. Try to shorten the time between the dissociation and cell sorting and add blebbistatin in the culture medium to reduce cell death.

5. Make sure that jagged-1 peptide is added directly into Matrigel but not culture medium. The colony formation efficiency is usually very low (~1%), so we recommend to use 10,000 single cells per 20 μL Matrigel.

6. When performing whole mount staining in organoids, it is advised to let the organoids sink to the bottom instead of centrifugation in order to keep the structure intact.

7. If possible, try to use low retention tips as the organoids can easily attach to the tip surface.

Acknowledgments

This work was supported by grants from the National Natural Science Foundation of China (31330049, 91519310) and the 973 Program (2013CB933700) to YGC.

References

1. van der Flier LG, Clevers H (2009) Stem cells, self-renewal, and differentiation in the intestinal epithelium. Annu Rev Physiol 71:241–260. https://doi.org/10.1146/annurev.physiol.010908.163145

2. Clevers H (2013) The intestinal crypt, a prototype stem cell compartment. Cell 154 (2):274–284. https://doi.org/10.1016/j.cell.2013.07.004

3. Barker N (2014) Adult intestinal stem cells: critical drivers of epithelial homeostasis and regeneration. Nat Rev Mol Cell Biol 15 (1):19–33. https://doi.org/10.1038/nrm3721

4. Barker N, van Es JH, Kuipers J, Kujala P, van den Born M, Cozijnsen M, Haegebarth A, Korving J, Begthel H, Peters PJ, Clevers H (2007) Identification of stem cells in small

intestine and colon by marker gene Lgr5. Nature 449(7165):1003–1007. https://doi.org/10.1038/nature06196

5. Sato T, Vries RG, Snippert HJ, van de Wetering M, Barker N, Stange DE, van Es JH, Abo A, Kujala P, Peters PJ, Clevers H (2009) Single Lgr5 stem cells build crypt-villus structures in vitro without a mesenchymal niche. Nature 459(7244):262–265. https://doi.org/10.1038/nature07935

6. Sato T, Clevers H (2013) Growing self-organizing mini-guts from a single intestinal stem cell: mechanism and applications. Science 340(6137):1190–1194. https://doi.org/10.1126/science.1234852

7. Straight AF, Cheung A, Limouze J, Chen I, Westwood NJ, Sellers JR, Mitchison TJ (2003) Dissecting temporal and spatial control of cytokinesis with a myosin II inhibitor. Science 299(5613):1743–1747. https://doi.org/10.1126/science.1081412

8. Chen GK, Hou ZG, Gulbranson DR, Thomson JA (2010) Actin-myosin contractility is responsible for the reduced viability of dissociated human embryonic stem cells. Cell Stem Cell 7(2):240–248. https://doi.org/10.1016/j.stem.2010.06.017

9. Ohgushi M, Matsumura M, Eiraku M, Murakami K, Aramaki T, Nishiyama A, Muguruma K, Nakano T, Suga H, Ueno M, Ishizaki T, Suemori H, Narumiya S, Niwa H, Sasai Y (2010) Molecular pathway and cell state responsible for dissociation-induced apoptosis in human pluripotent stem cells. Cell Stem Cell 7(2):225–239. https://doi.org/10.1016/j.stem.2010.06.018

10. Walker A, Su H, Conti MA, Harb N, Adelstein RS, Sato N (2010) Non-muscle myosin II regulates survival threshold of pluripotent stem cells. Nat Commun 1:71. doi:https://doi.org/10.1038/ncomms1074

11. Zhao B, Qi Z, Li Y, Wang C, Fu W, Chen YG (2015) The non-muscle-myosin-II heavy chain Myh9 mediates colitis-induced epithelium injury by restricting Lgr5+ stem cells. Nat Commun 6:7166. https://doi.org/10.1038/ncomms8166

12. Watanabe K, Ueno M, Kamiya D, Nishiyama A, Matsumura M, Wataya T, Takahashi JB, Nishikawa S, Nishikawa S, Muguruma K, Sasai Y (2007) A ROCK inhibitor permits survival of dissociated human embryonic stem cells. Nat Biotechnol 25(6):681–686. https://doi.org/10.1038/nbt1310

Isolation and Culture of Adult Intestinal, Gastric, and Liver Organoids for Cre-recombinase-Mediated Gene Deletion

Dustin J. Flanagan, Renate H.M. Schwab, Bang M. Tran,
Toby J. Phesse, and Elizabeth Vincan

Abstract

The discovery of Lgr5 as a marker of adult stem cells meant that stem cell populations could be purified and studied in isolation. Importantly, when cultured under the appropriate conditions these stem cells form organoids in tissue culture that retain many features of the tissue of origin. The organoid cultures are accessible to genetic and biochemical manipulation, bridging the gap between in vivo mouse models and conventional tissue culture. Here we describe robust protocols to establish organoids from gastrointestinal tissues (stomach, intestine, liver) and Cre-recombinase mediated gene manipulation in vitro.

Keywords: Intestinal organoids, Gastric organoids, Liver organoids, Cre-recombinase, 3D organoid culture, Lgr5

1 Introduction

The ability to grow gastrointestinal epithelial cells and liver hepatocytes in tissue culture as mini-tissue organoids has revolutionized our understanding of the stem cells that maintain homeostasis of these tissues and the response to injury. Coupled with the power of transgenic mouse models that allow tissue and cell specific manipulation of genes, organoid-based technology has advanced our understanding of the molecular mechanisms that lead to human disease in these tissues, such as cancer and infection with pathogens. Although for many decades human and mouse continuous cell lines have been the bedrock for interrogating the molecular mechanisms of cell and tissue function, it is now recognized that many processes cannot be adequately modeled in continuous cell lines or short-term primary cultures [i.e., two-dimensional (2D) culture]. Almost all cells in vivo are surrounded by other cells and extracellular matrix in a 3D fashion, and thus 2D cultures do not adequately represent this important aspect of their biology. 2D cultures also adhere to the plastic flasks they are grown in and thus not only represent a poor model of the in vivo 3D context, but also have significant changes in their biology and thus they do not closely resemble the living tissue they are meant to model [1]. This caveat is not confined to processes that are complex and multifaceted, such as organogenesis for example, as cell-intrinsic behavior, particularly of non-transformed cells, also requires a 3D context. Thus, much of what we understand about mammalian tissue biology is derived from in vivo mouse models. However, this too presents issues due to access to internal organs for analysis and the cost of generating transgenic mouse models and maintaining mouse colonies. The seminal discovery of Lgr5 (leucine-rich-repeat-containing G-protein-coupled receptor 5) as a specific marker of adult stem-cells [2, 3] has facilitated successful isolation and cultivation of these stem cells. Lgr5 stem cells generate mini-tissue organoids that faithfully recapitulate the characteristics of the intact tissue of origin [4–6]. Thus, organoids hold great promise for translational research, bridging an important gap between 2D cultures and in vivo mouse/human models. They provide more physiological relevance than 2D culture systems, and are easier to generate and more amenable to genome editing and biochemical manipulation than in vivo models. Here we describe robust protocols for the isolation and culture of adult mouse intestine, stomach, and liver organoids and their application to Cre-recombinase mediated gene deletion as a means to study gene function. Organoids are established from compound mice that carry transgenes for a tissue or cell type specific, inducible Cre-recombinase and the gene of interest that is flanked by DNA sequences (called LoxP sites) that are recognized by the Cre-recombinase enzyme. Once the organoids are established, Cre-mediated recombination to delete the intervening DNA sequence between the two LoxP sites is induced in vitro, thus manipulating the gene of interest. For the gastrointestinal epithelium this allows the study of deleting or activating a gene of interest in a model that faithfully recapitulates the cell types and organization of the intact epithelium [7]. For the liver, differentiation of the expanding organoid cultures faithfully mimics the functional characteristics of adult hepatocytes [5, 8].

2 Materials

Breed mice that harbor the relevant alleles or transgenes. Diligently follow ethical guidelines as outlined by your institution or governing body. Follow aseptic techniques throughout the procedure from harvesting the tissue until the organoids are harvested for further analysis using standard cellular and molecular techniques. Prepare and store tissue culture reagents in volume aliquots necessary to make the final medium or at a concentration for easy dilution to working concentrations. Use molecular biology and tissue culture grade reagents and materials.

2.1 Tissue Dissection

1. Female or male mice
2. Autoclaved forceps and dissection scissors, sterile scalpel
3. Phosphate buffered saline (PBS), pH 7.3, no magnesium or calcium (Invitrogen). Store at 4 °C, use ice cold.
4. Orbital nutator
5. Bench top centrifuge

2.2 Organoid Culture and Cre-recombinase Induction

1. Sterile 15 mL and 50 mL polypropylene centrifuge tubes (Greiner)
2. Cell strainers (Becton Dickinson), 70 μM and 100 μM pore size
3. Petri dish(Greiner), small (6 cm) and large (10 cm)
4. Chelation buffer: PBS with 2.5 mM EDTA (stomach), 2.0 mM EDTA (intestine). Make freshly by adding EDTA pH 8.0 to PBS and keep on ice.
5. Collagenase–dispase enzyme digestion: Add 12.5 mg collagenase type XI (Sigma) and 12.5 mg dispase (Gibco) to 100 mL of DMEM supplemented with 1 % (w/v) fetal bovine serum.
6. Advanced DMEM/F12 base medium (ADF): Advanced DMEM/F12 (Invitrogen) supplemented with 10 mM HEPES (N-2-hydroxyl piperazine-N'-2-ethane sulfonic acid), 2 mM L-glutamine, 100 U/mL penicillin and 100 μg/mL streptomycin, and 0.1 % (w/v) bovine serum albumin. Make 500 mL at a time and store at −20 °C in 50 mL tubes with 48.5 mL/tube.
7. Organoid culture medium base (ADF+): ADF supplemented with N2 and B27 (Invitrogen).
8. Organoid expansion culture medium for intestine (ADF-I): ADF+ supplemented with 50 ng/mL epithelial growth factor (EGF, Peprotech), 500 ng/mL R-spondin (R&D) (see **Note 1**), and 100 ng/mL Noggin.
9. Organoid expansion culture medium for stomach (ADF-S): ADF+ supplemented with 50 ng/mL epithelial growth factor (EGF, Peprotech), 500 ng/mL R-spondin (R&D) (see **Note 1**), 100 ng/mL Noggin, 50 ng/mL human fibroblast growth factor-10 (FGF-10, Peprotech), 100 ng/mL murine Wnt3a (Peprotech) (see **Note 2**), 10 nM [Leu^{15}]-Gastrin (Sigma-Aldrich), and 1 μM N-acetylcysteine (Sigma-Aldrich).
10. Organoid expansion culture medium for liver (ADF-L). ADF+ supplemented with 50 ng/mL epithelial growth factor (EGF, Peprotech), 500 ng/mL R-spondin (R&D) (see **Note 1**), 100 ng/mL Noggin, 100 ng/mL recombinant murine Wnt3a (Peprotech) (see **Note 2**), 50 ng/mL recombinant human fibroblast growth factor-10 (FGF-10, Peprotech), 50 ng/mL human growth factor (HGF, Peprotech), 10 nM

[Leu^{15}]-Gastrin (Sigma-Aldrich), 10 mM nicotinamide (Sigma-Aldrich), and 1 μM N-acetylcysteine (Sigma-Aldrich).

11. Organoid culture medium for liver differentiation (ADF-LD): ADF+ supplemented with 100 ng/mL recombinant human fibroblast growth factor-10 (FGF-10, Peprotech), 10 nM [Leu^{15}]-gastrin (Sigma-Aldrich), 10 mM nicotinamide (Sigma-Aldrich), 1 μM N-acetylcysteine (Sigma-Aldrich), 50 nM glycated human albumin (Sigma, A8301) and 10 μM DAPT (GSI-IX, LY-374973, N-[N-(3,5-difluorophenacetyl)-L-alanyl]-S-phenylglycine t-butyl ester). This medium is used for the first 9–12 days of differentiation and then replaced with ADF-LD supplemented with 3 μM dexamethasone (Sigma-Aldrich).

12. Matrigel basement membrane matrix, growth factor reduced, phenol red-free (BD).

13. 24-well adherent tissue culture dish (Greiner)

14. β-napthoflavone (BNF): Make a 180 μM working stock solution of 5–6 Benzoflavone in dimethylsulfoxide (DMSO). Store at −20 °C.

15. 4-Hydroxytamoxifen (4OHT): Make a 250 μM working stock solution of (Z)-4-(1-[4-(dimethylaminoethoxy)phenyl]-2-phenyl-1-butenyl)phenol, (Z)-4-OHT in 100 % ethanol. Store at −20 °C.

3 Methods

Thaw Matrigel on ice and keep it on ice throughout the procedure until culture incubation as it will solidify irreversibly at room temperature. Keep tissue and reagents on ice and work on ice as much as possible. For example, place petri dish onto a bed of ice during dissection. Place 24-well tissue culture plates into the 37 °C incubator to warm up, and keep them in there until ready to plate cultures. Prepare chelation buffer, digestion buffer, and organoid culture medium fresh each time. The procedures for stomach and liver organoid culture are variations of the protocol used for intestinal epithelial crypt organoid culture.

3.1 Intestinal Epithelial Crypt Isolation and Tissue Culture

1. Euthanize the mice using an appropriate ethically approved method, and then isolate the small intestine as an entire organ by standard surgical procedures.

2. Place the intestine in a 10 cm petri dish with ice-cold PBS. Using surgical scissors, cut the intestine longitudinally (see **Note 3**) and rinse it clear of luminal contents in petri dish with PBS on ice. Transfer the clean tissue to a 5 mL sterile tube containing PBS and take to tissue culture on ice. Continue procedure in a Class I or Class II hood.

3. Carefully scrape villi from the small intestine using a glass cover slip (see **Note 4**) and rinse the tissue free of villi with PBS in a petri dish on ice.

4. Over a 50 mL sterile tube, cut the intestine with dissection scissors into small (~5 mm) pieces and add 20 mL chelation buffer (2 mM EDTA in PBS) and incubate at 4 °C for 30 min.

5. Allow the tissue to settle to the bottom and discard the EDTA/PBS. Add 20 mL PBS to the tissue pieces.

6. Release the crypts from the EDTA treated tissue by either pipetting up and down using a 10 mL serological pipette or shaking the 50 mL tube.

7. Crypts are separated from the remaining tissue by allowing the large pieces of tissue to settle at the bottom and transferring the supernatant to a new tube (keep the supernatant and label it as fraction 1). **Step 6** is typically repeated 3–4 times (labeling the supernatant as fraction 2 and so on). Transfer a drop of each fraction onto a clean petri dish and check under an inverted microscope. Pool the crypt-rich fractions and pellet by centrifuging at 1200 rpm/306 rcf for 5 min. Discard the supernatant PBS.

8. Resuspend the crypts in 10 mL ADF and pass the crypts through a 70 μM cell strainer into a 50 mL centrifuge tube. Wash the cell strainer with 5 mL ADF and transfer the flow-through to a 15 mL centrifuge tube (see **Note 5**).

9. Wash the crypts 2–3 times in ADF and centrifugation at 600 rpm/76 rcf for 2 min (see **Note 6**).

10. Count the crypts by transferring 20 μL of the crypt suspension to a petri dish during washes. On the final centrifugation, discard as much of the ADF as possible (see **Note 7**).

11. Resuspend the crypts at 2000 crypts/mL of Matrigel on ice and seed plates by adding 50 μL of the crypt/Matrigel suspension per well in a pre-warmed 24-well tissue culture plate (see **Note 8**).

12. Allow the Matrigel to set by placing the tissue culture plate into the 37 °C incubator for 10–15 min.

13. Add 500 μL of ADF-I culture medium to each well carefully along the side of the well to avoid dislodging the Matrigel, and incubate at 37 °C, 10 %CO_2, humidified incubator. Typically, the crypts form cysts within a day and start to bud new crypts after 2–3 days after culture.

14. Add fresh EGF, R-spondin and Noggin every other day, and change the whole culture medium twice per week once the organoids are established (Fig. 1 shows established organoids).

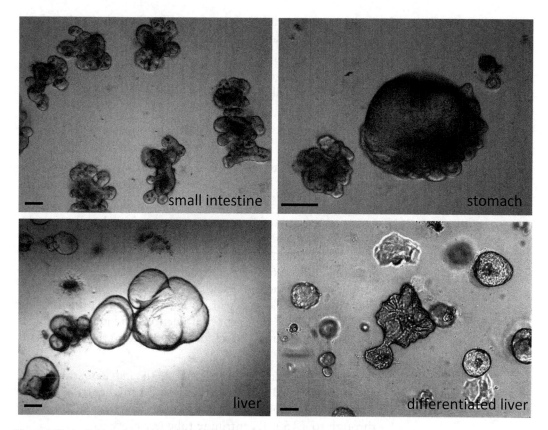

Fig. 1 Differential interference contrast (DIC) images of organoids established from intestinal crypts, gastric glands, and liver ductal structures, and differentiated liver hepatocyte organoids. Scale bar is 200 μM

15. After 7–10 days culture the organoids need to be passaged. Discard culture medium and mechanically disperse the organoids by pipetting the organoids/Matrigel with 1000 μL pipette tip. Transfer the dissociated organoids/Matrigel into a 15 mL centrifuge tube and dilute it with 2 mL ADF. Centrifuge at 600 rpm/76 rcf for 2–3 min. This will separate ADF, Matrigel and organoids into discrete layers, enabling the Matrigel to be removed from organoid pieces. Remove as much supernatant as possible without disturbing organoid pellet. Repeat the wash with ADF and mechanically disperse the pellet with 200 μL pipette tip until the organoids are dispersed and the released dead cells washed away.

16. Re-seed and incubate by repeating **steps 10–14**.

3.2 Antral Stomach Gland Isolation and Tissue Culture

1. Euthanize the mice using an appropriate ethically approved method, and then isolate the stomach as an entire organ by standard surgical procedures.

2. Transfer the stomach with forceps to a flat surface and use scissors to cut stomach along greater curvature and wash

twice in cold PBS to remove mucus and stomach contents (see **Note 9**). The following steps are performed in a Class I or Class II tissue culture cabinet.

3. Transfer stomach with forceps to petri dish on ice, gently flatten it using forceps and dissect out antral stomach with scalpel.

4. Transfer antral stomach with forceps to 50 mL tube containing 20 mL of chelation buffer and place on orbital nutator for 1.5 h at 4 °C. Use this time to prepare culture reagents.

5. Whilst securing the antral stomach to the side of tube with forceps, pour off chelation buffer and add 10 mL of ice-cold PBS to the tube.

6. Release the antral glands from the underlying stromal tissue by shaking the tube vigorously ~12–15 times (see **Note 10**).

7. Transfer 20 μL of gland suspension to a slide or petri dish and check under microscope for quality and amount of glands. Quality (intact) antral stomach glands will resemble a U-shaped sock. If there is less than ten glands in the 20 μL droplet, repeat **step 6**.

8. Discard underlying stromal layer and centrifuge PBS/gland suspension 1500 rpm/450 rcf for 5 min. Carefully discard supernatant with pipette to avoid disturbing the pellet.

9. Resuspend pellet in 1 mL ice-cold ADF and pass the suspension through a 100 μM cell strainer and collect the flow through in a new 50 mL centrifuge tube (see **Note 5**). Remove 20 μL of ADF/gland suspension and count the number of glands on slide/dish under microscope.

10. Transfer the glands to be cultured into a new tube and centrifuge ADF/gland suspension 1500 rpm/450 rcf for 5 min. On this final centrifugation, discard as much of the ADF as possible (see **Note 7**).

11. Resuspend the glands at 1000–2000 glands/mL of Matrigel on ice and seed plates by adding 50 μL of the gland/Matrigel suspension per well in a pre-warmed 24-well tissue culture plate (see **Note 8**).

12. Allow the Matrigel to set by placing the tissue culture plate into the 37 °C incubator for 10–15 min.

13. Add 500 μL of ADF-S culture medium to each well carefully along the side of the well to avoid dislodging the Matrigel, and incubate at 37 °C, 10 %CO_2, humidified incubator. Typically the organoids will begin to bud around 5–7 days following plating.

14. Discard medium and replace with 500 μL fresh ADF-S every other day (Fig. 1 shows established organoids).

15. After 7–10 days culture the organoids need to be passaged (*see* **Note 11**). To passage the gastric organoids, follow **step 15** in Sect. 3.1 (*see* **Note 12**) and re-seed following **steps 10 to 12**, Sect. 3.2.

3.3 Establishing Organoids from Liver Ductal Tissue

1. Euthanize the mice using an appropriate ethically approved method, and then isolate the liver as an entire organ by standard surgical procedures.
2. Rinse the tissue in PBS or DMEM and transport to a tissue culture cabinet on ice. The following steps are performed in a Class I or Class II tissue culture cabinet.
3. Place the tissue in a 10 cm petri dish on ice so that it is immersed in DMEM. Cut the tissue into small pieces and transfer it to a 50 mL centrifuge tube.
4. Clean the tissue pieces by pipetting up-down with a 10 mL serological pipette to remove some of the attached fat (*see* **Note 13**). Allow the tissue to settle and discard the supernatant.
5. Add sufficient collagenase–dispase medium to immerse the tissue and incubate at 37 °C in water bath.
6. Every 20 min, pipette up-down with a 10 mL serological pipette, allow the tissue pieces to settle to the bottom and collect fractions by transferring the supernatant to a new tube. Add fresh collagenase–dispase media and return the tube to the water bath to continue digestion.
7. After collecting each fraction, check the digest by transferring a drop onto a slide or petri dish. The aim is to isolate ductal structures. During the first hour the supernatant fraction only contains hepatocytes and very few ductal cells. These fractions are discarded.
8. Continue digestion by repeating **steps 4–6** until the whole tissue is disrupted (*see* **Note 14**).
9. Pool the fractions containing the ductal structures by centrifuging 300 rpm/18 rcf for 2–3 min (*see* **Note 6**).
10. It might be necessary to incubate again with collagenase–dispase to further disrupt the ducts. In that case, incubate for 10 min intervals at 37 °C with fresh collagenase–dispase and check after pipetting up and down.
11. Once the collagenase–dispase digestion is complete, repeat **step 8** and resuspend the ductal structures in 10 mL ADF.
12. Count and seed the ducts in Matrigel following **steps 9–12**, Sect. 3.1 and plate 30–50 ductal structures/well.
13. Add 500 μL of ADF-L culture medium to each well carefully along the side of the well to avoid dislodging the Matrigel, and incubate at 37 °C, 10 %CO_2, humidified incubator (*see* **Note 15**). Typically, the ductal cells form cysts within 1–2 days and begin to bleb/bud 3–4 days after culture.

14. Discard medium and replace with 500 μL fresh ADF-L every other day (Fig. 1 shows established organoids).

15. After 7–10 days culture the organoids need to be passaged (*see* **Note 11**). To passage the ductal organoids, follow **step 15** in Sect. 3.1 (*see* **Note 12**) and re-seed following **steps 9** to **12**, Sect. 3.1.

3.4 Differentiating Liver Ductal Oragnoids into Hepatocytes

1. Seed the ductal organoids into the required number of wells for an experiment following **steps 9** to **12**, Sect. 3.1, noting that once the organoids are differentiated to hepatocytes, they cannot be expanded further.

2. Add 500 μL of ADF-L culture medium to each well carefully along the side of the well to avoid dislodging the Matrigel, and incubate at 37 °C, 10 %CO_2, humidified incubator (*see* **Note 15**).

3. Two days after seeding, change the medium to ADF-LD. Replace with fresh ADF-LD every other day for the next 7–9 days (*see* **Note 16**).

4. After 7–9 days incubation in ADF-LD, change medium to ADF-LD supplemented with 3 μM dexamethasone (Sigma-Aldrich) and continue incubation for another 3 days to further differentiate towards an adult hepatocyte phenotype (*see* Fig. 1 for differentiated liver hepatocytes).

3.5 Cre-recombinase Mediated Gene Deletion

1. Seed the organoids while in an expansion phase in organoid expansion culture medium (ADF-I, ADF-S, or ADF-L depending on the organoids) in the appropriate number of wells required for the experiment.

2. Once organoids are established, typically 3–4 days after seeding, induce the Cre-recombinase enzyme by adding BNF (at 360 nM final concentration for intestinal organoids) or 4OHT (500 nM and 100 nM final concentration for intestinal and gastric organoids, respectively) depending on the Cre-recombinase transgene the compound mice harbor for 16–48 h at 37 °C, 10 %CO_2 (*see* **Note 17**).

3. Following Cre-induction, process organoid cultures for cellular and molecular analyses as required for the experiment.

4. To harvest the organoids for protein, DNA, or RNA extraction, the organoids are harvested in the same way as for passaging and the pellet is resuspended in the appropriate lysis buffer (*see* **Note 18**).

4 Notes

1. Recombinant R-spondin can be replaced by conditioned medium harvested from R-spondin producing cells, which is added at 20 % (v/v).
2. Recombinant Wnt 3a can be replaced by conditioned medium harvested from Wnt3a producing cells, which is added at 50 % (v/v).
3. Aim to not stretch the intestine while cutting it longitudinally.
4. A dissecting microscope can help with this step to better visualize scraping off of the villi without scraping off the crypts as well.
5. Passing the suspension through a cell strainer is essential as it removes larger pieces of tissue that has not dissociated properly.
6. The tube is centrifuged with lower speed (600 rpm/76 rcf for 2 min), so that single cells do not pellet.
7. To remove all of the supernatant ADF, take off as much as possible without disturbing the very lose pellet, then leave the tube on ice for several minutes and remove the remaining ADF with a filter tip. This is to ensure that the Matrigel is not diluted by ADF wash.
8. Keep the Matrigel suspension on ice while dispensing to the tissue culture plate and resuspend the suspension after 2–3 transfers to ensure the Matrigel suspension remains homogeneous.
9. Can use foam or cork board to pin out stomach with epidemiological pins to facilitate removing antral stomach.
10. Following the chelation buffer step, if vigorous shaking is not suitable (physical strain etc.), you can alternatively chop the antral stomach into small pieces on petri dish (1–2 mm) and place them into a 50 mL tube with 10 mL of cold PBS and using a serological 10 mL pipette, pipetting up and down 10–15 times, to dissociate glands from stromal tissue. This method often requires multiple (2–3) rounds of pipetting. Be sure to check/count how many glands are released following each round of pipetting.
11. The rate at which organoids grow and mature/bud will vary depending on how densely they are plated. However, if organoids are plated too densely this can be inhibitory to proper gastric organoid growth and maturation.
12. *Optional step*: pass through a 100 μM cell strainer when passaging the organoids.
13. This is a gentle wash to avoid destroying the tissue.

14. Typically it is necessary to incubate for about 2 h until ductal structures start to appear and then for another 2 h to release the ductal structures, i.e., a total of approximately 4 h.

15. Wnt3a and Noggin are not necessary for long term survival but help with initiating and establishing the organoid cultures, thus complete ADF-L is only added for the first 3 days and Wnt3a and Noggin are omitted after this time.

16. After 7–9 days, the ductal organoids should be differentiating to a hepatocytes phenotype and begin to express markers of adult hepatocytes such as albumin and cyp3a (cytochrome P450/family 3/subfamily A) [5, 8].

17. The agent used to induce the Cre-recombinase activity of the enzymes depends on which transgene is expressed. For example, 4OHT is used for *Villin-Cre^{ERT2}* while BNF is used for *AhCre* [7]. Parallel organoid culture are also treated with the carrier (DMSO or ethanol).

18. If the organoids are to be processed for immunohistochemical analysis, primary and secondary antibodies need to be pre-absorbed on Matrigel coated wells overnight at 4 °C. To coat the wells, add the minimum volume of Matrigel required to cover a 96-well flat bottom tissue culture tray, allow the Matrigel to set at 37 °C, then add the antibody [7].

References

1. Pampaloni F, Reynaud EG, Stelzer EH (2007) The third dimension bridges the gap between cell culture and live tissue. Nat Rev Mol Cell Biol 8(10):839–845
2. Barker N, Bartfeld S, Clevers H (2010) Tissue-resident adult stem cell populations of rapidly self-renewing organs. Cell Stem Cell 7(6):656–670
3. Barker N et al (2007) Identification of stem cells in small intestine and colon by marker gene Lgr5. Nature 449(7165):1003–1007
4. Barker N et al (2010) Lgr5(+ve) stem cells drive self-renewal in the stomach and build long-lived gastric units in vitro. Cell Stem Cell 6(1):25–36
5. Huch M et al (2013) In vitro expansion of single Lgr5+ liver stem cells induced by Wnt-driven regeneration. Nature 494(7436):247–250
6. Sato T et al (2009) Single Lgr5 stem cells build crypt-villus structures in vitro without a mesenchymal niche. Nature 459(7244):262–265
7. Flanagan DJ et al (2015) Frizzled7 functions as a Wnt receptor in intestinal epithelial Lgr5(+) stem cells. Stem Cell Reports 4(5):759–767
8. Huch M et al (2015) Long-term culture of genome-stable bipotent stem cells from adult human liver. Cell 160(1–2):299–312

The Three-Dimensional Culture of Epithelial Organoids Derived from Embryonic Chicken Intestine

Malgorzata Pierzchalska, Malgorzata Panek, Malgorzata Czyrnek, and Maja Grabacka

Abstract

The intestinal epithelium isolated from chicken embryos in last 3 days of development can be used to establish the 3D culture of intestinal organoids. When fragments of epithelial tissue released by incubation with EGTA (2.5 mM, 2 h) are embedded in Matrigel matrix on cell culture inserts the formation of empty spheres covered by epithelial cells is observed in first 24 h of culture. The growth and survival of organoids are supported by the addition of R-spondin 1, Noggin, and prostaglandin E_2 to the culture medium. The organoids are accompanied by myofibroblasts which become visible in the next 2 days of culture. The intestinal enteroids (free of myofibroblasts) can be obtained from adult chicken intestine.

Keywords: Intestinal epithelium, Three-dimensional cell culture, Epithelial organoids, Chicken embryo, Intestinal myofibroblasts

1 Introduction

For many decades avian embryos not only have been a remarkably useful model for cell biology studies [1] but also have been used as important source of many types of cell and tissue cultures. For example, chicken embryo fibroblasts are the cells, which were investigated in many pioneering and breakthrough research in cell [2] and cancer biology [3]. The numerous types of chicken cells, including embryonic and adult stem cells are also currently cultivated and studied [4]. There are two main reasons why *Gallus gallus* is the model animal whose development, immunology, gut function, and cell biology are relatively frequently studied in many laboratories worldwide. Firstly, the chicken embryos are easily available in all countries throughout the whole year. Secondly, poultry is the main source of meat in most human populations, regardless to religious restrictions. Therefore, the studies concerning many aspects of avian physiology are interesting, not

---

The original version of this chapter was revised. The correction to this chapter is available at DOI 10.1007/7651_2018_169

only for zoologists but also for veterinarian, and (as birds are the source of important zoonosis) from the medical point of view. Moreover, the gain of knowledge from scientific activity, which is connected with the use of chicken model systems, can be potentially applied by the profitable branches of husbandry and food industry.

In the last decade, there has been significant progress in the development of epithelial intestinal cells primary culture leading to the establishment of "mini-guts" model systems of various origin [5]. Although gut epithelium is a tissue of the profound proliferation potential and shows the fastest turnover rate in vertebrates, its cells practically cannot be efficiently propagated in the routine two-dimensional cell culture systems. This constraint limited substantially the modeling of physiological and pathological intestinal processes in vitro [6]. For many years technical challenges were an important obstacle for carrying out an in vitro research on intestinal immunity and gut–microbiota interactions and nutrition. Such studies were mainly performed on transformed or immortalized cell lines or on cells originated from colorectal and gastric tumors, which did not recapitulate the situation in the healthy or pathologically changed intestine. In case of mammalian laboratory species (mice and rats) and humans these difficulties were successfully overcome throughout the last few years due to employing the methods of three-dimensional culture of intestinal enteroids [7] or organoids [8, 9] embedded in extracellular matrix.

We have successfully introduced the method of intestinal organoid culture originated from intestinal tissue fragments isolated from 18- to 20-day-old chicken embryos and embedded in Matrigel [10]. The structures comprise both epithelial and mesenchymal cells. According to the nomenclature proposed in a recently published review by Finkbeier and Spence, they should be called intestinal organoids (in contrast to enteroids, which are built of epithelial cells only and derived from isolated intestinal epithelium or crypts or single intestinal epithelial cells) [11]. We use them to study various aspects of intestinal innate immunology, Toll-like receptor activation in intestinal epithelial cells, and probiotic bacteria influence on gut epithelial regeneration and maturation. Recently, we also established organoid culture from adult birds' intestine.

2 Materials

2.1 Reagents and Media

(a) PBS w/o Ca^{2+} and Mg^{2+}.
(b) PBS w Ca^{2+} and Mg^{2+}.
(c) Isolation solution—PBS w/o Ca^{2+} and Mg^{2+}, 0.5 % D-glucose (10 %, stock solution sterile, cell culture grade Sigma-Aldrich), EGTA (Sigma-Aldrich, 2.5 mM, diluted from the

0.25 M stock solution made in cell culture water and sterilized by filtration through 0.22 μm filter).

(d) Washing solution—Dulbecco's Modified Eagle's Medium (DMEM) with antibiotic–antimitotic solution—ZellShield™ (Minerva Biolabs, Germany).

(e) Cell culture medium components—DMEM/Ham's F12 1:1 mixture with GlutaMAX, antibiotic-antimitotic solution—ZellShield™, Insulin-transferrin-selenium reagent—ITS Premix (BD Biosciences, USA), Human recombinant R-spondin 1 (250 ng/ml, R&D Systems, USA), Human recombinant Noggin (25 ng/ml, R&D Systems, USA), EGF (25 ng/ml, PeproTech, Germany), PGE_2 (5 μg/ml, Cayman Chemicals, USA).

(f) BD Matrigel Basement Membrane Matrix (BD Biosciences).

(g) Chicken Serum (Pan-Biotech, Germany).

2.2 Cultureware and Equipment

(a) BD falcon cell culture inserts and 12-well companion plates (transparent, 0.4 μm pore size, 0.9 cm² area, BD Biosciences, USA).

(b) Plastic pipette tips, 1000 and 200 μl, "cell saver" type (with thick, smooth opening).

(c) Plastic tubes (15 and 50 ml), 0.5 ml conical Eppendorf tubes with high clarity.

(d) Falcon™ Cell Strainers (100 μm).

(e) Sterile scissors, small scissors, forceps, and needle pointed forceps.

(f) Tube rotator.

(g) Ice container.

3 Methods

1. Rinse and wipe four eggs with 19-day-old chicken embryos with 70 % ethanol and leave under the hood until shells are dry. Crack the shell, open it with the scissors, take out an embryo and decapitate. Place it in the cold petri dish on cooling blocks (*see* **Note 1**).

2. Isolate small intestines, wash them in a petri dish filled with ice-cold PBS containing Mg^{2+} and Ca^{2+} [PBS (Ca^{2+}, Mg^{2+})].

3. As quick as possible separate the intestine from loose mesenchymal tissue fragments and larger blood vessels, transfer to a new dish with cold PBS (Ca^{2+}, Mg^{2+}), cut into approximately 2-cm-long fragments, and dissect longitudinally with forceps to expose the inner epithelial lining (*see* **Note 2**).

4. Relocate the open pieces of intestine to a precooled 15-ml tube filled with 10 ml of cold PBS (lacking Mg^{2+} and Ca^{2+}), but containing 2.5 mM EGTA and 0.5 % glucose (isolation medium), and incubate with rotation at 4 °C for 15 min.

5. Allow the tube to stand for 2 min on ice to sediment tissue fragments. Discard carefully as much supernatant as possible (only about 0.5 ml of isolation solution should remain in the bottom of the tube). Add the fresh portion of isolation solution (10 ml) and incubate with shaking for 45 min.

6. Repeat **step 5**.

7. Place the 50 ml tube on ice, put 100-μm nylon cell strainer on it and filter the solution. Collect the tissue fragments from the sieve and transfer them into cold petri dish, add 5 ml of cold DMEM, mix vigorously with two broad ends of sterile 1 ml plastic tips. Collect the released tissue fragments and filter them through the same cell strainer into the 50 ml tube. Wash twice in cold DMEM by centrifugation ($1000 \times g$ for 5 min at 4 °C).

8. Resuspend the pellet containing liberated fragments of epithelium and single cells in 1.2 ml of DMEM and distribute 100 μl of cell suspension into 0.5 ml. Centrifuge Eppendorf tubes ($1000 \times g$ for 5 min at 4 °C) and place them on ice. Add 50 μl of cell culture medium to each tube. Use the cold cell saver pipette tips to resuspend the tissue fragments (*see* **Note 3**).

9. Prepare 50 μl of Matrigel in 0.5 ml Eppendorf tubes by transferring them from the freezer to the 4 °C (it needs to stay in the fridge for at least 5 h, optimally overnight to liquefy). Take the tubes from the fridge and place them on ice in the laminar flow cabinet.

10. Using cold "cell saver" tip (alternatively the end of normal tip can be cut off) transfer the cell suspension into equal volume of Matrigel solution. Mix twice by aspiration but so slowly and smoothly that no bubbles are created and pour on the center of 12-well plate cell culture insert. Force the layer to cover the whole area by gently rotating the insert. Add 750 ml of medium to the lower chamber and transfer the plate into the cell culture incubator (*see* **Note 4**).

11. Repeat **step 10** until all inserts are covered. Allow 50 % Matrigel to solidified—it usually takes about half an hour. Add gently (by touching the wall not the surface of insert with the tip) 250 μl of pre-warmed cell culture medium to all upper chambers.

12. During following week change the medium every other day and observe the formation of the organoids (*see* **Notes 5–7**). The spheroids are formed during the first 24 h of culture. Their diameter increase during the first 3 days reaching the diameter

Fig. 1 The growth dynamics of chicken embryo organoids. The changes in culture appearance were assessed by time-lapse recording with low magnification objective (5×) of the Zeiss Axio Observer.Z1 inverted microscope equipped with incubation chamber with CO_2 influx. The *numbers* in the *left upper corner* of all photos represent recording time (d.hh)

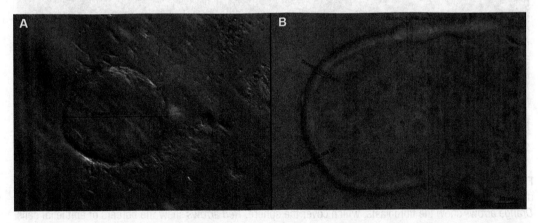

Fig. 2 The morphology and size of chicken embryo organoids in 4-day-old culture. Note the myofibroblasts around the large organoid (**a**) and the uneven thickness of the wall built of epithelial cells (**b**)

of about 250–1000 μm (Fig. 1). Most of organoids have the morphology of empty spheres, some also fuse and create structures of various shapes, resemble the elongated or branched tubes (Fig. 2). The apoptotic cells are visible in the inner lumens of spheroids. The epithelial wall is about 10–15 μm thick. From the second day the outgrowth of fibroblasts is usually observed around the majority of large epithelial spheroids (Figs. 2 and 3, *see* **Note 8**).

13. To obtain the organoids free of Matrigel (for RNA or protein isolation) it is enough to exchange the media for cold PBS and incubate for 1 h on ice. Than the cells can be collected by aspiration and washed by centrifugation (*see* **Note 9**).

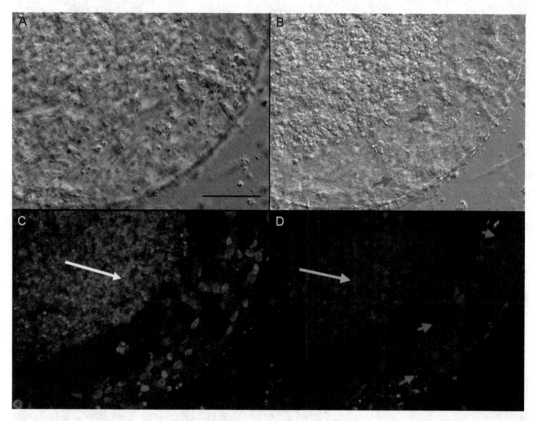

Fig. 3 The surface of large organoid from 4-day-old culture. The same organoid was observed with the differential interference optics focused above the outer surface (**a**) or focused on the epithelial cell layer (**b**). *Orange arrows* show the fibroblasts, which cover the sphere. *Red arrows* show the borders of epithelial cells. The results of cytoimmunofluorescence staining with anti-E-cadherin antibody reveal the presence of epithelial marker in the spheroid wall (**d**—the *green fluorescence*) and counterstaining with DNA stain, Hoechst 33258 shows the presence of the epithelial apoptotic cells and cells debris in the organoid lumen (**c**—the *blue fluorescence*)

4 Notes

1. Keep all the tissue culture plastic ware (e.g., dishes, pipette tips, tubes) ice cold. The tips that are used for Matrigel pipetting should be precooled in the freezer ($-20\ ^\circ C$).

2. During all steps of tissue isolation and culture establishment avoid warming the suspension and prolonging the procedures. The less time they take, the better.

3. The most difficult step is to adjust properly the amount of materials which should be used to prepare the culture in one insert. Counting the tissue fragments in Bürker's chamber (helpful with culture from adult animals' intestinal crypts) does not solve the problem, because the fragments are of different size and viability. We usually observe the size of the pellet in 0.5 ml Eppendorf tube—it should be visible, but not very big. Too much of initial material can spoil the outcome of the culture. Morphology of culture depends on the initial density. When the small amount of tissue fragments is used, the separated organoids are formed. With more materials at the beginning, the organoids form more elongated, branched tubular structure and tend to modify and contract extracellular matrix more easily. We roughly estimate that the four chicken embryos produce material sufficient to start the culture in one 12-well plate (12 inserts). But tissue yield and viability can vary from preparation to preparation and we usually observe the first insert under the inverted microscope to assess if we use proper initial density.

4. The volume of Matrigel–tissue suspension can be increased to 200 μl without affecting the organoids viability. Also the 6-well plates and larger inserts can be used. When the mixture is poured on the surface of normal culture dish the organoids rarely form and do not grow in size (or the growing is very much limited).

5. The small amount of chicken serum (up to 0.25 %) can be added to the culture medium. In such case, the growth of organoids is accelerated (Fig. 4). A higher percentage of serum blocks the tissue survival and organoids formation [10].

6. The human R-spondin-1 and Noggin can be omitted as the culture medium components although the organoids growth is slower without stimulation by these factors. In this case, the PGE_2 can be used in the higher concentration of 10 μg/ml.

7. It is very important not to use old or frequently frozen and thawed medium and media components. The complete media cannot be kept in the fridge for longer than 1 week. We routinely divided the media needed for particular culture and kept them in small volumes (5 ml) in the freezer. Only the amount, which is needed in a particular day is thawed and warmed before use. The medium poured on the Matrigel layer has to be prewarmed to

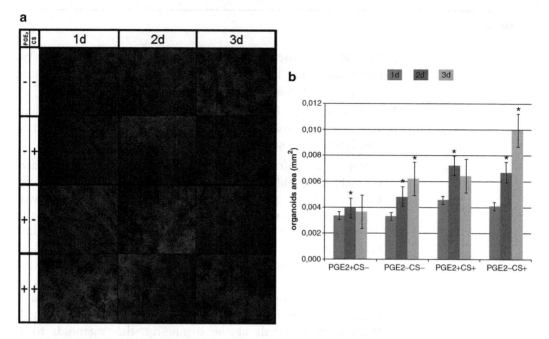

Fig. 4 The chicken serum and PGE$_2$ influence on the dynamic of organoids growth. Organoids were cultured in medium with or without PGE$_2$ (5 μg/ml) and with or without chicken serum (0.25 %, CS) and photographed on 1, 2, and 3 days (1d, 2d, 3d) from isolation (**a**). The area of at least 50 organoids was calculated with the image analysis tool (**b**). The bars represent the mean value (± standard deviation). The values statistically different from one another are marked by the same letters ($p < 0.5$, ANOVA test and contrast analysis regarding the media components)

37 °C. Cold medium might dissolve the Matrigel matrix upon contact.

8. In case of chicken embryonic tissue, we have never been able to obtain the pure enteroids—without myofibroblasts "contamination." The fibroblasts are strongly attached to the organoids surfaces and are almost not present in the area free of epithelial organoids (Fig. 3). When the same cell suspension to those used to establish organoid culture is plated directly on the plastic culture dish and incubated in DMEM supplemented with FCS (10–20 %) the fibroblast culture is not initiated. The organoid cultures can be also obtained from adult chicken intestine. In this case the outgrowth of fibroblasts is not usually observed (Fig. 5). Therefore, we assume that the myofibroblasts in embryonic organoid cultures originate from epithelial-mesenchymal transition process due to more plasticity of embryonic intestinal stem cells.

9. The released organoids are quite "sticky" and it is important not to let them remain attached to the walls of the pipette tips. The use of 1 ml or 200 μl "cell saver" tips are convenient but does not eliminate entirely the problem.

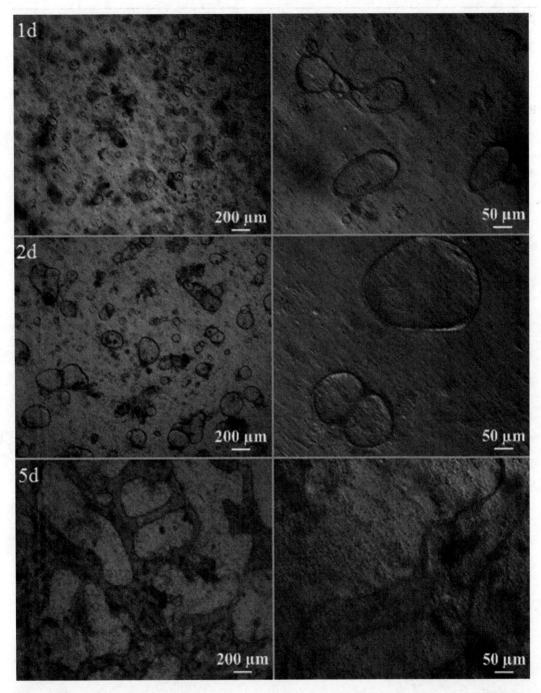

Fig. 5 The morphology of adult chicken organoid cultures. The organoids originated from adult intestine formed in Matrigel the structures very similar to those from embryonic tissue; however, in this case myofibroblasts outgrowth was not observed. The cultures were photographed on the first, second and fifth day of culture (1d, 2d, 5d) with differential interference contrast optics

Acknowledgment

The authors acknowledge the financial support of Polish National Science Center (grant no. **2013/09/B/NZ9/00285** to M.P.)

References

1. Kain KH, Miller JWI, Jones-Paris CR et al (2014) The chick embryo as an expanding experimental model for cancer and cardiovascular research. Dev Dyn 243:216–228. doi:10.1002/dvdy.24093
2. Brown S, Levinson W, Spudich JA (1976) Cytoskeletal elements of chick embryo fibroblasts revealed by detergent extraction. J Supramol Struct 5:119–130. doi:10.1002/jss.400050203
3. Stehelin D, Varmus HE, Bishop JM, Vogt PK (1976) DNA related to the transforming gene (s) of avian sarcoma viruses is present in normal avian DNA. Nature 260:170–173. doi:10.1038/260170a0
4. Intarapat S, Stern CD (2013) Chick stem cells: current progress and future prospects. Stem Cell Res 11:1378–1392. doi:10.1016/j.scr.2013.09.005
5. Sato T, Clevers H (2013) Growing self-organizing mini-guts from a single intestinal stem cell: mechanism and applications. Science 340:1190–1194. doi:10.1126/science.1234852
6. Chopra DP, Dombkowski AA, Stemmer PM, Parker GC (2010) Intestinal epithelial cells in vitro. Stem Cells Dev 19:131–142. doi:10.1089/scd.2009.0109
7. Sato T, Vries RG, Snippert HJ et al (2009) Single Lgr5 stem cells build crypt-villus structures in vitro without a mesenchymal niche. Nature 459:262–265. doi:10.1038/nature07935
8. Lahar N, Lei NY, Wang J et al (2011) Intestinal subepithelial myofibroblasts support in vitro and in vivo growth of human small intestinal epithelium. PLoS One 6:e26898. doi:10.1371/journal.pone.0026898
9. Spence JR, Mayhew CN, Rankin SA et al (2011) Directed differentiation of human pluripotent stem cells into intestinal tissue in vitro. Nature 469:105–109. doi:10.1038/nature09691
10. Pierzchalska M, Grabacka M, Michalik M et al (2012) Prostaglandin E2 supports growth of chicken embryo intestinal organoids in Matrigel matrix. Biotechniques 52:307–315. doi:10.2144/0000113851
11. Finkbeiner SR, Spence JR (2013) A gutsy task: generating intestinal tissue from human pluripotent stem cells. Dig Dis Sci. doi:10.1007/s10620-013-2620-2

New Trends and Perspectives in the Function of Non-neuronal Acetylcholine in Crypt–Villus Organoids in Mice

Toshio Takahashi

Abstract

Acetylcholine (ACh) is a neurotransmitter that is present in central, parasympathetic, and neuromuscular synapses of mammals. However, non-neuronal ACh is also predicted to function as a local cell signaling molecule. The physiological significance of the presence of non-neuronal ACh in the intestine remains unclear. Here, experiments using cultured crypt–villus organoids that lack nerve and immune cells led us to suggest that endogenous ACh is synthesized in the intestinal epithelium to evoke growth and differentiation of the organoids through activation of muscarinic ACh receptors (mAChRs). Extracts of cultured organoids exhibited a noticeable capacity for ACh synthesis that was sensitive to a potent inhibitor of choline acetyltransferase. Treatment of organoids with carbachol downregulated growth of organoids and expression of marker genes for each epithelial cell type. On the other hand, mAChR antagonists enhanced growth and differentiation of Lgr5-positive stem cells. Collectively, our data provide evidence that endogenous ACh released from mouse intestinal epithelium maintains the homeostasis of intestinal epithelial cell growth and differentiation via mAChRs.

Keywords: Non-neuronal cholinergic system, Intestine, Somatic stem cell, Growth and differentiation, Lgr5, Marker genes, Crypt–villus organoid

1 Introduction

Acetylcholine (ACh) functions as a classical neurotransmitter at synapses in the central and parasympathetic nervous systems and at neuromuscular junctions. In 1978, Sastry and Sadavongvivad [1] first observed expression of the cholinergic system outside the nervous system, demonstrating the presence of a non-neuronal cholinergic system. Moreover, ACh is also found in prokaryotes and eukaryotes including bacteria and plants [2–5]. Thus, in addition to the classical role in the nervous system, an emerging concept is that functional ACh is synthesized by and released from non-neuronal cells [5]. However, little is known about non-neuronal ACh, including how ACh may control physiological function in the intestine.

The structure and function of the small intestine vary along the crypt–villus (Fig. 1). The intestinal epithelium is a dynamic

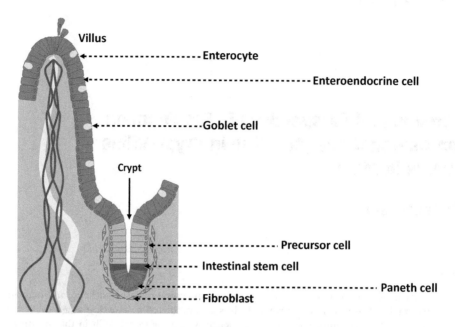

Fig. 1 A schematic illustration of the crypt–villus structure in the small intestine

structure that is constantly undergoing self-renewal as a result of continuous growth and differentiation. The intestinal epithelium consists of four lineages of differentiated cells: goblet cells, enteroendocrine cells, Paneth cells, and enterocytes [6]. In mice, cells are newly generated in crypts and are lost within 5 days following apoptosis at the tips of the villi. Self-renewing stem cells reside near the crypt bottom. Recently, Barker et al. [7] showed that Lgr5-positive cells are intestinal stem cells, which are interspersed between Paneth cells. Next, the same group established long-term culture conditions under which single crypts undergo multiple crypt fission events while simultaneously generating villus-like epithelial domains containing all types of differentiated cells [8]. However, Lgr5-positive stem cells cannot generate nerve and immune cells [8].

Currently, separation of the two groups of ACh-producing cells, neuronal ACh in the enteric nervous system and non-neuronal ACh in the intestinal epithelium, and detection of the function of non-neuronal ACh itself are difficult. Recently, we used mouse crypt–villus organoids and clearly showed that ACh is synthesized in non-neuronal cells in the intestinal epithelium and is involved in both the proliferation and differentiation of Lgr5-positive stem cells in the small intestine via muscarinic ACh receptors [9]. Thus, the crypt–villus organoid is a good assay system for studying the role of non-neuronal ACh in cell proliferation and differentiation. Here, we describe crypt isolation and crypt–villus organoid culture and methods for investigation of the growth and differentiation of crypt–villus organoids.

2 Materials

2.1 Crypt Isolation and Culture

1. C57/BL6: A standard wild-type strain of *Mus musculus* (*see* **Note 1**).
2. *Lgr5-EGFP-ires-CreERT2 mice*: The mice were purchased from Jackson Laboratory.
3. Matrigel (BD Biosciences).
4. Pen Strep (Penicillin: 10,000 units/mL, Streptomycin: 10,000 μg/mL) (Gibco).
5. Gentamicin sulfate solution (10 mg/mL) (Nacalai Tesque).
6. Advanced DMEM/F-12 (500 mL) (Gibco).
7. Sorbitol (Nacalai Tesque).
8. Fetal bovine serum (FBS) (Funakoshi).
9. Phosphate-buffered saline (PBS), pH 7.4 (Gibco).
10. 0.5 M EDTA (Nacalai Tesque).
11. PBS + ABx: PBS (500 mL) + Pen Strep (5 mL) + Gentamicin sulfate solution (2.5 mL).
12. Complete DMEM: Advanced DMEM/F-12 (500 mL) + Pen Strep (5 mL) + Gentamicin sulfate solution (2.5 mL) + FBS (5 mL).
13. Sorbitol DMEM: Sorbitol (1 g)/Complete DMEM (50 mL).
14. 10 % bovine serum albumin (BSA) in PBS (SIGMA).
15. Mouse epidermal growth factor (EGF) (20 μg/mL) (PEPROTECH): Add 100 μL water to one tube (100 μg) and dissolve completely in 5 mL PBS. Aliquot 250 μL of the stock solution into prechilled 1.5 mL tubes and store at −20 °C.
16. Recombinant mouse R-Spondin 1 (250 μg/mL) (R&D SYSTEMS): Add 200 μL PBS to one tube (50 μg). Aliquot 40 μL of the stock solution into prechilled 1.5 mL tubes and store at −20 °C.
17. Recombinant mouse Noggin (20 μg/mL) (R&D SYSTEMS): Add 1.25 mL 0.1 % BSA/PBS to one tube (25 μg). Aliquot 200 μL of the stock solution into prechilled 1.5 mL tubes and store at −20 °C.
18. Culture medium for crypt–villus organoids (Table 1).

2.2 PCR

1. cDNA synthesis kit (Invitrogen).
2. GeneAmp PCR System 9700 (Applied Biosystems).
3. CFX96™ Real-Time System (BIO-RAD).
4. SsoAdvanced™ Universal SYBR Green Supermix (BIO-RAD).

Table 1
Culture medium; upper table for 1–12 well(s) and lower table for 13–24 wells

Medium/well	Unit	1	2	3	4	5	6	7	8	9	10	11	12
P/S	μL	2.5	5	7.5	10	12.5	15	17.5	20	22.5	25	27.5	30
GM	μL	1.25	2.5	3.75	5	6.25	7.5	8.75	10	11.25	12.5	13.75	15
EGF (20 μg/mL)	μL	0.5	1	1.5	2	2.5	3	3.5	4	4.5	5	5.5	6
Noggin (20 μg/mL)	μL	2.5	5	7.5	10	12.5	15	17.5	20	22.5	25	27.5	30
mRspl (250 μg/mL)	μL	1	2	3	4	5	6	7	8	9	10	11	12
L-glutamine	μL	5	10	15	20	25	30	35	40	45	50	55	60
Advanced DMEM	mL	0.5	1	1.5	2	2.5	3	3.5	4	4.5	5	5.5	6

Medium/well	Unit	13	14	15	16	17	18	19	20	21	22	23	24
P/S	μL	32.5	35	37.5	40	42.5	45	47.5	50	52.5	55	57.5	60
GM	μL	16.25	17.5	18.75	20	21.25	22.5	23.75	25	26.25	27.5	28.75	30
EGF (20 μg/mL)	μL	6.5	7	7.5	8	8.5	9	9.5	10	10.5	11	11.5	12
Noggin (20 μg/mL)	μL	32.5	35	37.5	40	42.5	45	47.5	50	52.5	55	57.5	60
mRspl (250 μg/mL)	μL	13	14	15	16	17	18	19	20	21	22	23	24
L-glutamine	μL	65	70	75	80	85	90	95	100	105	110	115	120
Advanced DMEM	mL	6.5	7	7.5	8	8.5	9	9.5	10	10.5	11	11.5	12

2.3 Fluorescent Immunohistochemistry

1. 4 % paraformaldehyde (Nacalai Tesque). Store at 4 °C.
2. PBS containing 0.2 % polyoxyethylene p-t-octylphenyl ether (TX-100) (PBST).
3. Blocking solution: 1 % BSA (SIGMA) in PBST. Store at 4 °C.
4. Primary antibodies.
5. Fluorescent secondary antibodies.
6. Hoechst 33342 (AnaSpec).
7. Mowiol 4-88 (SIGMA): 0.2 M Tris–HCl, pH 8.5 (see **Note 2**).
8. Confocal immunofluorescence microscopy (FV1000; Olympus).

3 Methods

3.1 Crypt Isolation and Crypt–Villus Organoid Culture

Before carrying out crypt isolation and crypt–villus organoid culture, frozen Matrigel should be dissolved at 4 °C and kept on ice.

1. Isolate the small intestine, open the tissue longitudinally, and wash with cold PBS-ABx.
2. Cut the tissue into 5 × 5-mm pieces and wash with cold PBS-ABx. Collect the fragments into a 50-mL falcon tube.
3. Wash the fragments again with 30 mL cold PBS-ABx.
4. Incubate the washed tissue fragments in PBS-ABx containing 2 mM EDTA for 30 min on ice.
5. During the incubation time, place a 24-well plate into a 37 °C incubator (Panasonic).
6. After removal of the EDTA solution, shake the tissue fragments vigorously in cold PBS-ABx (30–40 times) (see **Note 3**).
7. Pass the resulting suspension through a 70-μm cell strainer (BD Biosciences) to remove residual villous material.
8. Centrifuge the isolated crypts at 390 × g for 3 min at 4 °C, and then discard the supernatant.
9. Resuspend the pellet in 20 mL sorbitol DMEM with pipetting and divide the solution into two 10-mL aliquots in two 15-mL falcon tubes.
10. Centrifuge the isolated crypts at 80 × g for 3 min at 4 °C, and then discard the supernatant (see **Note 4**).
11. Add 10 mL sorbitol DMEM to each tube. After suspension, centrifuge the isolated crypts at 80 × g for 3 min at 4 °C again.
12. After discarding the supernatant, add 10 mL sorbitol DMEM and centrifuge the isolated crypts at 80 × g for 3 min at 4 °C again.
13. After discarding the supernatant, add 10 mL complete DMEM and incubate on ice for 1 min (see **Note 5**).

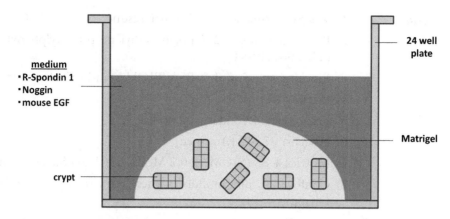

Fig. 2 A schematic illustration of three-dimensional culture of crypts and crypt–villus organoids

14. After 1 min, collect each 10-mL aliquot (total 20 mL) of supernatant and pass through a 70-μm cell strainer.
15. The final fraction should consist of essentially pure crypts. Count the purified crypts and pellet with centrifugation at 290 × *g* for 3 min at 4 °C in Complete DMEM.
16. Mix a total of 100 crypts (*see* **Note 6**) with 40 μL Matrigel and plate in 24-well plates (IWAKI) pre-warmed at 37 °C.
17. Incubate the 24-well plate for 15 min at 37 °C.
18. After polymerization of Matrigel, add 500 μL culture medium containing growth factors (20 ng/mL EGF, 500 ng/mL R-spondin 1, and 100 ng/mL Noggin) (Fig. 2). Begin crypt culture at 37 °C.

3.2 Subculture of Crypt–Villus Organoids

1. Discard old culture medium.
2. Wash three times with 1 mL cold PBS.
3. Add 1 mL BD Cell Recovery Solution (BD Biosciences).
4. Mechanically peel off the Matrigel and collect in a 15-mL Sumilon tube (SUMITOMO BAKELITE CO., LTD.).
5. Incubate for 15 min on ice with shaking.
6. Centrifuge at 200 × *g* for 1 min at 4 °C and carefully discard the supernatant.
7. Add 1 mL cold PBS and mechanically dissociate into single-crypt domains.
8. Centrifuge at 195 × *g* for 1 min at 4 °C and carefully discard the supernatant.
9. Add 40 μL fresh Matrigel, mix, and plate in a 24-well plate pre-warmed at 37 °C.
10. Incubate the plate for 15 min at 37 °C.
11. After polymerization of Matrigel, add 500 μL culture medium containing growth factors. Begin subculture of crypt–villus organoids in a 37 °C incubator (*see* **Note 7**).

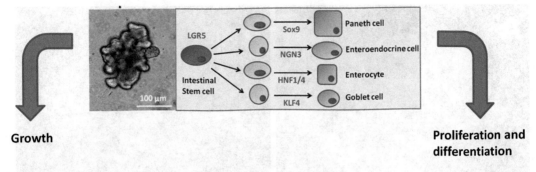

Fig. 3 Evaluation of organoid growth and differentiation with an assay system using marker genes (*Lgr5*, *Sox9*, *Ngn3*, *Hnf1*, and *Klf4*) to analyze the function of non-neuronal ACh

3.3 RT-PCR and Quantitative RT-PCR

1. Isolate crypt–villus organoids from the Matrigel (as in **steps 1–5**, Section 3.2).
2. Extract total RNA from the cultured organoids.
3. Synthesize cDNA using a suitable cDNA synthesis kit, using 3 μg total RNA as a template.
4. Use cDNA as a template for RT-PCR and quantitative RT-PCR. Amplify the intestinal stem cell marker gene and epithelial cell marker genes by PCR with designed PCR primers (Fig. 3). RT-PCR conditions: 3 min of initial denaturation at 94 °C; followed by 35 cycles of 94 °C for 30 s, 55 °C for 30 s, and 72 °C for 1 min; and then one final extension step of 72 °C for 5 min. Amplify glyceraldehyde 3-phosphate dehydrogenase (*GAPDH*) as an internal control.
5. Perform quantitative RT-PCR for intestinal stem cell and epithelial cell marker genes with the SYBR Green master mixture, according to the manufacture's procedure in triplicate. Quantitative RT-PCR conditions: 30 s of polymerase activation and DNA denaturation; followed by 45 cycles of 95 °C for 10 s and 55 °C for 30 s; then 65 °C for 5 s; and then 95 °C for 50 s for melt-curve analysis. Amplify *GAPDH* as an internal control.

3.4 Whole-Mount Immunohisto-chemistry

We visualized chromogranin A (enteroendocrine cells), mucin-2- (goblet cells), and lysozyme (Paneth cells) in whole mounts of organoids (Fig. 4).

1. Isolate crypt–villus organoids from the Matrigel (as in **steps 1–5**, Section 3.2).
2. Fix in 4 % paraformaldehyde for 60 min at room temperature (RT).
3. Wash three times for 15 min each in PBS.
4. Incubate in 0.2 % TX-100 solution for 20 min.
5. Wash three times for 15 min each in PBS.

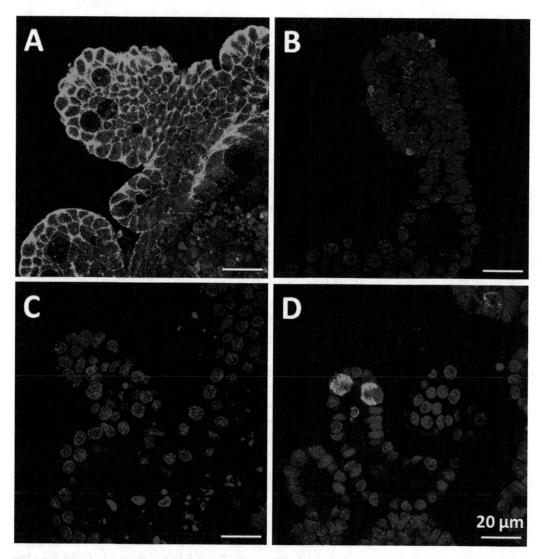

Fig. 4 Composition of crypt–villus organoids. (**a–d**) Visualization of (**a**) E-cadherin (*green*, adherens junctions), (**b**) chromogranin A (*red*, enteroendocrine cells), (**c**) mucin-2 (*red*, goblet cells), and (**d**) lysozyme (*green*, Paneth cells) in whole mounts of an organoid. Bars represent 20 μm

6. Block the samples with 1 mL blocking solution for 20 min.
7. Add 1 mL primary antibody diluted with blocking solution (1:1000) and incubate overnight at 4 °C.
8. Wash three times for 15 min each in PBS.
9. Add 1 mL secondary antibody and Hoechst diluted with blocking solution (1:1000 each) and incubate for 60 min at RT.
10. Wash three times for 15 min each in PBS.
11. Mount with Mowiol.

Table 2
List of drugs used in this study

Drug	Agonist or antagonist
Carbachol	ACh receptor agonist
Atropine	Muscarinic ACh receptor antagonist
Pirenzepine	M1 antagonist
AF-DX-116	M2 antagonist
4-DAMP	M3 antagonist
Tropicamide	M4 antagonist

3.5 Pharmacological Assay

Crypt–villus organoids contain a non-neuronal cholinergic system [9]. A pharmacological assay was performed with mechanically dissociated single-crypt domains (see Section 3.2), which were allowed to grow into organoids in the presence or absence of drugs (Table 2). The culture medium with or without drugs was replaced every other day. The extracts of the cultured organoids exhibit a noticeable capacity for ACh synthesis that is sensitive to a potent inhibitor of choline acetyltransferase [9].

3.5.1 Organoid Growth

1. Observe organoids with an inverted microscope and take a picture every day (1–7 days).
2. To assay growth, analyze organoid size with ImageJ 1.41 (NIH).

3.5.2 Cell Proliferation and Differentiation

The effects of the drugs on marker gene expression in intestinal stem cells (*Lgr5*) and the epithelium (*Sox9*, *Ngn3*, *Hnf1*, and *Klf4*; differentiation markers for Paneth cells, enteroendocrine cells, enterocytes, and goblet cells, respectively) (Fig. 3) were analyzed with RT-PCR and quantitative RT-PCR in treated and untreated (control) organoids.

1. After treatment for 7 days, extract total RNA from organoids (as in **step 1**, Section 3.3).
2. Synthesize cDNA (as in **step 2**, Section 3.3).
3. Perform RT-PCR.
4. Separate RT-PCR products with 1.5 % agarose gel electrophoresis.
5. Perform quantitative RT-PCR.
6. For relative quantification of gene expression, use the comparative C_T method.

3.6 Biological Assay

3.6.1 Lgr5-EGFP-ires-CreERT2 Organoids

1. Isolate crypts from *Lgr5-EGFP-ires-CreERT2* mice and establish the culture (*see* Section 3.1).
2. Incubate with carbachol (10^{-4} M) and atropine (10^{-5} M) for 3 days.
3. Dissociate into a suspension of single cells with TrypLE express (Invitrogen) for 30 min at 37 °C.
4. Count Lgr5-GFP-positive cells. Repeat three independent times.

3.6.2 Flow Cytometry Analysis

1. Dissociate the carbachol-treated, atropine-treated, and untreated control organoids by treating with TrypLE express for 30 min at 37 °C.
2. Pass through a 20-μm cell strainer (BD Biosciences).
3. Sort with BD FACSVerse (BD Biosciences) (*see* **Note 8**).
4. Analyze with BD FACSUITE (BD Biosciences).

4 Notes

1. This study was approved by the Committee of the Suntory Inc., and all animals were maintained in accordance with the guidelines of this committee for the care and use of laboratory animals. Mice used for small intestinal crypt isolation were 8–12 weeks of age.
2. Add 2.4 g Mowiol 4–88 to 6 g glycerol. Stir to mix. Add 6 mL H_2O and incubate for several hours at room temperature. Add 12 mL of 0.2 M Tris–HCl (pH 8.5) and heat to 50 °C for 10 min with occasional mixing. After the Mowiol dissolves, clarify by centrifugation at $5000 \times g$ for 15 min. Aliquot 500 μL of the stock solution into prechilled 1.5 mL tubes and store at −20 °C.
3. After suspension, aliquot 25 μL of the solution into 6-cm petri dishes. Observe the crypts in the aliquot with an inverted microscope (Nikon). The following steps should be carried out quickly.
4. As the pellet is easily broken, 2 mL of the supernatant should be kept in the tube.
5. Wait for 1 min. This is important because of the purified crypts gained.
6. Aliquot 25 μL of the solution into a 6-cm petri dish to examine two to three different aliquots and count the number of crypts. Calculate the average number of crypts counted (M) with the following formula:

$$[100 \times 25/M \times 25 \,(\mu L)]/well$$

7. Crypt–villus organoids that are cultured for >6 months retain their characteristics [9].
  8. Viable cells are gated by staining for 7-amino-actinomycin D (BD Biosciences). Mean fluorescence values of 2500 events per sample are read via flow cytometry.

Acknowledgement

This work was supported by a Grant-in-Aid for Scientific Research (C) (Grant number 26440184) from the Japan Society for the Promotion of Science (JSPS).

References

1. Sastry BV, Sadavongvivad C (1978) Cholinergic systems in non-nervous tissues. Pharmacol Rev 30:65–132
2. Horiuchi Y, Kimura R, Kato N, Fujii T, Seki M, Endo T et al (2003) Evolutional study on acetylcholine expression. Life Sci 72:1745–1756
3. Wessler I, Kilbinger H, Bittinger F, Kirkpatrick CJ (2001) The biological role of non-neuronal acetylcholine in plants and humans. Jpn J Pharmacol 85:2–10
4. Takahashi T, Hamaue N (2010) Molecular characterization of Hydra acetylcholinesterase and its catalytic activity. FEBS Lett 584:511–516
5. Wessler I, Kirkpatrick CJ (2008) Acetylcholine beyond neurons: the non-neuronal cholinergic system in humans. Br J Pharmacol 154:1558–1571
6. Cheng H, Leblond CP (1974) Origin, differentiation and renewal of the four main epithelial cell types in the mouse small intestine. V. Unitarian theory of the origin of the four epithelial cell types. Am J Anat 141:537–561
7. Barker N, van Es JH, Kuipers J, Kujala P, van den Born M, Cozijnsen M et al (2007) Identification of stem cells in small intestine and colon by marker gene Lgr5. Nature 449:1003–1007
8. Sato T, Vries RG, Snippert HJ, van de Wetering M, Barker N, Stange DE et al (2009) Single Lgr5 stem cells build crypt-villus structures in vitro without a mesenchymal niche. Nature 459:262–265
9. Takahashi T, Ohnishi H, Sugiura Y, Honda K, Suematsu M, Kawasaki T et al (2014) Non-neuronal acetylcholine as an endogenous regulator of proliferation and differentiation of Lgr5-positive stem cells in mice. FEBS J 281:4672–4690

Derivation of Intestinal Organoids from Human Induced Pluripotent Stem Cells for Use as an Infection System

Jessica L. Forbester, Nicholas Hannan, Ludovic Vallier, and Gordon Dougan

Abstract

Intestinal human organoids (iHOs) provide an effective system for studying the intestinal epithelium and its interaction with various stimuli. By using combinations of different signaling factors, human induced pluripotent stem cells (hIPSCs) can be driven to differentiate down the intestinal lineage. Here, we describe the process for this differentiation, including the derivation of hindgut from hIPSCs, embedding hindgut into a pro-intestinal culture system and passaging the resulting iHOs. We then describe how to carry out microinjections to introduce bacteria to the apical side of the intestinal epithelial cells (IECs).

Keywords Organoids, hIPSCs, Intestinal epithelium, Differentiation, Microinjection, Host–pathogen interactions

1 Introduction

Intestinal organoids provide a self-renewing, self-organizing, 3D in vitro model system for the intestinal epithelium. Intestinal organoids are rapidly becoming an essential tool for basic biology [1], including their use as a system for dissecting host–pathogen interactions [2–4]. Intestinal organoids also have multiple potential clinical applications, such as for diagnostic tests and large-scale drug screens [5]. Primary intestinal human organoids can be generated from adult intestinal stem cells isolated from intestinal biopsies [6]. An alternative route for generating intestinal human organoids (iHOs) is to start with human induced pluripotent stem cells (hIPSCs) and drive the hIPSCs to differentiate down the intestinal lineage. When embedded into an extracellular matrix, the iHOs organize into structures that resemble the architecture of the intestinal epithelium, exhibiting markers of a range of intestinal epithelial cell (IEC) types [7, 8].

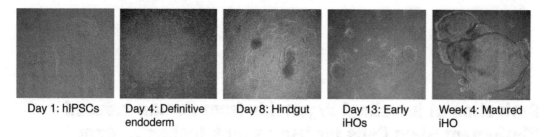

Day 1: hIPSCs | Day 4: Definitive endoderm | Day 8: Hindgut | Day 13: Early iHOs | Week 4: Matured iHO

Fig. 1 Differentiation of hIPSCs to iHOs. Time course showing the morphological changes accompanying the directed differentiation of hIPSCs to iHOs. Using defined concentrations of combinations of Activin A, FGF, BMP-4, LY294002, and CHIR99021, definitive endoderm is formed at differentiation day 4. After 8 days, patterning of this definitive endoderm with defined concentrations of CHIR99021 and retinoic acid results in the formation of hindgut. After moving this hindgut into a 3D intestinal culture system consisting of a supporting matrix of Matrigel and medium supplemented with pro-intestinal proliferation factors R-Spondin 1, Noggin, EGF, CHIR99021, and prostaglandin E2, small spheroids begin to form (Day 13). At this stage there may be some contaminating cell types, but after 4 weeks of sustained passaging, iHOs should be matured and contaminating cell types should be lost. Images were taken at ×100 magnification

hIPSCs provide an attractive progenitor cell type for iHO derivation because these cells are self-renewing, can be genetically manipulated, and are becoming more readily available to researchers through large iPSC banks [9]. In addition, hIPSCs are not lineage committed, so can be driven to differentiate into a range of different cell types [10–12].

Here, we provide protocols for the two-dimensional directed differentiation of hIPSCs to hindgut in a culture dish. We illustrate that by using different combinations of growth factors we can direct the differentiation of hIPSCs to endoderm, and then pattern that endoderm to hindgut (Fig. 1). This differentiation induces visible changes in the morphology of the starting colonies. We provide details on the processes for embedding hindgut into a three-dimensional, pro-intestinal culture system, and for passaging the iHOs as they mature. The resulting iHOs display features of the intestinal epithelium, such as tight junctions between epithelial cells, microvilli and both secretory and adsorptive IECs (Fig. 2). The ability to cryopreserve and indefinitely expand these iHOs makes them a tractable modeling system. To introduce pathogens into the luminal cavity of the iHO, we demonstrate the use of a microinjection system, which enables efficient delivery of bacteria to the apical side of the IECs. By using phenol red as a marker we are able to easily mark infected iHOs (Fig. 3). This provides an effective system for studying host–pathogen interactions at the intestinal interface.

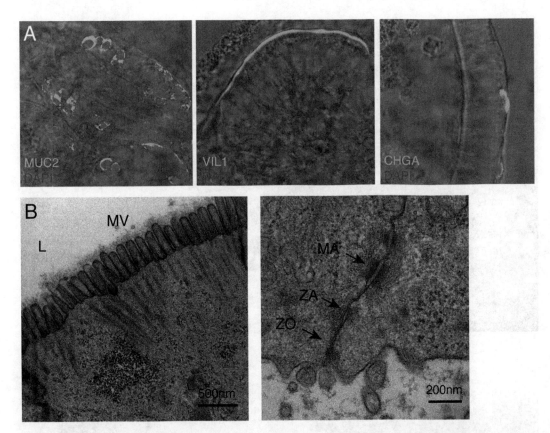

Fig. 2 iHOs developed from hIPSCs display morphology characteristic of the intestinal epithelium. After 6 weeks in culture, iHOs display markers of differentiated IECs including the Goblet cell marker MUC2, Enteroendocrine cell marker CHGA and enterocyte marker VIL1 (A). Using transmission electron microscopy (TEM), structures such as microvilli (MV) on the luminal (L) surface of the IECs, and structures closely resembling tight junctions between cells (ZO, zonula occludens; ZA, zonula adherens; MA, maculae adherens), should be visible (B). Phenotyping of iHOs generated from different hIPSC lines is important, as there can be some line-to-line variation

2 Materials

Prepare all solutions at room temperature and work under aseptic conditions at all times. Reconstitute growth factors in Dulbecco's phosphate-buffered saline (DPBS) (No calcium or magnesium), 0.1 % bovine serum albumin (BSA) or dimethyl sulfoxide (DMSO), as per manufacturer's instructions for each individual product. Store growth factors in aliquots of appropriate volumes at −20 °C. Store all cell culture medium at 4 °C; do not keep prepared media for longer than 6 weeks. Add all growth factors fresh to cell culture media immediately before use.

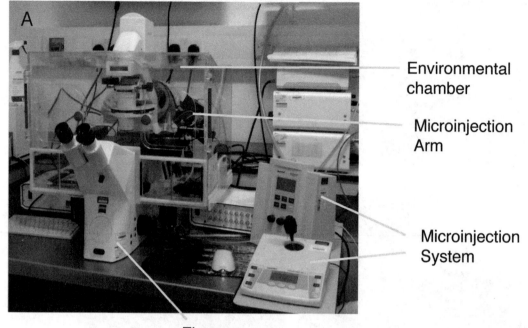

Fig. 3 Microinjection of iHOs. (**A**) shows the setup of a microinjection system, which allows for injection of the iHOs in a controlled environment (37 °C, 5 % CO_2). Using the Eppendorf TransferMan NK2-FemtoJet express system fitted with microcapillaries (**B**), bacterial inoculum can be delivered into the luminal cavity of the iHOs. By mixing bacterial inoculums with phenol red, injected iHOs can be marked for downstream processing (**C**). Images were taken at ×100 magnification

2.1 Stem Cell Growth and Passaging

1. Chemically defined medium supplemented with polyvinyl alcohol (CDM PVA)—Add 250 ml Iscove's Modified Dulbecco's Medium (IMDM) and 250 ml Ham's F-12 Nutrient Mix, GlutaMAX™ Supplement to a 500 ml sterile vacuum filter unit. Prepare PVA solution by weighing 10 g PVA powder and adding to 200 ml Water for Embryo Transfer. Mix on heated magnetic stirrer set to 100 °C (*see* **Note 1**). Allow to cool to room temperature and add 10 ml of resulting PVA

solution to the filter unit. Add 20 μl 1-thioglycerol, 5 ml concentrated lipids, and transferrin at a final concentration of 15 μg/ml. Reconstitute insulin at 10 mg/ml in Water for Embryo Transfer (*see* **Note 2**) and add 350 μl to the filter unit. Optional: add 5 ml penicillin–streptomycin (10,000 U/ml). Filter-sterilize.

2. MEF Media for plate coating—Add 450 ml of Dulbecco's Modified Eagle Medium (DMEM), 50 ml fetal bovine serum (FBS), 5 ml L-glutamine (200 mM) to a 500 ml sterile vacuum filter unit. Optional: add 5 ml penicillin–streptomycin (10,000 U/ml). Filter-sterilize.

3. Gelatin for plate coating—Weigh 500 mg gelatin powder (porcine) and add to 500 ml Water for Embryo Transfer. Place solution in a 60 °C water bath for 30 min, swirling regularly to ensure gelatin is dissolved. Once gelatin is completely dissolved transfer solution to a 500 ml sterile vacuum filter unit and filter-sterilize. Store at 4 °C and use within 2 months.

4. Dispase—Weigh 500 mg of dispase powder and add to 500 ml Advanced DMEM/F12. Swirl to mix, and once powder is fully dissolved transfer solution to a 500 ml sterile vacuum filter unit and filter-sterilize. Aliquot and store at −20 °C for up to 6 months.

5. Collagenase—Weigh out 500 mg of collagenase IV powder and add to 400 ml Advanced DMEM/F12. Add 100 ml Knockout™ Serum Replacement, 5 ml L-glutamine (200 mM), and 3.5 μl β-Mercaptoethanol and swirl to mix. Once powder is fully dissolved transfer solution to a 500 ml sterile vacuum filter unit and filter-sterilize. Aliquot and store at −20 °C for up to 6 months.

2.2 Organoid Differentiation and Passaging

1. RPMI/B27 medium—Add 500 ml of RPMI Medium 1640 with GlutaMAX™ supplement, 10 ml B27 Supplement (50×, serum free) and 5 ml nonessential amino acids to a 500 ml sterile vacuum filter unit. Optional: add 5 ml penicillin–streptomycin (10,000 U/ml). Filter-sterilize.

2. iHO Base growth medium—Add 500 ml of Advanced DMEM/F12 supplement, 10 ml B27 Supplement (50×, serum free), 5 ml N2 Supplement (100×, serum free), 5 ml HEPES (1 M) and 5 ml nonessential amino acids (100×) to a 500 ml sterile vacuum filter unit. Optional: add 5 ml penicillin–streptomycin (10,000 U/ml). Filter-sterilize.

3 Methods

Carry out all procedures at room temperature (unless otherwise specified) and work under aseptic conditions at all times.

3.1 Growing hIPSCs in CDM PVA Culture System

1. Prepare maintenance plates for CDM-PVA culture at least 24 h before passaging hIPSCs. Coat 10 cm plate with 10 ml 0.1 % gelatin and incubate at 37 °C for 30 min. Remove gelatin and add 12 ml MEF media. Incubate at 37 °C for at least 24 h (*see* **Note 3**). Pre-coated plates may be stored for up to 5 days at 37 °C.

2. Take selected hIPSC line growing on CDM PVA, aspirate media and add 3 ml collagenase and 3 ml dispase per 10 cm plate. Incubate at 37 °C for 20–50 min (*see* **Note 4**). If undifferentiated colonies are still attached, incubate for a further 5–10 min. Whilst hIPSC colonies and enzymes are incubating, add 10 ml CDM-PVA supplemented with 10 ng/ml Activin A and 12 ng/ml basic fibroblast growth factor (bFGF) (*see* **Note 5**) to pre-washed, pre-coated maintenance plate.

3. Harvest floating colonies into a 15 ml falcon and add 6 ml CDM-PVA media to inactivate enzymes. Centrifuge at $115 \times g$ for 1 min (*see* **Note 6**).

4. Aspirate media and gently resuspend pelleted cells in 10 ml CDM PVA, centrifuge at $115 \times g$ for 1 min. Remove supernatant and resuspend in 1 ml CDM PVA supplemented with 10 ng/ml Activin A and 12 ng/ml bFGF.

5. Using a P1000 pipette, break colonies by firmly pipetting 3–4 times, collecting and dispensing the cellular solution (*see* **Note 7**). Allow the larger aggregations to settle and move the medium sized aggregates to a new 15 ml falcon. Add another 1 ml of CDM PVA supplemented with 10 ng/ml Activin A and 12 ng/ml bFGF and repeat breaking process. Move the medium sized aggregates to the new falcon, and discard leftover small and large fragments.

6. Add 300–500 μl of hIPSC solution to the preprepared 10 ml of CDM PVA supplemented with 10 ng/ml Activin A and 12 ng/ml bFGF on the maintenance plate, depending on cell density. Rock the plate gently to evenly disperse clumps of cells and check density under a microscope. If density is too low, add more hIPSC solution until desired density is achieved (*see* **Note 8**).

7. Place freshly passaged cells into an incubator at 37 °C, 5 % CO_2. Gently rock the plate again once in place in the incubator to make sure colonies are evenly spaced.

8. Replace media with 12 ml fresh CDM PVA supplemented with 10 ng/ml Activin A and 12 ng/ml bFGF every 24 h (*see* **Note 9**). hIPSCs should be passaged every 4–7 days.

3.2 Differentiation: From hIPSC to Hindgut

The differentiation protocol requires 10 days of media changes to hIPSCs growing on a 10 cm plate. Over these 10 days the morphology of the colonies of the plate should change dramatically, moving from hIPSC colonies to definitive endoderm to hindgut (Fig. 1).

Day 0: Split hIPSCs into CDM PVA maintenance conditions on a 10 cm plate as described above.

Day 1: Replace media with 12 ml fresh CDM PVA supplemented with 10 ng/ml Activin A and 12 ng/ml FGF.

Day 2: Begin differentiation to definitive endoderm by replacing media with CDM PVA supplemented with 100 ng/ml Activin A, 100 ng/ml bFGF, 10 ng/ml BMP-4, 10 μM phosphoinositol 3-kinase inhibitor LY294002, and 3 μM GSK3 inhibitor CHIR99021 (*see* **Note 10**).

Day 3: Replace media with 12 ml CDM PVA supplemented with 100 ng/ml Activin A, 100 ng/ml bFGF, 10 ng/ml BMP-4, and 10 μM LY294002.

Day 4: By day 4, the formation of definitive endoderm should be visible (Fig. 1). Replace media with 12 ml RPMI/B-27 medium supplemented with 100 ng/ml Activin A and 100 ng/ml bFGF.

Day 5: Replace media with 12 ml RPMI/B-27 supplemented with 50 ng/ml Activin A.

Days 6–9: Begin patterning definitive endoderm to hindgut by replacing media with 12 ml RPMI/B-27 supplemented with 6 μM CHIR99021 and 3 μM retinoic acid. By day 8–9 3D structures of hindgut should be visible, covering the 10 cm plate (Fig. 1).

Day 10: embed hindgut in Matrigel supporting matrix (*see* **Note 11**):

1. Remove media from differentiation plate and wash once with DPBS (No Ca or Mg).

2. Add 5 ml collagenase, incubate at 37 °C. Check every 5 min under microscope. If hindgut starts to dissociate from plate proceed to next step. If no dissociation of hindgut is observed, after 20 min proceed to next step.

3. Add 5 ml of iHO base growth media to inactivate enzymes, and gently detach hindgut by scraping the plate with a cell scraper.

4. Pipette the hindgut suspension to a 15 ml falcon and centrifuge at $250 \times g$ for 1 min. Aspirate supernatant.

5. Resuspend the pelleted hindgut in 10 ml iHO base growth media, and break into smaller pieces by pipetting firmly with a 10 ml serological pipette. Centrifuge at $100 \times g$ for 1 min. Repeat wash step two more times.

6. Resuspend fragments of hindgut in a small volume of iHO base growth media (300–500 μl dependent on pellet size). Add ~150 μl of this hindgut suspension to 1.5 ml Matrigel (Growth factor reduced, phenol red free) (*see* **Note 12**) supplemented with 500 ng ml^{-1} R-spondin 1, 100 ng/ml Noggin, 100 ng/ml epidermal growth factor (EGF), 3 μM CHIR99021, 10 μM Y-27632 dihydrochloride monohydrate, and 2.5 μM prostaglandin E_2.

7. Spot 60 μl of Matrigel-hindgut suspension into the center of 1 well of a 24 well plate (*see* **Note 13**), allow Matrigel-hindgut droplet to set for 1 min and check density under microscope. Add more hindgut solution to Matrigel in increments until desired concentration is achieved. Spot out the rest of the Matrigel-hindgut suspension to fill a 24-well plate.

8. Incubate at 37 °C for at least 15 min. Overlay each spot of Matrigel with 800 μl iHO base growth medium supplemented with 500 ng/ml R-spondin 1, 100 ng/ml Noggin, 100 ng/ml EGF, 3 μM CHIR99021, 10 μM Y-27632, and 2.5 μM prostaglandin E_2.

9. Change iHO base growth media every 2–3 days, omitting Y-27632, which is only required when splitting/seeding. After initial seeding into Matrigel allow iHOs 7 days to develop before splitting. By day 3–4 distinct spheres should be visible in the culture (Fig. 1).

3.3 Maintenance and Passaging of Organoids

iHOs require at least 1 month of routine passaging after seeding to facilitate maturation, with splitting required every 4–7 days (*see* **Note 14**).

1. Aspirate media from iHOs and add 500 μl Cell Recovery Solution per well. Incubate at 4 °C for 40–50 min. Gently move the iHO solution to a 15 ml falcon, without breaking iHOs, and allow iHOs to settle. Aspirate supernatant.

2. Resuspend iHOs in 10 ml iHO base growth media and centrifuge at 115 × *g* for 2 min (*see* **Note 15**).

3. Remove supernatant and resuspend in 200–500 μl base growth medium. Using a P1000 pipette, break iHOs by vigorously pipetting 10–20 times, collecting and dispensing the iHO solution. Take 100 μl of the iHO solution and add to desired volume of Matrigel supplemented with 500 ng/ml R-spondin 1, 100 ng/ml Noggin, 100 ng/ml EGF, 3 μM CHIR99021, 10 μM Y-27632, and 2.5 μM prostaglandin E_2

4. Place a 60 μl droplet of Matrigel into the center of one well of a 24-well plate, leave to solidify for 1 min, and then check the density under a microscope. Add more iHO solution in 100 μl

increments to the Matrigel, until the desired concentration is achieved. Plate out the remaining iHO and Matrigel solution and incubate at 37 °C for at least 15 min (see **Note 16**).

5. Overlay set Matrigel with 800 μl iHO base growth media supplemented with 500 ng/ml R-spondin 1, 100 ng/ml Noggin, 100 ng/ml EGF, 3 μM CHIR99021, 10 μM Y-27632, and 2.5 μM prostaglandin E_2, changing iHO base growth media every 2–3 days, with the omission of Y-27632.

6. iHOs may be cryopreserved for long-term storage by removing iHOs from their Matrigel Matrix as described above in **step 1**, washing in iHO base growth media and resuspending in Recovery Cell Culture freezing medium. Move iHOs in freezing medium to cryogenic vials and place in a cell freezing chamber at −80 °C O/N. Move cryogenic vials to liquid nitrogen for long-term storage.

3.4 Microinjection of iHOs with Salmonella Typhimurium

The apical side of the iHO IECs is located on the inside of the spherical iHO structure. Microinjection provides a solution for introducing stimuli to this luminal surface. To perform microinjections, an Eppendorf TransferMan NK2-FemtoJet express system can be fitted to a standard fluorescence microscope with an environmental chamber, to allow all injections to be carried out at 37 °C and 5 % CO_2. Injections are performed with bacterial inoculums mixed with phenol red, to mark infected organoids for downstream processing.

1. After at least 6 weeks in culture, 4–5 days prior to infection, iHOs can be prepared for microinjection by passaging as described above up to **step 3**.

2. Plate Matrigel-iHO suspension into glass bottom microinjection dishes, but rather than plating droplets, the Matrigel-iHO suspension is spread into a thin layer in a circle in the center of the microinjection dish, and incubated at 37 °C for at least 15 min. Once solidified, this layer is covered with 3 ml iHO base growth medium (see **Note 17**), supplemented with 500 ng/ml R-spondin 1, 100 ng/ml Noggin, 100 ng/ml EGF, 3 μM CHIR99021, 10 μM Y-27632, and 2.5 μM prostaglandin E_2, changing supplemented iHO base growth media every 2–3 days, with the omission of Y-27632.

3. The day prior to microinjections, set up Salmonella Typhimurium cultures in Luria–Bertani broth to grow at 37 °C overnight with shaking. The following morning, dilute cultures in DPBS to an optical density at 600 nm (OD600) of 2 and then mix 1:1 with phenol red (see **Note 18**).

4. Load microinjection dish onto microscope stage once chamber is at the required temperature and remove the dish lid. If it is not already, rotate the injection arm back away from the

microscope stage. Load the Eppendorf Transferman microcapillaries (see **Note 19**) using Eppendorf microloaders with 10 μl of the bacterial/phenol red inoculum, and carefully attach to microinjection arm (see **Note 20**). See Fig. 3 for suggested set up of injection equipment.

5. Set the injector to the "course" setting and slowly move the arm up by rotating the joystick anticlockwise, whilst it is still pulled back to the right. Very slowly (manually) lower the arm down so that the needle is hovering about 1–2 cm above the microinjection dish. Change the injector control to the "fine" setting. Carefully rotate the needle down to just above the media by twisting the joystick clockwise. Stop when the needle is just above the surface of the media.

6. Focus a 10× objective up through the iHOs and find the tip of the needle. Once the tip is in focus, move the objective back down towards the iHOs and move the needle down in tandem, keeping focus on the needle. When the needle hits the surface of the media set this as "position 1" on the microinjector system by holding down the "Pos 1" button until it beeps.

7. Focus back down to the same plane as the iHOs, and move the stage to find an iHO suitable for injection. Move the stage so this iHO is just to the left of the "Pos 1" needle position. Ensuring that the injector is still set to fine, bring the needle down very slowly towards the iHO, by rotating the joystick clockwise. The iHO will start to move slightly as the needle gets closer.

8. When the needle and the iHO are in the same plane use the joystick to move the needle fairly quickly towards the iHO, inserting it into the side. Once the needle has broken into the lumen of the iHO press the "inject" button on the injector (see **Note 21**). Once the inoculum is inserted move the needle back out of the iHO. Press "Pos 1" once (see **Note 22**) to move the needle back to its original position and repeat protocol from **step 7** to **8** until required number of iHOs are injected. Injected iHOs can then be processed for subsequent phenotyping.

4 Notes

1. PVA is not soluble in culture media, so solution must be made, boiled and cooled, before adding separately.
2. Aliquot insulin and store at −20 °C for future use.
3. Both of these medias are toxic to cells so they must be removed from plates and washed with DPBS (no Ca, no Mg) immediately before use, without allowing time for plates to dry out.

4. Check plate regularly during this time, begin next step as hIPSC colonies start to lift off from plate. Do not leave unattended during this stage as hIPSC colonies should be harvested as soon as they begin to dissociate from plate.

5. Add Activin A and bFGF from frozen aliquots freshly to media just before use.

6. If cells do not pellet increase speed and/or time of spin, however the hIPSC colonies should be treated gently and therefore it is better to spin slowly and for a shorter duration where possible.

7. Do not pipette too harshly or too many times, as this will result in single cells.

8. For maintenance conditions, colony density can be lower, for differentiation it is advisable to increase density approximately two- to fourfold.

9. Media must be changed daily or stem cells will start to differentiate or lose viability, due to rapid FGF degradation at physiologically relevant temperatures. hIPSCs therefore need a new source of FGF every 24 h.

10. As with hIPSC maintenance media, growth factors must be added daily to media from frozen aliquots.

11. Put Matrigel on ice at 4 °C O/N so that it liquefies. Matrigel starts to gel rapidly at temperatures from 22 °C to 35 °C, so work quickly and keep Matrigel on ice throughout protocol. We have found it is not necessary to prechill pipettes as long as you work quickly, and that sometimes prechilling can result in iHO-Matrigel droplets not setting correctly. We recommend using Corning® Matrigel® Growth Factor Reduced (GFR) Basement Membrane Matrix, Phenol Red-Free, as this is the only extracellular matrix we have used for this protocol.

12. We recommend obtaining R-Spondin 1, Noggin and EGF recombinant proteins from R&D Systems.

13. We recommend using a plate heater, or pre-warming plates before seeding so that Matrigel droplets set quickly and do not run. If you work quickly you do not need to chill pipettes, but be aware that Matrigel solidifies rapidly at room temperature.

14. Split iHOs if luminal cavities start to fill up with dead cells (i.e., luminal cavity appears black under microscope); if the Matrigel starts to disintegrate; if the iHOs grow too large and bud out from the Matrigel; the culture is too dense and the media starts to discolor rapidly; if none of the above occur split on day 7.

15. At this point, if there is a lot of contaminating material for example, we sometimes find it useful to move iHO suspension

16. into a 6-well plate and use an in-hood imaging system to select the best iHOs for further passaging, pipetting them into a fresh base growth media in a 15 ml falcon.

16. This 15-min incubation is very important, if media is added before the Matrigel is properly set the droplets will start to break down and the iHOs will not be properly encased.

17. At this point iHO Base Growth Media must not contain penicillin–streptomycin.

18. We recommend using a high concentration of bacteria; we found lower concentrations were insufficient to generate a response from iHO IECs, and it was difficult to subsequently locate internalized bacteria via microscopy. Inoculums may have to be optimized for different bacterial strains.

19. We have found microcapillaries with a 6 μm inner diameter and a 25° tip angle work well.

20. These microinjection capillaries are very fine and can break easily. Handle with care at all times. Do not move the capillaries through the Matrigel to search for iHOs to inject, always bring needle out of the Matrigel before scanning plate as needle may break and the Matrigel will be disturbed.

21. If the microinjector stops delivering inoculum, the "clean" function can be used to clear any blockage in the microcapillary.

22. Do not hold down "Pos 1" at this point, as it will reset the position, tap the button only once, fairly sharply.

Acknowledgments

This work was supported by the Medical Research Council and the Wellcome Trust. We thank David Goulding for providing TEM images.

References

1. Fatehullah A, Tan SH, Barker N (2016) Organoids as an in vitro model of human development and disease. Nat Cell Biol 18(3):246–254
2. Finkbeiner SR, Zeng XL, Utama B, Atmar RL, Shroyer NF, Estes MK (2012) Stem cell-derived human intestinal organoids as an infection model for rotaviruses. MBio 3(4), e00159-12, doi:ARTN e00159.10.1128/mBio.00159-12
3. Zhang YG, Wu S, Xia Y, Sun J (2014) Salmonella-infected crypt-derived intestinal organoid culture system for host-bacterial interactions. Physiol Rep 2(9), e12147, doi:10.14814/phy2.12147
4. Leslie JL, Huang S, Opp JS, Nagy MS, Kobayashi M, Young VB, Spence JR (2014) Persistence and toxin production by Clostridium difficile within human intestinal organoids results in disruption of epithelial paracellular barrier function. Infect Immun 83:138–145. doi:10.1128/IAI.02561-14
5. Date S, Sato T (2015) Mini-gut organoids: reconstitution of the stem cell niche. Annu Rev Cell Dev Biol 31:269–289. doi:10.1146/annurev-cellbio-100814-125218
6. Jung P, Sato T, Merlos-Suarez A, Barriga FM, Iglesias M, Rossell D, Auer H, Gallardo M, Blasco MA, Sancho E, Clevers H, Batlle E

(2011) Isolation and in vitro expansion of human colonic stem cells. Nat Med 17 (10):1225–1227. doi:10.1038/nm.2470

7. Forbester JL, Goulding D, Vallier L, Hannan N, Hale C, Pickard D, Mukhopadhyay S, Dougan G (2015) Interaction of Salmonella enterica Serovar Typhimurium with intestinal organoids derived from human induced pluripotent stem cells. Infect Immun 83 (7):2926–2934. doi:10.1128/IAI.00161-15

8. Hannan NR, Fordham RP, Syed YA, Moignard V, Berry A, Bautista R, Hanley NA, Jensen KB, Vallier L (2013) Generation of multipotent foregut stem cells from human pluripotent stem cells. Stem Cell Reports 1(4):293–306. doi:10.1016/j.stemcr.2013.09.003

9. McKernan R, Watt FM (2013) What is the point of large-scale collections of human induced pluripotent stem cells? Nat Biotechnol 31(10):875–877. doi:10.1038/nbt.2710

10. Birket MJ, Ribeiro MC, Verkerk AO, Ward D, Leitoguinho AR, den Hartogh SC, Orlova VV, Devalla HD, Schwach V, Bellin M, Passier R, Mummery CL (2015) Expansion and patterning of cardiovascular progenitors derived from human pluripotent stem cells. Nat Biotechnol 33(9):970–979. doi:10.1038/nbt.3271

11. Lancaster MA, Renner M, Martin CA, Wenzel D, Bicknell LS, Hurles ME, Homfray T, Penninger JM, Jackson AP, Knoblich JA (2013) Cerebral organoids model human brain development and microcephaly. Nature 501 (7467):373–379. doi:10.1038/nature12517

12. Hale C, Yeung A, Goulding D, Pickard D, Alasoo K, Powrie F, Dougan G, Mukhopadhyay S (2015) Induced pluripotent stem cell derived macrophages as a cellular system to study Salmonella and other pathogens. PLoS One 10(5):e0124307, doi:ARTN e0124307. 10.1371/journal.pone.0124307

Murine Colonic Organoid Culture System and Downstream Assay Applications

Yang-Yi Fan, Laurie A. Davidson, and Robert S. Chapkin

Abstract

Colonic organoids, three-dimensional colonic crypts grown in vitro that show realistic microanatomy, have many potential applications for studying physiology, developmental biology, and pathophysiology of intestinal diseases including inflammatory bowel disease and colorectal cancer. Here, we describe detailed protocols for mouse colonic crypt isolation, organoid culture, and downstream applications. Specific culture strategies including growth factor enriched Matrigel and Wnt and R-spondin conditioned media serve as key factors for enhancing the growth and cost efficiency of colonic organoid cultures.

Keywords: Niche-independent crypt expansion, Stem cells, In vitro primary culture

1 Introduction

The ability to grow intestinal tissue-derived cultures in vitro, broadly referred to as organoid culture, has allowed the gastrointestinal research field to move away from transformed cell lines to a more physiologically relevant "mini-gut" model. For example, the unique three-dimensional culture system allows the crypt, the basic functional unit of the intestine, to expand to generate organoids, comprising multiple crypt-like domains surrounding a central lumen lined by an epithelium, similar to the basic crypt physiology in vivo. Significant research has highlighted the use of organoid systems to characterize the adult stem cell niche and gut morphogenesis. This organoid primary culture system is being applied to study intestinal physiology and pathophysiology with great expectations for translational applications, including regenerative medicine [1, 2]. Interestingly, despite the fact that small intestinal cancer is fairly rare compared to colorectal cancer, most studies utilize organoids derived from small intestine. Few studies have been conducted in colonic crypt-derived organoids, or colonoids. This may be due to the fact that colonic organoids are more difficult to culture and maintain.

Building on the protocol developed in the Clevers lab [3, 4], our laboratory has refined the culture system with unique features suited to colon tissue, resulting in higher growth efficiency and lower culture maintenance costs. Furthermore, several downstream applications are included in this chapter to provide insight into potential usage of this unique in vitro culture system.

2 Materials

Most ingredients for organoid culture should be prepared as stock solutions, aliquoted (*see* **Note 1**) and kept at −20 °C or −80 °C for prolonged storage.

2.1 Commercially Available Materials

Advanced DMEM/F-12 medium (ADF) (Gibco #12634-010); DMEM, high glucose with GlutaMAX (Gibco #10569-010); non-essential amino acids (Gibco #11140-050); Zeocin (Gibco #R25001); GlutaMAX (Gibco 35050-061); Ca/Mg free Dulbecco's PBS (Gibco #14190-144); growth factor reduced basement membrane matrix (Matrigel) (Corning #356231); recombinant mouse epidermal growth factor (EGF) (Gibco #PMG8041, working stock [100 μg/mL in PBS + 0.5 % FBS]); LDN-193189 (Cellagen Technology #C5361-2s, working stock [0.2 mM in PBS + 0.5 % FBS]), (*see* **Note 2**); N2 supplement, 100×, (Gibco #17502-048); B27 supplement minus vitamin A, 50×, (Gibco #12587-010); N-acetyl-L-cysteine (Sigma #A9165, working stock [400 μM in water]); Y-27632 (Sigma #Y0503, working stock [10 mM in ADF]) (*see* **Note 3**). ADF+: ADF medium, supplemented with 2 mM GlutaMAX, 100 units/mL penicillin–streptomycin, and 10 μM HEPES.

2.2 Wnt Conditioned Medium Generation

1. L-Wnt3A cell medium: DMEM high glucose with GlutaMAX supplemented with 10 % FBS, 100 units/mL penicillin–streptomycin, and 1× nonessential amino acids.

2. Grow L-Wnt3A cells (ATCC #CRL-2647) (*see* **Note 4**) in one T-75 flask. Add Zeocin to the culture medium at a final concentration of 125 μg/mL.

3. When confluent, split cells into two T-175 flasks. Add fresh Zeocin to the culture medium at a final concentration of 125 μg/mL.

4. When confluent, split each flask into four T-175 flasks with 25 mL medium per flask, *without* addition of Zeocin.

5. After 96 h of culture, collect media and centrifuge at 200 × g for 5 min to remove cells and debris. Collect the supernatant and filter through 0.22 μm filter. Note that the cells will become over-confluent during the growth period.

6. This is the first batch of Wnt conditioned medium. Aliquot and store at −80 °C.

7. (Optional): Fresh medium can be added to the same flasks, and the culture continued (without Zeocin) for another 5 (batch 2) or 10 days (batch 3). Collect the conditioned medium as described in **step 5** (*see* **Note 5**).

2.3 R-Spondin Conditioned Medium Generation (See Note 6)

1. Initial growth medium: DMEM high glucose with GlutaMAX supplemented with 10 % FBS and 100 units/mL penicillin–streptomycin.

2. Grow RSPO1 cells (HA-R-Spondin1-Fc, Trevigen #3710-001-01) in one T-175 flask in initial growth medium. Add fresh Zeocin to the culture medium at a final concentration of 125 µg/mL.

3. When confluent, split cells into six T-175 flasks, culture in initial growth medium *without* addition of Zeocin.

4. When confluent, aspirate initial growth medium, change to ADF+ medium (50 mL/flask). Continue incubation for 7 days (*see* **Note 7**).

5. Collect media and centrifuge at 200 × *g* for 5 min to remove cells and debris. Collect the supernatant and filter through 0.22 µm filter. Aliquot and store at −80 °C.

3 Methods

3.1 Isolation of Colonic Crypts for Organoid Culture

1. Thaw Matrigel *on ice*, keep at 4 °C. Prechill 200 and 1000 µL pipet tips in a 4 °C refrigerator the day before crypt isolation (*see* **Note 8**).

2. Prepare complete organoid medium by adding the following growth factors to ADF+ medium at the final concentrations given: EGF [50 ng/mL], LDN-193189 [0.2 µM], N2 supplement [1×], B27 supplement [1×], N-acetylcysteine [1 µM], R-Spondin conditioned medium [10× dilution] and Wnt conditioned medium [2× dilution]. 500 µL of complete media is needed per organoid well.

3. Prepare growth factor enriched Matrigel (GF-Matrigel) by adding the following growth factors to Matrigel (*see* **Note 9**): EGF [50 ng/mL], LDN-193189 [0.2 µM], N2 supplement [1×], B27 supplement [1×], N-acetylcysteine [1 µM], R-Spondin conditioned medium [10× dilution], and Y-27632 [10 µM]. Keep on ice. 50 µL of growth factor enriched Matrigel is needed per well of organoids to be plated. Due to the viscosity of Matrigel, significant loss occurs in the pipet tips. Therefore, make allowance for approximately 15 % extra volume for plating.

4. Make fresh 20 mM EDTA in Ca^{2+}/Mg^{2+} free PBS, adjust to pH 7.4. Warm to 37 °C in a water bath.

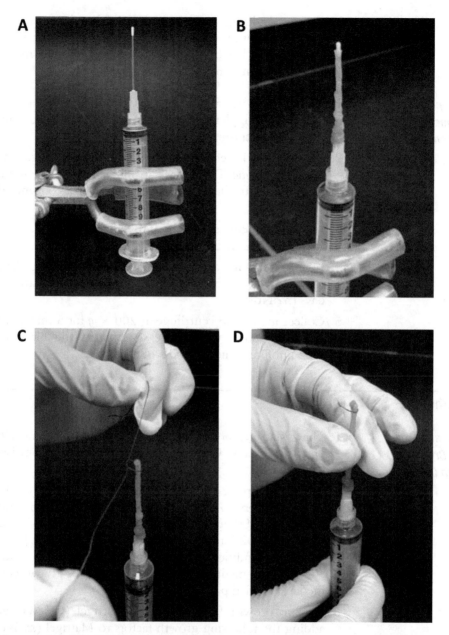

Fig. 1 Colon eversion station for colonic crypt isolation. (**a**) Disposable gavage needle on a 10 mL syringe. Note the white tip which will help hold the colon in place. (**b**) Mouse colon situated on the needle. (**c**) Thread is tied tightly around the colon, just under the white tip of the plastic needle. (**d**) By gently pulling upward, the colon everts on itself and can be turned completely inside out, remaining tied to the needle

5. Set up a colon eversion station by positioning a disposable 10 mL syringe upright in a clamp and attach a flexible disposable feeding needle (such as Soloman Scientific #FTP-20-38) onto the syringe (Fig. 1a).

6. Fill a 5 mL syringe with cold PBS and attach a reusable animal feeding needle (such as Popper & Sons, #7922).

7. For each tissue, prepare three 50 mL conical tubes containing 30 mL cold PBS (*see* **Note 10**) supplementing one tube with 100 units/mL penicillin–streptomycin.

8. Euthanize the mouse with CO_2 asphyxiation followed by cervical dislocation.

9. Remove the colon and place in a cup of cold PBS.

10. Remove excess fat from the outside of the colon using a forceps.

11. Using the preloaded 5 mL syringe from **step 6**, perfuse the colon in order to flush out feces.

12. Gently thread the proximal end of the colon (wider end) onto the disposal gavage needle. Once the entire colon passes through the tip of the needle (Fig. 1b), tie the distal end onto the needle with a piece of thread (Fig. 1c) and cut off the extra length. Evert the tissue by placing thumb and index finger on each side of the colon (Fig. 1d) and gently pulling upward. The colon will easily slide over itself and evert. Be careful to not damage the now-exposed inner lining of the colon.

13. Place the tissue, attached to the gavage needle, into the conical tube containing 30 mL cold PBS + penicillin–streptomycin. Keep on ice.

14. Vortex this tube at 80 % of maximum, $5\times$, 5 s each, to remove remaining fecal material, swirling the tube to untangle the colon between vortexing cycles.

15. Use a forceps to transfer the colon/needle to another conical tube containing 30 mL cold PBS. Vortex at maximum speed $3\times$, 5 s each, again detangling colon after each cycle.

16. Transfer colon/needle to the pre-warmed 20 mM EDTA in PBS in a 50 mL conical tube. Incubate at 37 °C in a water bath for 35 min.

17. Meanwhile, warm a 24-well culture plate (such as Costar #3524) in a 37 °C cell culture incubator (*see* **Note 11**).

18. Following incubation, transfer the colon still on the gavage needle to a conical tube containing 30 mL cold PBS and vortex at maximum speed $8\times$, 5 s each to release crypts, untangling the colon after each cycle by swirling tube. (*The PBS should turn cloudy, from crypts released from the colon tissue*). Figure 2a shows an aliquot of freshly released crypts viewed under the microscope.

19. Remove colon/needle and discard. Add 0.6 mL FBS to the tube containing crypts to yield a final concentration of 2 % and spin down the crypts at $125 \times g$ for 3 min.

20. Aspirate solution and resuspend crypts with 10 mL cold ADF^+ and transfer to a 15 mL conical tube.

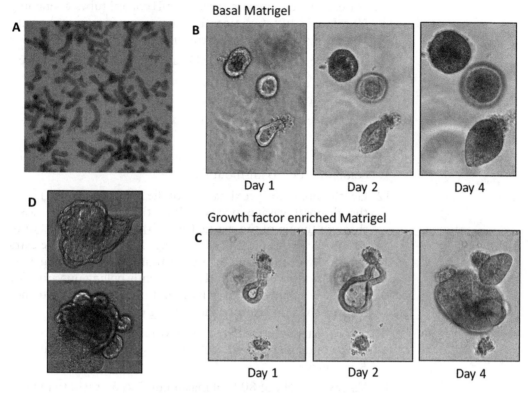

Fig. 2 Crypts and organoids at various stages during isolation and growth. Note that crypts and organoids are three dimensional and, hence, the single plane images shown can appear out of focus in some areas. (**a**) An aliquot of freshly detached crypts is show at ×10 magnification. As can be seen, the crypts are isolated as mainly intact units. (**b**) Growth of representative organoids in basal Matrigel (no growth factors added) and (**c**) GF enriched Matrigel (×20 magnification). By day 4, both basal and GF Matrigel groups show maturing organoids especially around the edges of the Matrigel mound, with the 100 % GF group exhibiting more highly developed organoids compared to the basal group. (**d**) Examples of highly developed organoids at day 6 of growth in the GF enriched group (×20 magnification)

21. Centrifuge at 70 × g for 2 min. This step helps remove single cells.

22. Remove supernatant and gently resuspend pellet with 10 mL ADF⁺ to repeat the wash.

23. Pipet a 10 μl aliquot onto a slide and count crypts in the entire 10 μl volume. (*Typical yield ~ 80–120,000 crypts from one mouse colon which is 80–120 crypts in the 10 μl aliquot*).

24. Transfer the required number of crypts (750 crypts/well to be plated) into a 15 mL conical tube, fill the tube with ADF⁺ to ~10 mL in order to resuspend the crypts well, and spin down at 100 × g for 3 min.

25. Thoroughly discard the supernatant and keep the crypts cold at all times. All buffer must be removed from the crypt pellet to avoid Matrigel becoming diluted. Thinning the Matrigel with

buffer will cause it to spread when plated, rather than forming a mound. If the pellet becomes loose and the remaining buffer cannot be removed, spin the tube again for 30 s at $70 \times g$. Carefully aspirate the remaining buffer with a P-200 tip.

26. Place the *prechilled* pipet tip boxes on ice and use the chilled tips to gently but thoroughly resuspend the crypt pellet in GF-Matrigel at the density of ~750 crypts/50 μL Matrigel. Avoid bubbles. *See* **Note 12**.

27. Using the prechilled pipet tips and holding the pipet perpendicular to the well, seed 50 μL of Matrigel–crypt mixture to the center of the wells of the pre-warmed 24-well plate, and incubate for 5–10 min in 37 °C incubator until solidified. (*The droplet of Matrigel should remain in the center of the well and not spread*).

28. During the Matrigel solidification step, prewarm the complete organoid medium to 37 °C.

29. Once the Matrigel is solidified, add 500 μL warm complete organoid medium to each well without touching the Matrigel mound. (*The medium should barely cover the Matrigel–crypt mound*).

30. Incubate at 37 °C in a CO_2 incubator. The plate is now ready for application of ex vivo treatments. Change complete medium every 2–3 days as needed. Representative images of organoid culture in basal vs. GF enriched Matrigel are shown in Figs. 2b, c and 3.

Basal Matrigel   **GF enriched Matrigel**

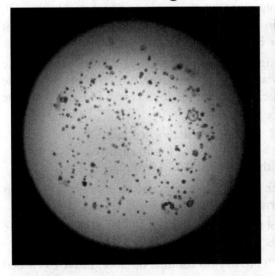
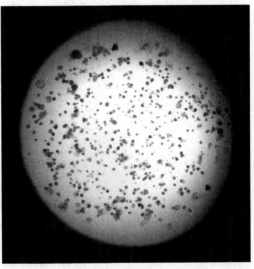

Fig. 3 Whole well organoid growth. The entire 50 μl Matrigel mound is shown 6 days after plating. Note that more vigorous growth occurs around the edges of the mound. Significantly increased organoid number and size can be seen in the growth factor enriched Matrigel at right (×4 magnification)

3.2 Organoid Harvest for Immunohistochemistry (IHC) Analysis

1. Place organoid culture plates on ice. Carefully aspirate and discard culture medium, add 0.5 mL ice cold ADF$^+$ and mechanically break up the Matrigel by pipetting up and down multiple times using a 1000 μL pipet. Transfer the dissociated Matrigel into a 15 mL conical tube and keep on ice. Typically, several wells of organoids (~5–8) are pooled for one embedded sample in order to obtain sufficient organoid pieces on each slide cut from the block.

2. Add additional cold ADF$^+$ to the tube to ~10 mL to dissociate and wash the organoids from the Matrigel.

3. Centrifuge at $100 \times g$ for 3 min.

4. Remove supernatant and wash once in cold PBS. Centrifuge at $100 \times g$ for 3 min.

5. Resuspend the pellet in 2 mL 4 % PFA in PBS for 10 min at room temperature for fixation. Stop reaction by adding 4 mL 0.1 M glycine in PBS.

6. Centrifuge at $100 \times g$ for 3 min. Remove supernatant and wash once in PBS.

7. Dehydrate the samples in a series of ethanol washes:

    Resuspend the pellet in 2 mL 30 % ethanol on ice for 10–20 min. Centrifuge at $100 \times g$ for 3 min. Repeat with 50 % ethanol and then 70 % ethanol. (*Can be stored at 4 °C in 70 % ethanol until ready to embed*).

8. After removal of 70 % ethanol, resuspend the organoids in 2 mL of 96 % ethanol then transfer the fixed, washed organoids to a flat bottom glass vial (VRW, #66011-041) using a plastic disposable pipet.

9. Add an equal volume of 0.1 % eosin in 96 % ethanol to the vial, which will aid in visualization of the organoids in subsequent steps. Let sit for 30 min at RT.

10. For the following washes, allow the organoids to settle to the bottom of the vial and aspirate the solution with the disposable pipet, tipping the vial at an angle so the organoids collect in the corner.

    Wash with 2 mL fresh 96 % ethanol, 10 min.
    Wash with 100 % ethanol, 3×, 10 min each.
    Finally, wash with n-butanol, 3×, 10 min each.

11. Heat a dry block to 58 °C. Place a glass dropper into the wells of heater to warm (FisherBrand #14-955-216)

12. Aspirate butanol from organoids and add liquid paraffin to the vial of organoids.

13. Keep vial at 58 °C to dry off residual butanol. Leave to embed for 30 min.

14. Perform two more changes of paraffin, 30 min each.

15. Using a warm glass dropper, suck out organoids in paraffin and transfer into a 1 × 1 cm metal embedding mold at room temperature.
16. Use a heated metal poker to push organoids to the center of the mold.
17. Briefly chill the mold on cold plate until the organoids adhere, then cover with a tissue cassette and add additional liquid paraffin so the mold is full and let set at room temperature.
18. The organoid blocks are now ready for sectioning onto slides for IHC applications. If using reporter mice, such as the Lgr5-EGFP-IRES-creER^2 transgenic line [5], co-localization of GFP-positive stem cells can be done with other targets of interest.

3.3 Single Cell Harvesting from Organoid Culture for Further Analyses (See Note 13)

1. Harvest organoids as described in Section 3.2, **steps 1–3**.
2. Carefully remove supernatant and wash again with 10 mL cold ADF$^+$. Spin down and aspirate media.
3. Resuspend organoids in 1 mL of 0.25 % Trypsin-EDTA (Gibco #25200-056) at 37 °C for 5 min.
4. After incubation, gently pass suspension through a 20-gauge needle 3× to aid dissociation. Incubate for 1 min at room temperature and pass through needle one more time.
5. Add 5 mL ice-cold 5 % FBS in ADF$^+$ to stop reaction. Spin down at 500 × g for 4 min.

   The cell pellet is now ready for resuspending in appropriate buffers/reagents for further analyses, such as fluorescence activated cell sorting.

4 Notes

1. Aliquot growth factors to avoid repeated freeze–thaw cycles. We typically do not freeze-thaw the growth factors more than 3 times.
2. LDN-193189 is a synthetic, highly potent bone morphogenic pathway inhibitor. It can be used as a more economical replacement for Noggin [6] while providing similar growth support for the organoid culture.
3. Y-27632 is a highly potent, cell-permeable, reversible, and selective ROCK inhibitor [7]. It is required for growing organoids from single colonic cells, but optional for organoid culture from crypts.
4. The L-Wnt3A cell line is commercially available from ATCC (#CRL-2647). A new cell line is now available which produces

Wnt-3A, R-spondin 3, and noggin (ATCC #CRL-3276). This cell line was not tested for this manuscript.

5. Multiple batches of Wnt-conditioned media can be collected from the cultures. In our experience, the 1st batch of conditioned medium (5 days culture) consistently provides superior growth support compared to media collected from 10 or 15 days cultures.

6. R-Spondin can be purified from the conditioned media using Protein A Agarose via the Fc tag on the protein. However, in our hands, conditioned media supports growth of colonic organoids equivalent to that of organoids grown with the purified protein.

7. RSPO expressing cells will become very confluent and will lift in patches but will continue to grow in suspension.

8. Prechilled pipet tips are used for pipetting the liquefied Matrigel, helping to avoid solidification of Matrigel during the plating procedures.

9. The addition of growth factors to the Matrigel, modified from Ahmad et al. [8], greatly enhances the viability and growth of organoids. Without addition of growth factors, the viability of colonic organoids 1 day after plating is ~10 % compared to 30–50 % with the addition of growth factors to the Matrigel. The organoids also grow larger and with more structural complexity.

10. Keep the volume to 30 mL as recommended. Subsequent vortexing steps do not work efficiently if the volume is over 40 mL.

11. Pre-warming of the culture plate will facilitate the solidification of Matrigel after plating the crypts.

12. Manipulation of Matrigel takes some practice. When resuspending crypts, take care not to push air into the Matrigel. If bubbles do occur within your Matrigel mound, however, the organoids will grow with no problem. If you are plating an entire 24-well plate, pipet half of the wells, place the plate in the incubator to solidify, then complete the plating in the rest of the wells. This helps avoid sedimentation of the crypts to the plate surface while the Matrigel is liquid, where they will attach and grow in two dimensions.

13. Organoid cultures can be digested to a single cell suspension, which can be used for flow cytometry [9], bioenergetic profiling [10], or other assays. In addition, harvested organoids can be incubated with Annexin V before single cell isolation, for apoptosis measurement by flow cytometry [11]. Proliferation assays can also be performed by treating the organoid culture 4 h prior to harvest with 10 μM EdU (Molecular

Probes Click-It kit #C10340) and following kit instructions after removing the organoids from the Matrigel (Section 3.2, **steps 1–4**). Use of a stem cell reporter line, such as Lgr5-EGFP-IRES-creER^2 will allow identification of GFP-positive stem cells in the assays described above.

Acknowledgement

This work was supported by funding from NIH R35CA197707 and P30ES023512. We would like to thank Drs. Hans Clevers, Nick Barker, and Toshiro Sato for initial guidance in organoid culture technique.

References

1. Sato T, Clevers H (2013) Growing self-organizing mini-guts from a single intestinal stem cell: mechanism and applications. Science 340:1190–1194
2. Zachos NC, Kovbasnjuk O, Foulke-Abel J, In J, Blutt SE, de Jonge HR et al (2016) Human enteroids/colonoids and intestinal organoids functionally recapitulate normal intestinal physiology and pathophysiology. J Biol Chem 291:3759–3766
3. Sato T, Vries RG, Snippert HJ, van de Wetering M, Barker N, Stange DE et al (2009) Single Lgr5 stem cells build crypt-villus structures in vitro without a mesenchymal niche. Nature 459:262–265
4. Sato T, Clevers H (2013) Primary mouse small intestinal epithelial cell cultures. Methods Mol Biol 945:319–328
5. Barker N, van Es JH, Kuipers J, Kujala P, van den Born M, Cozijnsen M et al (2007) Identification of stem cells in small intestine and colon by marker gene Lgr5. Nature 449:1003–1007
6. Zhao J, Li S, Trilok S, Tanaka M, Jokubaitis-Jameson V, Wang B et al (2014) Small molecule-directed specification of sclerotome-like chondroprogenitors and induction of a somitic chondrogenesis program from embryonic stem cells. Development 141:3848–3858
7. Ohata H, Ishiguro T, Aihara Y, Sato A, Sakai H, Sekine S et al (2012) Induction of the stem-like cell regulator CD44 by Rho kinase inhibition contributes to the maintenance of colon cancer-initiating cells. Cancer Res 72:5101–5110
8. Ahmad AA, Wang Y, Gracz AD, Sims CE, Magness ST, Allbritton NL (2014) Optimization of 3-D organotypic primary colonic cultures for organ-on-chip applications. J Biol Eng 8:9
9. Fan YY, Davidson LA, Callaway ES, Goldsby JS, Chapkin RS (2014) Differential effects of 2- and 3-series E-prostaglandins on in vitro expansion of Lgr5+ colonic stem cells. Carcinogenesis 35:606–612
10. Fan YY, Davidson LA, Callaway ES, Wright GA, Safe S, Chapkin RS (2015) A bioassay to measure energy metabolism in mouse colonic crypts, organoids, and sorted stem cells. Am J Physiol Gastrointest Liver Physiol 309:G1–G9
11. DeClercq V, McMurray DN, Chapkin RS (2015) Obesity promotes colonic stem cell expansion during cancer initiation. Cancer Lett 369:336–343

Intestinal Organoids as a Novel Tool to Study Microbes–Epithelium Interactions

Giulia Nigro, Melissa Hanson, Cindy Fevre, Marc Lecuit, and Philippe J. Sansonetti

Abstract

The gut, particularly the colon, is the host of approximately 1000 bacterial species, the so-called gut microbiota. The relationship between the gut microbiota and the host is symbiotic and mutualistic, influencing many aspects of the biology of the host. This homeostatic balance can be disrupted by enteric pathogens, such as *Shigella flexneri* or *Listeria monocytogenes*, which are able to invade the epithelial layer and consequently subvert physiological functions. To study the host–microbe interactions in vitro, the crypt culture model, known as intestinal organoids, is a powerful tool. Intestinal organoids provide a model in which to examine the response of the epithelium, particularly the response of intestinal stem cells, to the presence of bacteria. Furthermore, the organoid model enables the study of pathogens during the early steps of enteric pathogen invasion.

Here, we describe methods that we have established to study the cellular microbiology of symbiosis between the gut microbiota and host intestinal surface and secondly the disruption of host homeostasis due to an enteric pathogen.

Keywords: Intestinal organoids, Host–microbe interactions, *Listeria monocytogenes*, Bacterial products, Flow cytometry, Microinjection

1 Introduction

The microbiota, as well as bacterial products and bacterial metabolites, provides continuous stimuli to the entire intestinal epithelial layer, possibly indirectly affecting stem cells that might sense signals from neighboring cells responding to bacterial agonists. These signals may influence the survival of stem cells and therefore control both proliferation and regeneration of the whole epithelium or conversely, pathogens may subvert the physiological functions of stem cells [1].

In the past, cell lines have been extensively used to study not only the invasive capacity of a pathogen, but also to define the response of eukaryotic cells to said pathogen.

Most of the cell lines used in this kind of research are immortalized through a process called transformation, which can occur spontaneously or can be chemically or virally induced, acquiring the ability to divide indefinitely, and this does not reflect the situation in vivo. Instead, the use of intestinal primary cells provides the opportunity to work with variegate cell populations, thus more accurately representing what it is observed in vivo. Moreover, the newly developed organoid model, described by the group of Clevers, facilitates the cultivation of intestinal crypts, mimicking crypt–villus structures [2]. The organoids are three-dimensional structures, in which all four type of mature post mitotic epithelial cells are present. Furthermore, these epithelial cells are polarized, with the apical surface facing toward the internal lumen. From the central body of an organoid, there are budding structures, which recapitulate the crypt region in which the stem cells are located. The survival of the organoids is directly linked to the presence of stem cells in these budding structures.

The stimulation of organoids with (1) bacterial products, (2) dead bacteria, or (3) living bacteria, represents an initial step towards better understanding how commensal bacteria or pathogens interact directly or indirectly with the various cell types of the intestinal epithelium and the subsequent host responses.

The use of bacterial products is suitable when, for example, we need to observe a specific pathway, such as a signaling cascade activated through specific pattern recognition receptor (PRR), i.e., TLRs or NODs [3]. Moreover, bacteria may have evolved with modified microbial associated molecular patterns (MAMPs), such as modified LPS or PGN, so by using purified structures from these bacteria it is possible to compare the effect of these modified structures vs "normal" structures. Alternatively, the use of dead bacteria makes it possible to compare the effect of full bacterial structures and therefore analyze the effect on the epithelial cells while multiple bacterial structures are present. On the other hand, using living bacteria, such as commensals, facilitates the characterization of how released components, such as metabolites or effectors, influence the epithelial compartment. Moreover, using pathogens, such as the enteric pathogen *Listeria monocytogens*, may not only provide information regarding the effect on the host side but it could also facilitate the study of specific mechanisms during the invasion process, or enable the identification of specific cell targets [4, 5].

2 Materials

2.1 Crypt Extraction

1. Phosphate-buffered saline without Ca^{2+} and Mg^{2+} (PBS).
2. Bleach 0.3 % in PBS.
3. 10 mM EDTA in PBS.
4. BSA 0.1 % in PBS.
5. 70 μm cell strainer.

2.2 MAMPs Stimulation

6. Cell scraper.
7. Advanced DMEM/F12 (adDMEMF12).
8. Large-orifice pipet tips.

1. Complete medium: Advanced DMEM F12 (ThermoFisher), GlutaMAX ThermoFisher) 1×, Hepes (ThermoFisher) 10 mM, N-2 supplement (ThermoFisher) 1×, B-27 supplement (ThermoFisher) 1×, Recombinant Mouse EGF (R&D) 50 ng/mL, Recombinant Mouse Noggin (R&D) 100 ng/mL, Recombinant Mouse R-spondin1 (R&D) 500 ng/mL, penicillin 100 U/mL and streptomycin (ThermoFisher) 100 μg/mL (*see* **Note 1**).
2. Geltrex LDEV-free reduced Growth Factor Basement Membrane Matrix (ThermoFisher) (Matrix) (*see* **Note 2**).
3. 48-well plates.
4. MAMPs.
5. Dead bacteria (heat killed or fixed bacteria with PFA 4 %).

2.3 Internal Bacteria Delivery

1. Microscope with a 10× or 20× objective; ideally with an enclosure to maintain temperature and CO_2 conditions (*see* **Note 3**).
2. Eppendorf InjectMan and FemtoJet microinjection system (*see* **Note 4**).
3. Eppendorf TransferTip-RP injection microcapillaries.
4. Eppendorf 20 μL Microloader pipet tips.
5. Ibidi μ-Dish 35 mm, low wall, ibiTreat plates.

2.4 FACS

1. Cell recovery solution (Corning).
2. Dissociation buffer: HBSS, Dispase (Corning) 0.3 U/mL, DNase 0.8 U/μL (Sigma), Y-27632 10 μM (Sigma).
3. adDMEMF12.
4. Live/dead fixable dead cell staining kit (ThermoFisher).
5. Foxp3/Transcription Factor Staining Buffer Set (eBioscience) (*see* **Note 5**).
6. Antibody Ki-67 clone:SolA15 (eBioscience) e-Fluor660 (*see* **Note 6**).
7. PBS.

3 Methods

3.1 Crypt Extraction

Put all reagents on ice before starting.

1. Extract the small intestine from a 6–8-week-old mouse and cut it into 5–10 cm pieces.

2. Flush the intestine with 10 mL of ice-cold PBS (use a syringe fitted with a 200 μL pipette tip as a needle), 10 mL of ice-cold Bleach 0.3 % (see **Note 7**) and finish with 10 mL of ice-cold PBS.

3. Open the intestine longitudinally.

4. Remove the villi using a cell scraper (see **Note 8**) and cut pieces of 0.5–1 cm length.

5. Transfer the tissue in a 50 mL tube containing PBS, let the pieces settle down and remove the PBS.

6. Add 20 mL of 10 mM ice-cold EDTA.

7. Incubate for 30 min on ice (mixing every 10 min).

8. Let the piece settle down, remove the EDTA.

9. Add 10 mL of ice-cold BSA 0.1 % and vortex the tube (if possible regulate the vortex below maximal speed; when the speed is indicated use between 18 and 22 g) between 30 s and 1 min (see **Note 9**).

10. Recover the supernatant and observe it under an inverted microscope for the presence of villi or crypts (see **Note 10**). Repeat procedures 12 and 13 at least 2–3 times. Due to the high variability from extraction to extraction, the enriched crypts fraction could appear at different steps. If few crypts are released in the supernatant, another incubation with EDTA could be done.

11. Chose one or more crypt-enriched fractions, combine, and pour through a 70 μm cell strainer (to remove the remain villi).

12. Centrifuge at 100 × g at 4 °C for 5 min (see **Note 11**).

13. Resuspend in 10 mL ice-cold adDMEMF12.

14. Count the number of crypts present in 20 μL (multiply the obtained number by 500 to get the total number of crypts).

3.2 MAMPs/Dead Bacteria Stimulation

Examples of bacterial products include the microbe-associated molecular patterns (MAMPs), which are conserved bacterial motifs present in all bacteria classes, such as peptidoglycan (PGN) muramyl-dipeptide (MDP), muramyl-tri and tetrapeptide, lipopolysaccharide (LPS), flagellin, and lipoproteins (Pam3CSK). The use of MAMPs in the organoids culture can provide information regarding the effect of bacterial products on the intestinal cells' response. Use of fluorescent compounds (such as MDP-Rhodamine) will help to visualize the internalization of the bacterial components in the luminal portion of the organoids. In addition, this same technique can be used with dead bacteria in lieu of bacterial products.

1. Transfer the amount of needed crypts in another tube, centrifuge at 100 × g at 4 °C, 5 min

2. Remove as much supernatant as possible.
3. Add the MAMPs/dead bacteria to the crypts' pellet, taking into consideration the concentration for the final volume of the well (*see* **Notes 12** and **13**).
4. Resuspend the crypts with Matrix diluted 1:2 with complete medium in order to have a final concentration of 250 crypts/ 30 μL Matrix (*see* **Note 14**).
5. Distribute 30 μL of crypts/Matrix into a 48-well plate by making a drop in the middle of each well.
6. Incubate the plate at 37 °C with 5 % CO_2 for 10–15 min in order to let the Matrix solidify.
7. Add 300 μL of Complete medium per well.
8. After 3–4 days of culture, exchange medium with fresh Complete medium + MAMPs.

3.3 Internal Bacteria Delivery

For studies focusing on the mechanisms by which enteric pathogens cross the intestinal epithelium, it is useful to microinject the pathogen of interest into the center of an organoid (which is the physiological equivalent of the lumen) and to then study the host-pathogen interactions. We have developed a microinjection method, described here, to inject not only (live fluorescent) bacteria, such as *Listeria monocytogenes*, but also drugs, staining reagents etc. This microinjection requires specific equipment and as such, there is a significant cost barrier to installation. However, once implemented, microinjection is a robust and reliable method to deliver biologics to the interior of organoids. Here, we will detail a general method for the injection of bacteria into organoids.

1. Install the Eppendorf InjectMan and FemtoJet microinjection system on your microscope of choice (*see* **Note 15**).
2. After passaging the organoids of interest, resuspend in 100 % matrix and plate into Ibidi μ-Dish 35 mm plates. Use 50 μL of matrix *per* plate and seed at 50–100 crypts/50 μL (*see* **Note 16**).
3. On day 3–5 after passaging, organoids will be large enough to perform the injection procedure. On the day of your choice, turn on the microscope, establish a 37 °C and 5 % CO_2 environment, and set up one Ibidi plate of organoids on the microscope.
4. Turn on the InjectMan and FemtoJet and raise the capillary holder to its maximum Z-position. Swivel the capillary holder up and away from the Ibidi plate.
5. Prep your bacteria. After culturing, wash three times in PBS and resuspend in adDMEMF12 (*see* **Note 17**).
6. Using a 10 μL pipet with a Eppendorf 20 μL microloader pipet tip, load 10 μL of the bacteria into a TransferTip-RP injection

microcapillary. Then load the microcapillary into the capillary holder of the InjetMan (*see* **Note 18**).

7. Carefully lower the capillary holder + capillary down to the plate. Carefully turn the capillary so that the cantilevered portion is exactly level (*see* **Note 19**).

8. Hit the "INJECT" button of the InjectMan several times while observing the capillary through the microscope's eyepiece to be sure that the bacteria is being ejected when you hit inject (*see* **Note 20**).

9. Using the microscope eyepiece and transmitted light, turn the Ibidi plate until you find an organoid with three or more budding structures that is on the outer circumference of the matrix.

10. Slowly lower the capillary until it's on the same plane as the organoid of interest (and thus both are in focus).

11. Steadily and slowly move the capillary horizontally into the organoid and to the far side of the organoid's interior (*see* **Notes 21** and **22**).

12. Press the Inject button of the InjectMan to inject your material into the organoid. Wait 5–10 s before proceeding to the next step (*see* **Notes 23–25**).

13. Slowly extract the capillary from the organoid. Just before and just after crossing the organoid wall, pause for 2 s (*see* **Note 26**).

14. Move along the edge of the matrix to continue injecting organoids, alternatively turning the Ibidi plate and adjusting the position of the capillary (*see* **Note 27**).

3.4 FACS

In order to study the effect of the bacterial products or the bacteria on cells, FACS analysis can be used to measure several parameters such as viability, differentiation state, and rate of proliferation. Here we describe a method to isolate single cells starting from embedded organoids and to stain them intracellularly for Ki-67 in order to measure the rate of proliferation.

1. Remove the medium from the well where the organoids are plated.

2. Add 300 μL cell recovery solution and pipette smoothly 2–3 times using a 1000 μL pipette to detach the Matrix from the well.

3. Transfer to a 15 mL tube.

4. Incubate on ice for 10 min.

5. Centrifuge at $300 \times g$ at 4 °C for 5 min and remove the supernatant.

6. Add 2 mL of Dissociation buffer and incubate in a thermomixer at 37 °C with agitation at 1 g for 30 min maximum (*see* **Note 28**).
7. Add 3 mL of Advanced DMEM/F12 and centrifuge at 300 × g at 4 °C for 5 min, discard the supernatant and wash the cells with PBS.
8. Add the live/dead fluorescent reactive dye in 100 μL of PBS following the manufacturer's instructions and incubate on ice for 30 min and then wash the cells with PBS (*see* **Note 29**).
9. Add 300 μL of Permeabilization/Fixation buffer (from the Foxp3/transcription factor staining buffer set) to the cells following the manufacturer's instructions and incubate for 30 min at 4 °C.
10. Without washing add 600 μL of Permeabilization buffer following the manufacturer's instructions; centrifuge the samples at 300 × g for 5 min and discard the supernatant.
11. Add 100 μL of antibody diluted 1/200 in permeabilization buffer and incubate for 30 min on ice.
12. Add 400 μL of permeabilization buffer, centrifuge at 300 × g for 5 min, discard the supernatant and resuspend the cells with 100 μL of PBS and acquire the samples on a flow cytometer.

4 Notes

1. When experiments with living bacteria are made, the use of the antibiotics should be avoided unless the studied bacteria are resistant to them or in case the interest is to study dead bacteria.
2. We have used also Matrigel Matrix from Corning (Matrix). The use of the Geltrex is preferred to the Matrigel due to the absence of the gentamicin in this product. Moreover, if live-imaging will be performed, we suggest to use a phenol-red free compounds in order to avoid interferences with the images.
3. In our hands, organoids survive much longer and are healthier when their exposure to ambient conditions is limited. Therefore, an enclosure to enable the maintenance of 37 °C and 5 % CO_2 is very important to procure if at all possible.
4. When used together, the InjectMan and FemtoJet allow for the precise manipulation of the injection microcapillary in all three axes while also enabling you to standardize the amount of material injected by controlling the time and the pressure of the injection.
5. Any other kit or homemade buffer to fix and permeabilize can be used.

6. The choice of fluorophores will depend on the available instrument.

7. This step is not always necessary but it is strongly suggested if studying the effect of bacteria or bacterial products and particularly in conditions in which the antibiotics are not used. Gentamicin solution could be used instead of bleach.

8. This step helps to remove most of the villi and provides direct access to the crypts, but it has to be done very gently, otherwise the tissue can be broken. If necessary, it can be avoided by doing an additional incubation with EDTA.

9. We describe here a method that is an alternative to the previously described method of mechanical dissociation done by pipetting with 10 mL pipettes. The two methods are equivalent in term of yields but we have found the use of the vortex to be faster and to consume less materials.

10. We suggest to use for this step, and for the following ones, large-orifice pipet tips in order to not disrupt the crypt, particularly when working with small volumes.

11. If, in the pulled fractions, there are not only crypts but also isolated single cells or remaining villi structures, then repeat the centrifugation step a lower speed ($100 \times g$).

12. The addition of the MAMPs/dead bacteria at this step will allow the compounds to have access to the luminal compartment of the organoids because the crypts are still open at this step. The use of fluorescent compounds (such as MDP-Rhodamine) will help to visualize the localization of the bacterial components in the luminal portion of the organoids (Fig. 1).

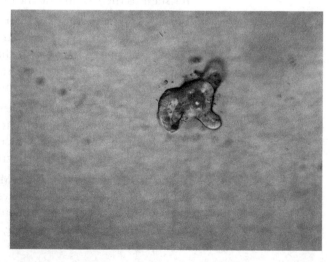

Fig. 1 Bright field and fluorescence image of a mouse intestinal organoid upon MDP-Rhodamine stimulation for 3 days (*red*: MDP-Rhodamine, *green*: Lgr5 Stem cells)

13. During their growth, bacteria release products, such as MAMPs or metabolites, that can influence epithelial cells. Using living bacteria could therefore be necessary to study their effect on the host cells. Embedding living bacteria in the matrix, particularly for long term exposure, could be deleterious to the organoids culture due to the high rate of multiplication of the bugs that could affect the matrix itself. Instead, to study long term effect of bacteria we highly recommend adding the bacteria to the culture medium and exchanging it every day with a fresh bacterial culture.

14. If the removal of the medium is not complete then do not use diluted matrix. In general we have observed a better growth of the organoids in matrix diluted with complete medium compared to undiluted matrix. For some specific applications, such as live microscopy, the use of pure matrix is still recommended to limit the movement of the organoids.

15. Remember to consider the handedness of the principle user of the microinjection system. Generally, you will find it most intuitive to set up the injection system so that the InjectMan is manipulated by the dominant hand and that the injection microcapillary enters the organoid from that respective side of the microscope (e.g., on the right side of the microscope for right-handed people).

16. Be sure to put the Matrigel droplet exactly in the center of the plate. This procedure is limited by some simple geometries; the wall of the plate cannot be too high, and the Matrigel cannot be too close to the edge of the plate; otherwise the capillary will be unable to access the organoids. We have optimized this protocol using the Ibidi μ-Dish 35 mm plates, but these specifications can be modified as necessary to fit a given project.

17. This step depends on the choice of bacteria. For *Listeria monocytogenes*, culture one colony of *Lm* overnight in BHI, then dilution 0.5 mL of the overnight culture in 9.5 mL of BHI for approximately 2 h at 37 °C until an OD600 of 0.8 is reach. Wash by three cycles of centrifugation and resuspension in PBS. Make the final resuspension in 100 μL adDMEMF12.

18. The cantilevered tip of the microcapillary is extremely fragile. Be very careful not to touch it.

19. To accomplish this step, first turn on the transmitted light of the microscope. This will help you see the cantilevered portion and you can "eyeball" how level the capillary tip is. Then, using a 10× or 20× objective, find the capillary through the microscope eyepiece. The tip should appear to be perfectly straight. If there is any "bend", then it is not parallel to the *x*-axis and needs to be adjusted accordingly.

20. If bacteria is not expelled after several injection button taps, then select the "clean" button. This will send a high pressure burst through the microcapillary and force the bacteria down to capillary tip.

21. There will be some initial resistance that you will feel upon trying to pierce into the organoid. That's to be expected. But in your effort to pierce the organoid, be aware that once the capillary breaks through, it can quite easily continue on to pierce the opposite organoid wall, resulting in a hole in which, upon injection, the bacteria will simply seep out of. It takes practice to avoid this, so we encourage you to practice with a fair number of organoids before proceeding to an experiment.

22. If the organoid folds in around the capillary, you can add water to the media, but no more than 1/5th the volume of media. This will stiffen the organoids due to increased osmotic pressure.

23. We suggest playing around with the settings of the InjectMan to find a good set of injection settings. In our hands, 100 psi and a 2 s injection time works well for the injection of *Listeria monocytogenes* into organoids.

24. If you have successfully injected into the interior of the organoid, the organoid will "plump out" just after infection. Imagine that the organoid is an inflated balloon and you have just blown a bit more air in. This is the same principle. Therefore, if it does not plump out, there are several reason for why it failed:

    (a) The organoid wall may have a break or a hole which will allow the injected material to escape.

    (b) The capillary tip is not actually in the interior of the organoid. Check the focus of the microscope and try again.

    (c) If the capillary is in the mass of dead cells often present in the interior of the organoid, it may be that the pressure of the injection is not sufficiently high enough to overcome the dead cell blockade. In this case, try to move the needle to an open space within the organoid.

    (d) Lastly, the capillary may simply be clogged. To diagnose this, extract the capillary from the organoid and Matrigel and hit inject while the capillary is out in the external milieu. If there is no material expelled, try using the "Clean" function to clear the clog. In the case of a severe clog, a replacement of the capillary needle may be required.

25. A simple way to confirm successful organoid injection is to include a fluorescent reagent and to check the organoid via a fluorescent microscope (Fig. 2).

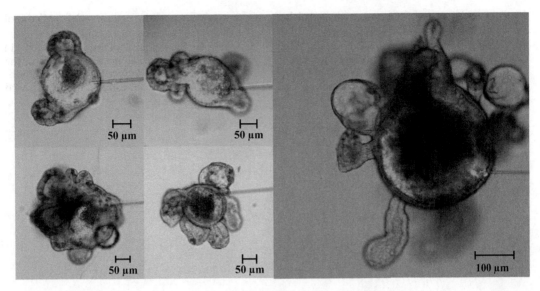

Fig. 2 Bright field and fluorescence confocal image of a mouse intestinal organoid injected with *Listeria innocua* GFP (*green*). The injection microcapillary can be seen to the right of the organoid

26. This step helps reduce the leakage of injected material out of the organoid as well as reducing the build-up of cell debris on the outside of the capillary.

27. Over time, you can expect a certain amount of cell debris build-up on the injection capillary. To avoid this, try to avoid touching the dead cells in the interior of the organoids. The capillary needs to be replaced when the cell debris affects the capillary's ability to pierce into the organoids.

28. During this incubation step, check a small aliquot to verify the dissociation of the organoids and to stop the incubation as soon as single cells are observed. Too long of incubations may increase the amount of dead cells.

29. In order to evaluate the viability of the cells we proceed to stain the cells using a kit for live/dead discrimination compatible with fixation, as this is a necessary step in the case of Ki-67 staining. In the case that the parameter in analysis does not need fixation, any other live/dead discrimination could be applied (i.e., propidium iodide, dapi). The amount of dead cells will be dependent on the size of the organoids: organoids after 2 days of culture will present less dead cells accumulated into the lumen as compared to organoids after 7 days of culture.

References

1. Sansonetti PJ (2004) War and peace at mucosal surfaces. Nat Rev Immunol 4:953–964

2. Sato T, Vries RG, Snippert HJ et al (2009) Single Lgr5 stem cells build crypt-villus

3. Nigro G, Rossi R, Commere P-H et al (2014) The cytosolic bacterial peptidoglycan sensor Nod2 affords stem cell protection and links microbes to gut epithelial regeneration. Cell Host Microbe 15:792–798

4. Bonazzi M, Lecuit M, Cossart P (2009) *Listeria monocytogenes* internalin and E-cadherin: from structure to pathogenesis. Cell Microbiol 11:693–702

5. Nikitas G, Deschamps C, Disson O et al (2011) Transcytosis of *Listeria monocytogenes* across the intestinal barrier upon specific targeting of goblet cell accessible E-cadherin. J Exp Med 208:2263–2277

structures in vitro without a mesenchymal niche. Nature 459:262–265

The Isolation, Culture, and Propagation of Murine Intestinal Enteroids for the Study of Dietary Lipid Metabolism

Diana Li, Hongli Dong, and Alison B. Kohan

Abstract

Since the initial report in 2009 by Sato and Clevers, primary enteroids have been of major interest in the fields of stem cell biology and gastrointestinal (GI) tract biology. More recently, we and others have made major inroads into the physiological relevance of these enteroid models and have shown that enteroids derived from the stomach, intestine, or colon recapitulate major functions of these tissues, namely, gastric acid secretion, lipid absorption and lipoprotein secretion, and ion transport. Here, we detail the isolation of stem cells from the small intestine and the culture and propagation of those stem cells into mature three-dimensional enteroids. We will also detail how we use enteroids to determine intestinal mechanisms behind dietary lipid absorption and lipoprotein secretion. The primary enteroid model is a powerful tool that significantly expands our ability to model GI tract function in vitro.

Keywords Chylomicron, Enteroids, Intestinal stem cell, Lipid absorption, Model of small intestine, Organoids, Primary enterocytes

Abbreviations

apo	Apolipoprotein
BSA	Bovine serum albumin
FFA	Free fatty acids
ISCs	Intestinal stem cells
LGR5	Leucine-rich repeat-containing G-protein-coupled receptor 5
MG	Monoacylglycerol
OA	Oleic acid
TAG	Triacylglycerol

1 Introduction

The intestine plays a crucial role in metabolic homeostasis through dietary fat absorption and secretion, the secretion of incretin hormones, and the actions of the gut immune system [1]. Despite the importance of studying these functions, it has been notoriously difficult to study the intestine because the tissue rapidly degrades during isolation and due to the lack of cell culture models [2, 3]. Primary enterocytes are short lived (~24 h); everted gut sacs cannot be transfected; Caco-2 cells are a monolayer colon cancer cell line that lacks essential biology of the small intestine. Overall, the lack of a culture model has been a significant roadblock to gaining mechanistic insights into the function of the small intestine as a metabolic organ.

Intact intestine is a complex tissue comprised of multiple cell types, including enterocytes (the absorptive cells of the intestine), enteroendocrine cells (which secrete incretin hormones), goblet cells (which secret mucus), and mast cells (which secrete immune modulators). There are two primary structures in the intestinal epithelium: the villus and crypt. The crypt is found at the base of the villus, and these structures contain the multipotent intestinal stem cells (ISCs), which express the transcription factor LGR5 and give rise to all the cells lining the intestinal epithelium [4–8]. In vivo, the intestinal epithelium is in a constant state of differentiation, renewal, and replacement, driven by these ISCs within the crypt niche.

The isolation and propagation of these stem cells were first established by Sato and Clevers in 2009 and have had a major impact on our field [4]. They showed that the crypt, when plated into 3D Matrigel and treated with growth factors, will differentiate into a 3D enteroid. In the enteroid culture, the LGR5+ stem cells within those crypts grow and differentiate into all the cell types normally found in the intestinal epithelium, including new stem cells. As the primary stem cell within an isolated crypt differentiates in response to growth factors, it forms a three-dimensional enteroid. Mature enteroids (by convention "enteroid" refers to mouse-derived cultures, whereas "organoids" refer to human-derived cultures) form at approximately day 10 in growth media, retain intestinal barrier function, and express amino acid transporters and intestine-specific stem cell markers [9–11]. These cells maintain their physiological orientation around a central lumen (the apical surface) and a basolateral surface facing media. The stem cells differentiate in culture by sloughing off cells into the luminal compartment followed by regeneration of crypt epithelium.

We have used these enteroids as a model for dietary fat absorption; they absorb micellar fat across their luminal surface and secrete a chylomicron-sized particle. We have also shown that enteroids can be generated from transgenic mice, and crypts isolated from various regions in the small intestine maintain their regional characteristics. Key advantages of this model are that enteroid cultures can be grown and cultured like transformed cells cultures because the stem cells continue to replenish the culture, and we can directly isolate all the secretions from the enteroids (including lipoproteins, which are normally secreted into the lymph) [9–11]. Enteroids are therefore not only powerful models of intestinal function but also represent a significant advance in our ability to determine intestinal mechanisms for dietary fat absorption and lipoprotein synthesis and secretion.

2 Materials

2.1 Animals

We use C57BL/6J wild-type (WT) mice, age 6–12 weeks (with best results using mice younger than 12 weeks old). Mice are housed 3–4 per cage in a temperature-controlled (21 ± 1 °C) vivarium on a 12-h light-dark cycle. Though the crypt isolation procedure thoroughly removes intestinal contents and presumably microbiota, it may be important to standardize cage bedding between different strains (if you are going to compare enteroid function between 2 difference strains). All animals receive free access to water and chow diet (LM-485 Mouse/Rat Sterilizable Diet, Harlan Laboratories), and there is no need to remove food prior to intestinal isolation. All animal procedures are performed in accordance with the University of Connecticut Internal Animal Care and Use Committee and in compliance with the National Institutes of Health Guide for the Care and Use of Laboratory Animals.

2.2 Enteroid Culture Buffers

1. Matrigel (Corning #356237).
2. Minigut culture media:
   Advanced DMEM/F12 (Invitrogen #12634-010).
   1% L-glutamine (Invitrogen #25030).
   1% pen/strep (Invitrogen #15140-148).
   1% Hepes 10 mM (Invitrogen #15630-106).
   N2 supplement 1:100 (Invitrogen #17502048).
   B27 supplement 1:50 (Invitrogen #17504044).
3. Enteroid growth media:
   500 μL/well minigut culture media (see above) with:
   1 μL rMouse-Noggin (R&D systems, #1967-NG-025/CF), stock concentration: 50 μg/mL in DPBS.
   1 μL rMouse R-spondin1 (R&D systems, #3474-RS-050), stock concentration: 250 μg/mL in DPBS.
   0.25 μL rMouseEGF (R&D systems, #2028-EG-200), stock concentration: 100 μg/mL in DPBS.
4. Wash buffer:
   DPBS (Corning Cell Gro #21-031-CV).
5. Shaking buffer:
   DPBS with:
   1% sucrose (Fisher Scientific #S5-3).
   1% sorbitol (Fisher Scientific #BP439).

6. Chelation buffer:

    DPBS with 2 mM EDTA.

7. Rho-kinase inhibitor:

    Y-27632 (Sigma-Aldrich #Y0503-1MG), stock concentration: 10 mM in DPBS.

2.3 Enteroid Treatment

1. Fatty acid bound to bovine serum albumin:

    OA (Nu-Chek prep #S-1120).
    Stock concentration: 4 mM in 0.1% BHT H_2O.
    Final concentration: 400 μM.
    BSA (Sigma-Aldrich Cat#A-7030).
    Stock concentration: 1 mM in 0.1% BHT H_2O.
    Final concentration: 100 μM.

2. Lipid Micelles.

    OA (Nu-Chek prep #S-1120).
    Working concentration: 6 mM.
    Final concentration: 0.6 mM.
    PC (Sigma-Aldrich #P2772).
    Working concentration: 2 mM.
    Final concentration: 0.2 mM.
    2-MG (Sigma-Aldrich #2787).
    Working concentration: 2 mM.
    Final concentration: 0.2 mM.
    Cholesterol (SupeLco, #47127-U).
    Working concentration: 0.5 mM.
    Final concentration: 0.05 mM.
    TC (Sigma-Aldrich #86339).
    Working concentration: 40 mM.
    Final concentration: 2 mM.
    BODIPY-C12 FFA (Molecular Probes #D3822).
    Working concentration: 2.5 mM in methanol.
    Final concentration: 10 μM.

2.4 Preparation of Fatty Acid Treatment

Though mature enteroids mimic the 3D architecture of the intestine, they do not contain pancreatic lipase, which is necessary to hydrolyze TAG to FFA and MG. To mimic dietary fat absorption, one approach we used was to treat with BSA-bound FFA. Fatty acid treatment was prepared as a 4:1 M ratio of oleic acid/fatty acid-free bovine serum albumin in 0.1% butylated-hydroxytoluene. Oleic acid (Nu-Check Prep) was prepared as 4 mM stock solutions in complex with 1 mM fatty acid-free bovine serum albumin (BSA). Enteroids not receiving the 400 μM OA:BSA complex were treated with an equivalent amount of BSA alone. Mature enteroids were dissociated from Matrigel by washing with ice-cold DPBS, followed by a 150 × g (Eppendorf centrifuge 5810R rotor A-4-81) spin for 10 min. After removing the supernatant, the intact enteroids were then placed in a 1.5 mL Eppendorf tube with 1 mL minigut culture media with 10 μM Rho-kinase inhibitor and final concentration of 400 μM OA:BSA or BSA alone.

2.5 Preparation of Lipid Micelle Treatment

Stock solutions of oleic acid (OA), phosphatidylcholine (PC), 2-monooleoylglycerol (2-MG), and cholesterol were prepared in chloroform. Stock solutions of sodium taurocholate were prepared in PBS. To prepare lipid micelles, stock solutions of OA, PC, 2-MG, and cholesterol were combined into a glass vial and dried under a stream of nitrogen. After the lipids were completely dried, warm sodium taurocholate (TC) was added to the mixture and vortexed. The mixture was brought to working concentration with warm PBS. The final lipid micelle preparation therefore consisted of PBS containing 6 mM OA, 2 mM PC, 2 mM 2-MG, 0.5 mM cholesterol, and 40 mM TC [12]. Intact enteroids were treated with 1 mL minigut culture media with 10 μM Rho-kinase inhibitor and lipid micelle mixture brought to final concentration. Lipid micelles/culture media were delivered to the lumen of enteroids as with the treatment of FFA. For treatment with BODIPY-C12 FFA (Molecular Probes, Inc., catalog #D3822), 10 μM BODIPY-FFA was incorporated into mixed micelles.

3 Methods

3.1 Isolation of Intestinal Stem Cells and Crypt Culture

1. Mice are killed via CO_2 inhalation or cervical dislocation.

2. Wet the fur with 70% EtOH. Make a midline incision to access body cavity. We find it helpful to make two small cuts perpendicular to the midline incision to allow better access to the body cavity.

3. Gently pull the entire small intestine from the body cavity and using scissors, cut the small intestine 3 cm distal to the stomach (to avoid Brunner's glands), and cut again 1 cm proximal to the colon.

4. Flush the small intestine twice with cold DPBS to remove intestinal contents (being careful not to dislodge enterocytes by using too much pressure). Flush from proximal to distal. We find that a curved syringe (Carner Supply, cat#8881412012) inserted into the proximal end of the cut intestine works best.

5. Cut the flushed intestine into three equal sections, duodenum, jejunum, ileum, and place each section into a fresh 15 mL tube in 5 mL of cold DPBS. Tissue should sit on ice for at least 1 h or can be placed at 4 °C overnight.

6. Decant PBS from each tube and add 5 mL of fresh cold DPBS. Cut tissue open longitudinally and mince (into ~1 cm pieces).

Place tubes on a rocker at 4 °C for 2–5 min (less time if stored overnight).

7. Remove tubes from rocker and allow intestine to settle at the bottom the tube ~1 min. Aspirate as much DPBS as possible, and add 5 mL chelation buffer. Place tubes on rocker for 30 min at 4 °C.

8. The tissue is again allowed to settle and chelation buffer aspirated and is replaced with 5 mL of shaking buffer and shaken by hand for 1–3 min to dissociate individual crypts from whole tissue. You should use a "martini shaking" motion; the goal is to release the crypts from the muscle layer of the intestine. After 1 min of shaking (approximately 2–3 shake per second), check for crypts by taking a 20 μL aliquot to a culture dish (or cover slip) and view under 20× magnification to determine adequate crypt dissociation from villus tips (see **Note 1**). Villus tips that are the first to break free of mucosal layer are significantly larger than both crypts and single enterocytes. Continue shaking for 30 s to 1 min intervals until crypts begin to come off. These have the best chance of surviving. If necessary, continue shaking in 60-s intervals.

9. After viewing another aliquot under 20× magnification to determine adequate crypt dissociation, filter the crypt/shaking buffer mixture through a 70 μm nylon mesh cell strainer (Fisher, Cat. # 22363548). Centrifuge crypts at $150 \times g$ for 10 min (Eppendorf centrifuge 5810R rotor A-4-81). Aspirate the supernatant, and wash twice with 5 mL of DPBS.

10. Take 20 μL aliquot from the 5 mL filtrate crypts mixture. Count crypts using 20× magnification microscope.

11. Finally, resuspend the crypt pellet at a concentration of 500 crypts/50 μL of depolymerized Matrigel (BD Biosciences, Franklin Lakes, NJ, USA) containing 1 μL of R-spondin 1 (250 μg/mL), 1 μL of Noggin (50 μg/mL), and 0.25 μL of EGF (100 μg/mL). The Matrigel/crypt suspension is then plated in prewarmed 24-well plates at a density of 50 μL/well and allowed to polymerize at 37 °C in a 5% CO_2 incubator.

12. 500 μL of minigut culture media is then added to the wells (be careful not dislodge the polymerized Matrigel).

13. Replace the media the following morning with enteroid growth media. Crypts should be closed at Day 1 (see **Note 2** and Fig. 1).

14. Media is then replaced every 3–4 days with enteroid growth media. As the enteroids mature, they will have the characteristic "daisy" pattern of a central mucus-filled lumen, surrounded by new crypt buds (Fig. 1). The cells will metabolize media more rapidly once mature (see **Note 3**). If media becomes yellow, change every other day rather than every 3 days.

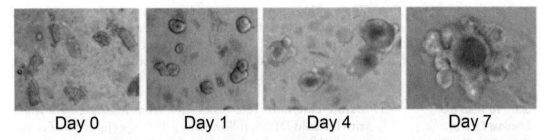

Fig. 1 Growth and differentiation of duodenal crypts into mature primary murine intestinal enteroids. Isolated duodenal crypts were plated in 3D culture in Matrigel at day 0. Crypts were provided complete growth media on day 1. Crypts form tight junctions by day 1. Crypts were passaged at day 4, and the formation of new crypts is observed. Crypts form mature enteroids with daisy-like shape by day 7. Images were taken at 20× magnification on a Zeiss Axiovert 40 °C inverted light microscope

3.2 Enteroid Maturation and Passaging

1. Enteroids will mature 7–10 days in culture and form 3D hollow balls (with the "daisy" appearance) (Fig. 1). We call these mature enteroids and use these for experiments or for passage.

2. To passage enteroids, aspirate the media, and then add ice-cold DPBS to dislodge and depolymerize the 3D Matrigel/enteroid culture. Place the cold DPBS/Matrigel/enteroid mixture in a 15 mL conical tube and gently pipet with a p200. The goal is to depolymerize the Matrigel while keeping enteroids intact.

3. Spin the mixture at $150 \times g$ (Eppendorf centrifuge 5810R rotor A-4-81) for 5 min in a conical tube. Wash an additional two times with cold DPBS.

4. After aspirating the DPBS, enteroids can be gently broken open by repeated pipetting (5–20 times) with a 27 gauge needle and syringe. This step is highly susceptible to individual variation; you should view an aliquot of the cells under 20× magnification (as in the isolation procedure) to monitor the level of enteroid breakage. You can break the enteroids into individual crypts (which will result in a significant reduction in viable material) or you break the enteroid into multi-crypt pieces. Generally, if your goal is to increase the volume of your culture, it is simpler to start from fresh intestine than it is to passage crypts. However, if your goal is to keep particular crypts (from a specific strain of mice or crypts that have been virally transformed), then it is best to split the enteroids into multi-crypt pieces since these are most viable.

5. As in the primary isolation, individual crypt pieces will reform mature enteroids in approximately 7–10 days, and alternatively, multi-crypt pieces will form mature enteroids in ~4 days. The crypts are placed back into Matrigel with growth factors (as in step 1.11), and after 30 min of polymerization, complete growth medium is added. The following day, replace the growth media with fresh media containing growth factors.

6. To directly compare enteroid lines to each other, we counted crypts and plated the same number of crypts at day 1, adhering to identical culture conditions throughout the experiment, and isolation on day 10. You should also isolate genomic DNA to compare the relative density of cells in each well.

3.3 To Treat the Luminal Face of Enteroids with Lipid

1. Mature enteroids were dissociated from Matrigel by washing with ice-cold DPBS, followed by a $150 \times g$ (Eppendorf centrifuge 5810R rotor A-4-81) spin for 10 min.

2. After removing the supernatant, the intact enteroids were then placed in a 1.5 mL Eppendorf tube with 1 mL of either fatty acid treatment or lipid micelle treatment media.

3. The enteroids were very gently opened by pipetting up and down with a p1000 pipette (Fig. 2), followed by incubation with the lids open in a 37 °C 5% CO_2 incubator for 2 h.

4. After 2 h, the enteroids were centrifuged at $150 \times g$ for 10 min and the supernatant collected.

5. Enteroids were washed again with 1 mL ice-cold DPBS followed by a spin at $150 \times g$ for 10 min; the enteroids were resuspended in 1 mL of minigut culture media with 10 μM Y27623 Rho-kinase inhibitor and placed back in the incubator for 4–6 h with the tube lids left open, allowing enteroids to close (Fig. 2).

6. The media and cell pellet were then collected via centrifugation at $150 \times g$ for 10 min.

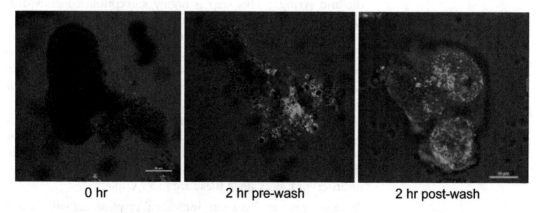

0 hr   2 hr pre-wash   2 hr post-wash

Fig. 2 Incorporation of BODIPY-C12 FFA into enteroid lumen during 2 h treatment with lipid micelles. At 0 h, mature enteroids are resuspended in treatment media containing BODIPY-labeled micelles. Mature enteroids are opened, and lumenal face of enteroids is exposed to micelles. After 2 h treatment, BODIPY label is incorporated into the mature enteroid lumens. At 2 h post-wash, the BODIPY label remains only in the lumenal compartment of enteroids

4 Notes

We have determined that there are several key reasons why crypt isolation and culture do not work.

1. You don't see any crypts (or conversely you see a lot of cell debris but few crypts):
   (a) We find that this is most common when new people learn the technique. Either you are not shaking the tissue enough to "pop" the crypts off or you are shaking too hard and the crypts are destroyed along with the generation of cell debris, which likely triggers apoptosis in the isolated crypts. Practice until you hit the sweet spot of a few minutes in chelation buffer and an appropriate amount of shaking. Do not plan to culture a huge number of wells until you have figured out what works in your own hands.
   (b) Relatedly, when passaging the enteroids, avoid over-pipetting and breaking the crypts down so much that they die.

2. The crypt isolation is great, and cells are alive and well and then do not close:
   (a) In this scenario, you happily leave the tissue-culture room having plated crypts in Matrigel with growth factors, and you come back to find that none of the crypts have formed their characteristic donut shape by day 1 (Fig. 1). You add media anyways, but by the next morning, nothing has changed. Occasionally, this occurs. We've determined that this is more likely to happen in very young (<6 week) and old (>12 week) mice and may also be due to a poor or over-chelation step. If you do not get enough crypts to come off during those first few minutes, it is possible that the longer incubation with chelation buffer makes the crypts more susceptible to apoptosis, and they never recover. Sometimes this happens when the tissue is shaken too vigorously. Sometimes this just happens without any other warning.

3. Over-plating:
   (a) Do not add too many crypts per well. They become overcrowded and will die quickly (day 2–3).

References

1. Mansbach CM 2nd, Gorelick F (2007) Development and physiological regulation of intestinal lipid absorption. II. Dietary lipid absorption, complex lipid synthesis, and the intracellular packaging and secretion of chylomicrons. Am J Physiol Gastrointest Liver Physiol 293:G645–G650
2. Date S, Sato T (2015) Mini-gut organoids: reconstitution of the stem cell niche. Annu Rev Cell Dev Biol 31:269–289
3. Wilson TH, Wiseman G (1954) The use of sacs of everted small intestine for the study of the transference of substances from the mucosal to the serosal surface. J Physiol 123:116–125
4. Sato T, Vries RG, Snippert HJ, van de Wetering M, Barker N, Stange DE, van Es JH, Abo A, Kujala P, Peters PJ et al (2009) Single Lgr5 stem cells build crypt-villus structures in vitro without a mesenchymal niche. Nature 459:262–265
5. VanDussen KL, Carulli AJ, Keeley TM, Patel SR, Puthoff BJ, Magness ST, Tran IT, Maillard I, Siebel C, Kolterud Å et al (2012) Notch signaling modulates proliferation and differentiation of intestinal crypt base columnar stem cells. Development 139:488–497
6. Murphy CL, Polak JM (2002) Differentiating embryonic stem cells: GAPDH, but neither HPRT nor beta-tubulin is suitable as an internal standard for measuring RNA levels. Tissue Eng 8:551–559
7. Fevr T, Robine S, Louvard D, Huelsken J (2007) Wnt/beta-catenin is essential for intestinal homeostasis and maintenance of intestinal stem cells. Mol Cell Biol 27:7551–7559
8. Tetteh PW, Basak O, Farin HF, Wiebrands K, Kretzschmar K, Begthel H, Van Den Born M, Korving J, De Sauvage F, Van Es JH et al (2016) Replacement of lost Lgr5-positive stem cells through plasticity of their enterocyte-lineage daughters. Cell Stem Cell 18:203–213
9. Levy E, Beaulieu JF, Delvin E, Seidman E, Yotov W, Basque JR, Ménard D (2000) Human crypt intestinal epithelial cells are capable of lipid production, apolipoprotein synthesis, and lipoprotein assembly. J Lipid Res 41:12–22
10. Foulke-Abel J, In J, Yin J, Zachos NC, Kovbasnjuk O, Estes MK, de Jonge H, Donowitz M (2016) Human enteroids as a model of upper small intestinal ion transport physiology and pathophysiology. Gastroenterology 150:638–649.e8
11. Mahe MM, Sundaram N, Watson CL, Shroyer NF, Helmrath MA (2015) Establishment of human epithelial enteroids and colonoids from whole tissue and biopsy. J Vis Exp 97: e52483–e52496
12. Chateau D, Pauquai T, Delers F, Rousset M, Chambaz J, Demignot S (2005) Lipid micelles stimulate the secretion of triglyceride-enriched apolipoprotein B48-containing lipoproteins by Caco-2 cells. J Cell Physiol 202:767–776

Oncogenic Transformation of Human-Derived Gastric Organoids

Nina Bertaux-Skeirik, Jomaris Centeno, Jian Gao, Joel Gabre, and Yana Zavros

Abstract

The culture of organoids has represented a significant advancement in the gastrointestinal research field. Previous research studies have described the oncogenic transformation of human intestinal and mouse gastric organoids. Here we detail the protocol for the oncogenic transformation and orthotopic transplantation of human-derived gastric organoids.

Keywords: Human fundic gastric organoids, CRISPR/Cas-9, Nucleofection, Organoid orthotopic transplantation

1 Introduction

Advances in our knowledge of stem cell biology have allowed the ability to grow cultures of gastrointestinal tissues in vitro. These organotypic models, known as organoids, are derived from either primary tissue or pluripotent stem cells and closely recapitulate the structure and cellular diversity of native tissue. Investigators thus far have been successful in developing organoids from gastrointestinal organs including stomach, intestine, colon, and pancreas [1–9]. Organoid models are rapidly emerging as a fundamental system used to study gastrointestinal development [1], physiology [3] and disease [4, 7–10].

Researchers strive to develop cancer models that recapitulate the cellular diversity characteristic of human tumors. To begin to address this challenge, researchers have developed three dimensional organotypic cancer models. For example, human colon organoids derived from normal tissue and adenoma have been engineered by clustered regularly interspaced short palindromic repeats (CRISPR)/Cas9 gene editing to introduce deletions of *APC*, *TP53*, and *SMAD4*, combined with point mutations in $KRAS^{G12V}$ and $PIK3CA^{E545K}$ to model colon cancer mutations in vitro [11]. Similarly, knockdown of the *Tgfbr2* gene, encoding the TGF-β receptor 2, in murine $Cdh1^{-/-}$; $Trp53^{-/-}$ gastric organoids confirmed tumor suppressor activity and exhibited increased metastasis after in vivo transplantation [12]. Moreover, oncogenic transformation of pancreatic and gastric organoids, as a result of expression of Kras carrying G12D mutation, loss of p53, or both also generated adenocarcinoma after in vivo transplantation [13]. A key limitation that remains is accurately capturing the tumor heterogeneity within the context of the tissue microenvironment. Here we detail the protocol for the oncogenic transformation of human-derived gastric organoids as an in vitro cancer model. Moreover, these organoids may be used for orthotopic transplantation as a model of gastric cancer in vivo.

2 Materials

2.1 Human-Derived Fundic Gastric Organoids (hFGOs)

1. Storage and washing buffer for gastric tissue, glands, and organoids: ice-cold Dulbecco's phosphate buffered saline (DPBS) without calcium or magnesium (Fisher, 21-031-CV) supplemented with antibiotics, 100 U/ml Penicillin/Streptomycin (Thermo Scientific, SV30010), 100 U/ml Kanamycin (Thermo Fisher, 15160054), 1× Gentamicin/Amphotericin (Thermo Fisher, R01510).

2. Incubation media: Advanced Dulbecco's modified Eagle medium/F12 medium (Life Technologies, 12634-010) supplemented with 2 mM GlutaMAX (Life Technologies, 25030-081), 100 U/ml Penicillin/Streptomycin (Thermo Scientific, SV30010), 10 mM HEPES Buffer (Sigma, H0887), Collagenase from *Clostridium histolyticum* at 1 mg/mL (Sigma, C9891), and Bovine Serum Albumin cell culture grade at 2 mg/mL (Sigma, A9418).

3. Conical tubes 50 mL (Fisher, 352070), 5 mL tubes (Fisher, 14-956-3C).

4. 12 well plate, cell culture treated (Midwest Scientific, 92012).

5. 200 µL wide tips (Thermo Fisher, 02-707-134).

6. Basement membrane for growth of organoids: Growth Factor Reduced, Phenol Red-free Matrigel™ (Fisher, CB-40230C).

7. Growth medium for organoids: Advanced Dulbecco's modified Eagle medium/F12 medium (Life Technologies, 12634-010) supplemented with 2 mM GlutaMAX (Life Technologies, 25030-081), 100U/ml Penicillin/Streptomycin (Thermo Scientific, SV30010), 100 U/ml Kanamycin (Thermo Fisher, 15160054), 1× Gentamicin/Amphotericin (Thermo Fisher, R01510), 10 mM HEPES Buffer (Sigma, H0887), 1 mM n-Acetylcysteine (Sigma, A9165), 1× N2 (Life Technologies, 17502-048), 1× B27 (Life Technologies, 12587-010), 50 % Wnt-conditioned medium, 20 % R-spondin-conditioned

medium supplemented with gastric growth factors including 100 ng/ml bone morphogenetic protein inhibitor, Noggin (PeproTech, 250-38), 1 nM gastrin (Tocris, 3006), 50 ng/ml epidermal grow factor (EGF) (PeproTech, 315-09), and 100 ng/ml fibroblast growth factor 10 (FGF10) (PeproTech, 100-26), 10 mM nicotinamide (Sigma, N0636), 10 μM Y-27632 ROCK inhibitor (Sigma, Y0503).

2.2 Nucleofection of Human Gastric Organoids

1. Passaging Solution: Accutase (Stemcell Technologies, 07920).
2. Storage medium (same medium as Growth medium described above).
3. $26^{3/8}$ G Syringe (Thermo Fisher, 309625).
4. Nucleofector Core unit (Lonza, AAF-1001B) and X Unit (Lonza, AAF-1001X).
5. Nucleofection Kit: P1 Primary cell kit (Lonza, V4XP-1024), Setting 1F (Code, CM-113).
6. Extra Long Tips (Thermo Fisher, 7281).

2.3 Tumor-Derived Gastric Organoids

1. Stripping buffer: Hank's Balanced Salt Solution (HBSS) (Thermo Fisher, 21021CV) with 2 mM EDTA, 25mmM HEPES Buffer (Sigma, H0887), and 10 % Fetal Calf Serum.
2. Incubation medium: RPMI with L-glutamine (Thermo Fisher, 10040CV), 100 U/ml Penicillin/Streptomycin (Thermo Scientific, SV30010), 100 U/ml Kanamycin (Thermo Fisher, 15160054), 1× Gentamicin/Amphotericin (Thermo Fisher, R01510), 10 mM HEPES Buffer (Sigma, H0887), supplemented with 1.5 mg/mL Collagenase from *Clostridium histolyticum* (Sigma, C9891), and 20 μg/mL Hyaluronidase from bovine testes (Sigma, H3884).
3. Growth medium: Same as for normal gastric organoids.

2.4 Orthotopic Transplantation of Human Gastric Organoids

1. For ulcer injury induction: Glacial acetic acid and glass capillary tubes.
2. For transplantation of organoids: $26^{3/8}$ G syringe (Thermo Fisher, 309625) and DPBS without Ca^{2+} and Mg^{2+}.
3. Suggested surgical tools: curved and straight scissors, forceps, sutures (Ethicon, G121), and wound clips (Becton Dickinson, 427631).
4. Other Supplies: Puralube eye ointment (Dechra, NDC17033-38), Buprenex, Insulin syringes (Thermo Fisher, 329461), Betadine solution swabstick (Thermo Fisher, NDC67618-153-01), alcohol sterile prep pads (Thermo Fisher, NDC 10819-5910-1), heating pad.

3 Methods

3.1 Isolation of Human Gastric Glands

Human Fundus was collected during sleeve gastrectomies (IRB protocol numbers: 2015-4869 and 2015-5537).

1. Thaw desired volume of Matrigel™ on ice or at 4 °C for 2 h prior to gland isolation (*see* **Note 1**).

2. Wash stomach tissue thoroughly in ice cold sterile DPBS without Ca^{2+} and Mg^{2+}, and lay flat on a sterile surface. Use sterile gauze to wipe away mucus on the epithelial side of the stomach tissue. Take a pair of blunt curved scissors and scrape the epithelial layer away from the muscle layer of the stomach. Collect the epithelial fragments and wash in ice cold sterile DPBS without Ca^{2+} and Mg^{2+} (*see* **Note 2**).

3. Cut the epithelium into small fragments in a sterile petri dish and wash with ice cold sterile DPBS without Ca^{2+} and Mg^{2+} until wash solution turns clear. Transfer the fragments into 50 mL of Incubation media supplemented with 1 mg/mL collagenase and 2 mg/mL BSA in a sterile round flask. Keep tissue stirring with an electric stir bar, and place in a water bath at 37 °C, supplemented with oxygen flow for 30 min.

4. In a sterile hood, dilute the incubation media containing the stomach glands with 50 mL of fresh incubation media without the collagenase or BSA to stop the digestion reaction. Collect the glands by filtering the fragments with 2 layers of gauze (*see* **Note 3**) and allow the glands settle in 50 mL conical tubes on ice for 10 min (*see* **Note 4**).

5. Remove 35 mL by vacuum pipet, and wash the remaining solution with another 35 mL cold incubation media, and let glands settle another 10 min. Remove 40 mL of the supernatant, leaving the glands in 10 mL of remaining media.

6. Using a pipette, gently transfer 5 mL to two 5 mL Fisher Tubes to collect the glands. Centrifuge the glands at a low speed, $65 \times g$ for 5 min. Carefully remove supernatant, and gently flick the tube to resuspend the glands. Add 4 mL ice-cold DPBS without Ca^{2+} and Mg^{2+} containing antibiotics (*see* **Note 5**) to wash the glands and spin again at $65 \times g$ for 5 min. Wash twice with DPBS without Ca^{2+} and Mg^{2+} supplemented with antibiotics.

7. Remove supernatant carefully, so as not to disturb the pellet of glands (*see* **Note 6**). Keeping the tube on ice, transfer desired volume of Matrigel™ to the glands (*see* **Note 7**) and mix carefully to avoid air bubbles.

8. Using a wide-tipped 200 μL pipet, pipet 60 μL of the glands in Matrigel™ directly onto the middle of the wells of the 12 well plate (*see* **Note 8**).

9. Incubate the plate at 37 °C for 15 min to allow Matrigel™ to solidify. After this incubation add 1 mL of growth media to each well. Allow organoids to grow for 7 days.

3.2 Nucleofection of Human Gastric Organoids

1. To prepare organoids for nucleofection, collect the organoids into ice-cold DPBS without Ca^{2+} and Mg^{2+} to wash away the Matrigel™, centrifuge the organoids at 195 × *g* (*see* **Note 9**).
2. Resuspend in 1 mL pre-warmed Accutase per 2–3 wells of organoids, and place at 37 °C for 10–12 min.
3. Using a $26^{3/8}$ G syringe, pass the organoids in the Accutase solution through the syringe 6–8 times to break the organoids into single cells (*see* **Note 10**).
4. Dilute the Accutase with equal volume DPBS and count the cells with a hemocytometer to determine maximum number of cuvettes that can be nucleofected (*see* **Note 11**). Centrifuge the cells at 195 × *g* for 5 min and resuspend in growth media in appropriate aliquots. Keep on ice until nucleofection.
5. Prepare nucleofection buffer according to the manufacturer's instructions by adding 82 μL of nucleofector solution to 18 μL of Supplement, then add 100 μL total of P1 buffer to 1×10^6 cells, and 2 μg of desired plasmid(s) with the cells and P1 buffer (*see* **Note 12**).
6. Transfer complete solution to nucleofection cuvette and place in appropriate holder in the *X* unit. Press start on the F1 program (Code: CM-113), and following nucleofection, set cuvette at room temperature for 5 min (*see* **Note 13**).
7. Dilute the P1 nucleofection buffer with 500 μL of organoid growth media and transfer cuvette to 37 °C for 30–45 min to allow cells to recover.
8. After this time, transfer cells to eppendorf tube, centrifuge at 400 × *g* for 5 min, and resuspend in desired volume of Matrigel™ (*see* **Note 14**).
9. Incubate the plate at 37 °C for 15 min to allow Matrigel™ to solidify, and provide appropriate volume of growth media. Change to fresh media every 4–5 days. Spheres should be visible after approximately 48–72 h, and will continue to grow for 14 days (*see* **Note 15**, Fig. 1).

3.3 Organoids Derived from Gastric Cancer Tissue

1. Wash tumor tissue in DPBS, and cut into small pieces.
2. Add 10 mL pre-warmed EDTA stripping buffer (HBSS, 5mmM EDTA, 25 mM HEPES, 10 % FCS), incubate for 10 min in a shaking incubator at 37 °C.
3. Change to fresh EDTA stripping buffer, and incubate for a further 5 min in a shaking incubator at 37 °C.
4. Wash the tissue with HBSS twice to remove stripping buffer.

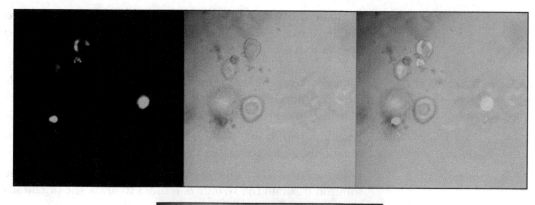

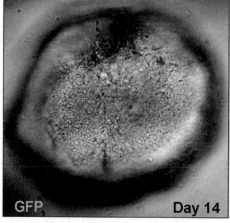

Fig. 1 Oncogenic transformation of human gastric organoids. Human gastric organoids transfected with GFP-expressing plasmids at day 3 and 14 following nucleofection

5. Add 10 ml pre-warmed incubation buffer (RPMI with 1.5 mg/mL Collagenase and 20 μg/mL Hyaluronidase), and mince the tissue well with a pair of scissors (*see* **Note 16**).
6. Incubate at 37 °C for 30 min in a shaking incubator.
7. Dilute the digest with 20 mL DPBS without Ca^{2+} and Mg^{2+}, and filter it through 70 μm filter.
8. Centrifuge the cells at 500 × *g* for 5 min.
9. Wash with DPBS without Ca^{2+} and Mg^{2+} at least once.
10. Centrifuge cells at 1200 rpm for 5 min, and resuspend in desired volume of Matrigel™. Spheres may be visible in 24–48 h and will continue to grow for 14 days.

3.4 Orthotopic Transplantation of Human Gastric Organoids

All mouse studies were approved by the University of Cincinnati Institutional Animal Care and Use Committee (IACUC) that maintains an American Association of Assessment and Accreditation of Laboratory Animal Care (AAALAC) facility.

1. Sterilize appropriate tools for surgery including: small straight and curved scissors, forceps, and glass capillary needle. Place 1 mL of glacial acetic acid in a sterile 1.5 mL tube.
2. Harvest organoids by washing in cold DPBS without Ca^{2+} and Mg^{2+} until Matrigel™ has been removed (centrifuge at $195 \times g$). Resuspend organoids in cold DPBS without Ca^{2+} and Mg^{2+} in order to inject approximately 50 µL per mouse stomach.
3. Anesthetize the mouse.
4. The stomach is then exteriorized through a midline abdominal laparotomy opening.
5. Place the capillary tube containing acetic acid on the serosal surface of the stomach as shown in Fig. 2a and hold for 25 s to induce the injury.

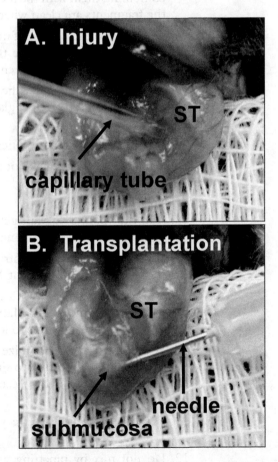

Fig. 2 Acetic acid-induced injury and organoid transplantation. (a) Acetic acid-induced injury to the serosal surface of the mouse stomach with capillary tube. (b) Orthotopic transplantation of human gastric organoids into submucosa of the mouse stomach

6. With a 26$^{3/8}$ G syringe inject 50 μL of organoids resuspended in cold DPBS without Ca^{2+} and Mg^{2+} adjacent to the injured area, being careful to insert the needle gently into the submucosa (*see* **Note 16**, Fig. 2b).

7. The stomach is then replaced into the abdominal cavity and the muscle and skin incisions sutured.

4 Notes

1. Thaw Matrigel™ in an ice bucket at 4 °C for 16 h if you are starting with a new bottle. Mix gently before using, and keep all materials on ice to prevent Matrigel™ from solidifying.

2. Make sure to cut into small fragments to ensure the best results, and make sure that the fragments are completely clean before combining them with the incubation buffer. You will know that the fragments are clean when the DPBS washes become clear.

3. You can also use a piece of mesh, but make sure to test the size of the openings to ensure that glands can move through the mesh before using on the entire volume.

4. You should be able to see the glands falling to the bottom of the tube, and you can give the glands more time to settle if you wish to increase your yield.

5. It is critical to add antibiotics to avoid contamination.

6. The slow centrifuge will not create a tight pellet, so pipet slowly and carefully so that the glands are not lost as you remove the supernatant.

7. Add the volume of Matrigel™ that is required, and then mix with a pipet of lesser volume to avoid bubbles.

8. Keep the tube with the Matrigel™ and organoids on ice during this and all steps to prevent Matrigel™ from solidifying.

9. You may need to wash the organoids 2 or 3 times, to completely remove the Matrigel™, and be careful when you remove supernatant not to lose any organoids after spinning.

10. You may need to optimize this step to determine that the organoids are digesting to single cells, and you can increase or decrease the time in the Accutase, as well as syringe more times or less times.

11. For each 100 μL cuvette you can transfect a maximum of 1×10^6 cells.

12. Do not mix by pipetting after adding the plasmid to avoid shearing, instead gently flick the tube. For CRISPR-Cas9 oncogenic transformation, you can order CRISPR plasmids from ORIGENE™ Technologies for gene of interest.

13. For gastric epithelial spheroids, the F1 program (code = CM-113) yields the optimum transfection efficiency; however, you may determine a different code or nucleofection buffer is best. You may need to order a 16 test strip cuvette to optimize your transfection, available by Lonza.

14. There may be some cell death at this step, but even if you do not see an obvious cell pellet, use a pipet to remove the supernatant slowly. Try to keep the cells dense when you plate back out into Matrigel™.

15. The organoids may be ready earlier than this time-point, so be careful to monitor the speed of growth so that you can harvest at the best time.

16. You should be able to see the end of the needle when you insert the needle parallel to the stomach serosa to inject the organoids.

Acknowledgements

This work was supported by NIH 1R01DK083402 grant and College of Medicine Bridge Funding Program (Zavros), NIH 5T32GM105526, Albert J. Ryan Fellowship, and Dean's Fellowship (Bertaux-Skeirik).

References

1. McCracken KW et al (2014) Modelling human development and disease in pluripotent stem-cell-derived gastric organoids. Nature 516:400–404
2. Sato T et al (2009) Single Lgr5 stem cells build crypt-villus structures in vitro without a mesenchymal niche. Nature 459:262–265
3. Schumacher MA et al (2015) The use of murine-derived fundic organoids in studies of gastric physiology. J Physiol (Lond) 593:1809–1827
4. Bertaux-Skeirik N et al (2015) CD44 plays a functional role in Helicobacter pylori-induced epithelial cell proliferation. PLoS Pathog 11, e1004663
5. Huch M et al (2013) In vitro expansion of single Lgr5+ liver stem cells induced by Wnt-driven regeneration. Nature 494:247–250
6. Huch M et al (2015) Long-term culture of genome-stable bipotent stem cells from adult human liver. Cell 160:299–312
7. Dedhia PH (2016) Organoid models of human gastrointestinal development and disease. Gastroenterology. doi:10.1053/j.gastro.2015.12.042
8. Sato T et al (2011) Long-term expansion of epithelial organoids from human colon, adenoma, adenocarcinoma, and Barrett's epithelium. Gastroenterology 141:1762–1772
9. Boj SF et al (2015) Organoid models of human and mouse ductal pancreatic cancer. Cell 160:324–328
10. Neal JT, Kuo CJ (2016) Organoids models for neoplastic transformation. Annu Rev Pathol 11:199–220
11. Matano M et al (2015) Modeling colorectal cancer using CRISPR-Cas9-mediated engineering of human intestinal organoids. Nat Med 21:256–262
12. Nadauld LD et al (2014) Metastatic tumor evolution and organoid modeling implicate TGFBR2 as a cancer driver in diffuse gastric cancer. Genome Biol 15:428
13. Li X et al (2014) Oncogenic transformation of diverse gastrointestinal tissues in primary organoid culture. Nat Med 20:769–777

Intestinal Crypt Organoid: Isolation of Intestinal Stem Cells, In Vitro Culture, and Optical Observation

Yun Chen, Chuan Li, Ya-Hui Tsai, and Sheng-Hong Tseng

Abstract

The isolation and culture of intestinal stem cells (ISCs) was first demonstrated in the very recent decade with the identification of ISC marker Lgr5. The growth of ISCs into crypt organoids provides an in vitro model for studying the mucosal physiology, intestinal cancer tumorigenesis, and intestinal regeneration. Here, we describe two different isolation protocols and demonstrate a fixation method that aids in the confocal observation of the organoids.

Keywords: Intestinal stem cell (ISC), Crypt organoid, Isolation, Confocal observation, Three-dimensional imaging

1 Introduction

Intestinal epithelial cell lines HIEC-6, Caco-2 or primary intestinal epithelial cells (IECs) were used to mimic the mucosal environment when the research came to focus on the biophysical functions and signaling pathway within intestinal epithelium [1–4]. However, the lack of a definite epithelial model or organoid environment had limited the research theme. The emerging development of regenerative medicine prompted the search for tissue-specific stem cells as well as the advances in the culture techniques for stem cells. The identification of ISC marker Lgr5 with the ISC isolation protocol was first came out in 2007 described by Baker et al. [5], leading to a number of studies among intestinal biophysiology and pathology, and intestinal regeneration [6–11].

With the research interests in intestinal transplantation and regeneration, we have performed the ISC isolation and culture from mice intestine as well as three-dimensional imaging of crypt organoids [12, 13]. Here presents the methods and technical tips from our experimental experiences.

Electronic supplementary material: The online version of this chapter (doi:10.1007/7651_2017_21) contains supplementary material, which is available to authorized users. Videos can also be accessed at http://link.springer.com/chapter/10.1007/7651_2017_21.

2 Materials

2.1 Isolation of Intestinal Stem Cells and Culture of Crypt Organoids

1. Mice: 4–6-week-old mice.
2. Scissors and forceps.
3. Cover glasses (size 24 × 50 mm).
4. 10 cm sterilized petri-dish.
5. 15 mL corning tube.
6. Gentamycin sulfate (Biological Industries, #03-035).
7. PBS: Dulbecco's Phosphate Buffered Saline (Life Technologies, #21600–010).
8. PBS-Ca^{2+}/Mg^{2+}: PBS added with 100 mg/L $CaCl_2$ and 100 mg/L $MgCl_2 \cdot 6H_2O$.
9. PBS-EDTA/EGTA: PBS added with 1 mM EDTA and 1 mM EGTA.
10. Cell strainer (70 μm) (BD Biosciences).
11. Laminin-rich Matrigel™ (BD Biosciences, #356234).
12. Crypt medium: Advanced DMEM/F12 (Thermo Fisher Scientific, #12634028) containing 2% fetal bovine serum, 2 mmol/L of L-glutamine, 100 U/mL of penicillin, 100 mg/mL of streptomycin, 100 ng/mL of Noggin, 500 ng/mL of R-spondin 1, 50 ng/mL of epidermal growth factor, 1 μmol/L of Jag-1, 10 μmol/L of Y-27632 [10].

2.2 Preparation of Crypt Organoids for Confocal Microscopic Observation

1. PBS.
2. 4% para-formaldehyde.
3. Triton X-100.
4. Normal goat serum.
5. Primary and secondary antibodies for fluorescence staining.
6. FocusClear (CelExplorer Labs Co., #FC-101).
7. Paper hole reinforcement stickers.
8. Circular cover glasses (size 12 mmΦ).
9. Mounting solution with DAPI (Abcam, #ab104139).
10. Microscopic Slide (size 76 × 26 mm).
11. Tape.
12. Confocal Microscopy.

3 Methods

The procedure mainly follows the method described in ref. [5] with little modification.

3.1 Isolation of Intestinal Stem Cells from Mice

3.1.1 Protocol A

1. Mice should be euthanized and sacrificed under the regulation and supervision of IACUC in the institute.
2. Harvest the small intestine tissue but exclude the first 1 cm (duodenum) and the last 1 cm close to cecum. Immediately soak the tissue in ice-old PBS.
3. Fill the 10 cm petri-dish with ice-cold PBS, and let the dish stand firmly on ice.
4. Transfer the intestine into the petri-dish and cut open the tissue first longitudinally and then cross cut into 0.6–0.8 cm fragments. Remove the fecal masses on the mucosa.
5. Transfer the tissue pieces into a 15 mL corning tube with 5 mL PBS containing gentamycin (0.5 mg/mL) and shake the tube on an orbital shaker at 100 rpm for 10 min at 4 °C to remove remnant feces and contaminants.
6. Discard the supernatant PBS and add 5 mL PBS-Ca^{2+}/Mg^{2+} into the tube. Shake the tube on an orbital shaker at 75 rpm for 20 min at 4 °C.
7. Discard the supernatant PBS-Ca^{2+}/Mg^{2+} and then add 5 mL PBS-EDTA/EGTA into the tube. Shake the tube on an orbital shaker at 75 rpm for 10 min at 4 °C.
8. Discard the supernatant PBS-EDTA/EGTA and then add 5 mL fresh PBS-EDTA/EGTA into the tube.
9. Vortex the tube intermittently for 20 s, then immediately harvest the supernatant as fraction 1 and store the fraction on ice.
10. Soak the intestine slices in 5 mL fresh PBS-EDTA/EGTA again, shake the tube on an orbital shaker at 75 rpm for 10 min at 4 °C.
11. Repeat **steps 9** and **10** for seven more times to get fraction 2–8.
12. Fractions 6–8 are filtered through the cell strainer and the flow-through is centrifuged at 900 rpm, 4 °C, 5 min.
13. Resuspend the cells in crypt medium and keep the cell suspensions on ice. Use small amount of aliquot (50 µL) for trypan blue staining and count the number of viable crypts under microscopy.
14. Mix 500–1000 crypts with 400 µL of Matrigel and plate the mixture for each well of the 12-well plate. Put the plate in the 37 °C incubator for 15 min.
15. Add 1 mL pre-warmed crypt medium onto the solidified Matrigel.
16. Put the plate in the 37 °C incubator.

3.1.2 Protocol B

1–3. Same as Protocol A.

4. Transfer the intestine into the petri-dish and cut open the tissue longitudinally and then cross cut into 4–5 cm fragments. Gently scrap off the villi and the fecal masses on the mucosa with a cover glass (size 24 × 50 mm). Then further cut the fragments into 0.6–0.8 cm pieces.

5. Transfer the tissue pieces into a 15 mL corning tube with 5 mL PBS containing gentamycin (0.5 mg/mL) and shake the tube on an orbital shaker at 100 rpm for 10 min at 4 °C to remove remnant feces and contaminants.

6. Discard the supernatant PBS and add 5 mL PBS-Ca^{2+}/Mg^{2+} into the tube. Shake the tube on an orbital shaker at 75 rpm for 20 min at 4 °C.

7. Discard the supernatant PBS-Ca^{2+}/Mg^{2+} and add 5 mL PBS into the tube. Remove the PBS to wash away the remnant PBS-Ca^{2+}/Mg^{2+}.

8. Add 5 mL PBS-EDTA/EGTA into the tube. Shake the tube on an orbital shaker at 75 rpm for 30 min at 4 °C (*see* **Note 1**).

9. Discard the supernatant PBS-EDTA/EGTA and then add 5 mL fresh PBS-EDTA/EGTA into the tube.

10. Vortex the tube intermittently for 20 s, then immediately harvest the supernatant as fraction a.

11. Soak the intestine slices in 5 mL fresh PBS-EDTA/EGTA again and repeat **step 10** for two more times to get fractions b and c.

12. Fractions b and c are filtered through the cell strainer and the flow-through is centrifuged at 900 rpm, 4 °C, 5 min.

13–16. Same as Protocol A (*see* Fig. 1).

3.2 Culture and Passage of Crypt Organoids

1. The crypt medium should be refreshed every 2–3 days in the first week after isolation and every other day from the second week.

2. To quantify the organoid forming efficiency, let the organoids grow to days 10–14. The number of organoids grow out of the number of seeded crypt is calculated as the efficiency (*see* **Note 2**).

3. When the organoids grow into the size as around 500 μm in diameter, passage is required to prevent the burst from the central body of the organoid (*see* **Notes 3** and **4**, Fig. 2).

4. To start the passage procedure, remove the culture medium without touching the matrigel. Directly add 1 mL ice-cold PBS onto the matrigel and stand for 1 min.

5. Use 1 mL micro-pipette to gently mix PBS and the Matrigel to let the gel dissociate and homogenize in the PBS.

Culture of Intestinal Crypt Organoid   219

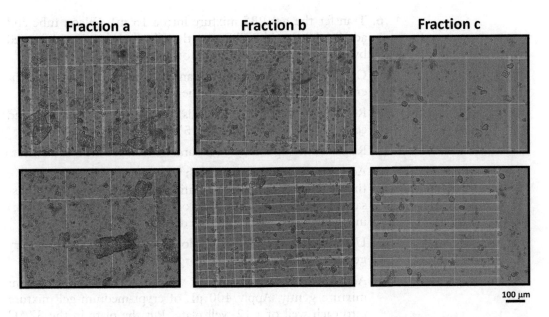

Fig. 1 Microscopic views of trypan blue-stained crypt fractions **a**, **b**, and **c** (obtained from method Section 3.1.2) on the hemocytometer

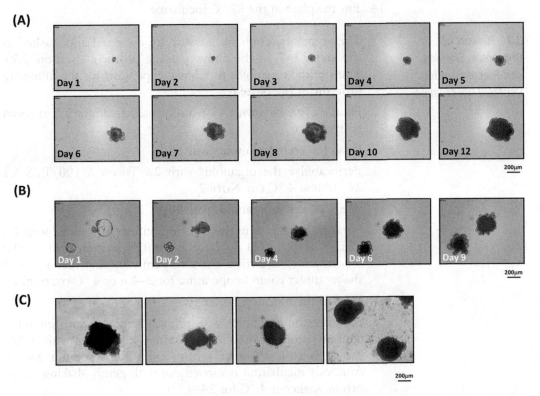

Fig. 2 Culture of crypt organoids. (**a**) The images of ISC and crypt organoids captured from Day 1 to Day 12 after isolation. (**b**) The growth of organoids after passage. (**c**) The ruptured and unhealthy crypt organoids

6. Transfer the gel-PBS mixture into a 15 mL Falcon tube and add 8 mL ice-cold PBS. Invert the tube for a few times followed by centrifugation at 900 rpm for 5 min at 4 °C.

7. Carefully aspirate the supernatant and gel layer while not bothering the crypt organoids in the bottom.

8. Resuspend the crypt organoids with 5 mL ice-cold PBS and centrifuge as **step 6** (*see* **Note 5**).

9. Remove the supernatant PBS and leave as few PBS as possible.

10. Add crypt medium to mix with the crypt organoids. Calculate the medium volume by estimating the well number after passage. For example, to amplify one well (from a 12-well plate) into four wells, use 400 μL medium.

11. Use 1 mL micro-pipette to do pipetting up and down very gently to dissociate the large organoid into smaller parts.

12. Add 1200 μL matrigel to mix with 400 μL of crypt-medium mixture gently. Apply 400 μL of crypt-medium-gel mixture into each well of a 12-well plate. Put the plate in the 37 °C incubator for 15 min.

13. Add 1 mL pre-warmed crypt medium onto the solidified Matrigel.

14. Put the plate in the 37 °C incubator.

3.3 Harvest of Organoids, Staining, and Preparation for Confocal Observation

1. The crypt organoids for harvest are collected and washed as the **steps 1–7** in the above passage protocol (Section 3.2). Transfer the organoids to 1.5 mL eppendorf for the following procedures (*see* **Note 6**).

2. Fix the organoids with 4% para-formaldehyde for 2 h at room temperature.

3. Wash the crypt organoids with PBS.

4. Permeabilize the organoids with 2% Triton X-100/PBS for 24–48 h at 4 °C (*see* **Note 7**).

5. Wash the crypt organoids with PBS.

6. Soak the organoids in the blocking buffer (PBS added with 2% Triton X-100, 10% goat serum, 0.02% sodium azide). The blocking step is carried out with gently shaking on an orbital shaker under room temperature for 2–4 h or 4 °C overnight.

7. Discard the blocking buffer.

8. Incubate the organoids with primary antibody with intended concentrations in the dilution buffer (PBS added with 0.25% Triton X-100, 1% goat serum, and 0.02% sodium azide). Antibody incubation is carried out with gently shaking on an orbital shaker at 4 °C for 24–48 h.

9. Wash the crypt organoids with dilution buffer for 3 × 10 min with gently shaking.
10. Incubate the organoids with fluorescence-conjugated secondary antibody diluted in the dilution buffer. The incubation is carried out by gently shaking on an orbital shaker at room temperature for 2–4 h. Keep off the light from this step on.
11. Wash the crypt organoids with dilution buffer for 3 × 10 min with gently shaking.
12. Discard the dilution buffer and rinse the organoids with PBS once.
13. Soak the organoids in FocusClear at 4 °C overnight (*see* **Note 7**).
14. Prepare the glass chamber for organoid fixation: stack 3–5 paper hold reinforcement sticks on a circular cover glass. Make sure to compact the stickers tightly without bubbles between.
15. Use 1 mL micro-pipette to transfer the organoids onto the circular cover glass. Then use 200 µL micro-pipette to aspirate the PBS around the organoids.
16. Add DAPI-mounting solution to cover the organoids and fill the chamber space. Try to leave the organoids in the center of the chamber.
17. Cover the chamber with another circular cover glass. Make sure that there are no bubbles in the chamber space, either around the organoids or the stickers. Surround the lateral side of the stickers with nail polish to prevent air get into the chamber. Place the chamber in the dark and let the mounting solution and nail polish solidify (*see* Fig. 3, **Note 8**).
18. Use tape to fix the sealed stands on a microscopic slide. Remember to keep organoids at the side distant from the microscopic slide.
19. Store the slides in the dark at 4 °C.
20. Before confocal microscopic observation, place the slides in the room temperature for 10 min. When doing observation, place the stands downward and slide upward so that the organoids are facing the objectives.
21. The image acquisition is done using confocal microscope with 40× or 63× oil objective. Recommended resolution is equal to or above 1024 × 1024 pixels.
22. For three-dimensional imaging, the image series along the Z-stack from the top to the bottom of the organoid are captured with a consistent Z-axis increment (usually found in the acquisition panel within the imaging software). Afterward, these z-stack images are processed with three-dimension reconstruction software (Supplementary **Videos 1** and **2**).

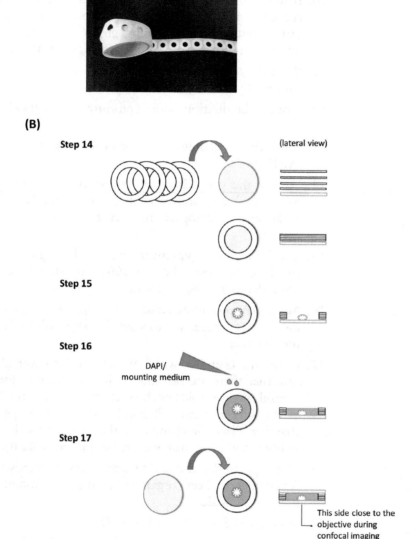

Fig. 3 Preparation of crypt organoids for confocal microscopic observation. (**a**) Paper hole reinforcement sticker. (**b**) Illustration of organoid fixation method described in Section 3.3, **steps 14–17**

3.4 Summary and Future Perspectives

Culture for ISCs is a delicately skillful procedure. To obtain healthy and long-termed survived ISCs, remember to pay high attention to the quality of growth factors and other reagents used in the whole procedure (*see* **Notes 9–12**).

The progress in tissue engineering and autologous stem cell therapy has made it possible to repair or replace tissue under injured or diseased conditions [14, 15]. For patients with irreversible intestinal diseases, especially to whom are not eligible for small bowel transplantation, an engineered small intestine seems to be a promising alternative as the ultimate treatment [16]. Recently, engineered small intestine has been reported in mice and rat model by transplanting crypt organoid units onto a polymer scaffold with digestive and absorptive functions [17–20]. We believe that robust culture and application of 3-D observation would be greatly helpful in the future research.

4 Notes

1. The main difference between isolation protocols in Sections 3.1.1 and 3.1.2 resides in the incubation condition and the subsequent yield of crypts. The EDTA-containing PBS helps to disrupt the tight junctions between cells. Our experimental testing showed that a continuous 30 min shaking (Section 3.1.2, **step 8**) in PBS-EDTA/EGTA made a higher yield of crypt number. With the efficient detachment of cells, abundant crypts are flushed out during the second and the third vortex as found in the fractions b and c in Section 3.1.2 (*see* Fig. 1). However, the villus scrap off step (Section 3.1.2, **step 4**) is a critical point that shall determine the resultant mucosal remnants in the desired fraction. By the adequate removal of villi layer, the crypt embedded in the mucosa can readily be ripped off without long and eight times of repeated shaking and vortex as in Section 3.1.1.

2. The quantification of the organoid forming efficiency (Section 3.2, **step 2**) is by counting the number of the crypt organoids grown in the matrigel after a period of culture. Originally, we counted the organoid number under the microscopy or within the microscopic images captured from the random views of each well. However, this is quite time-consuming and not so objective among different researchers. In addition, the diameters of organoid could vary a lot in the very same well with the same isolation and culture conditions. Therefore, we utilized digital analysis with ImageXpress Micro XLS Widefield High-Content Analysis System (Molecular Devices). The size of organoids and the number of organoids per well can be quantified using a customized module used TRITC autofluorescence intensity for organoid identification. Once acquired, images are stitched and processed through MetaXpress analysis software (*see* Fig. 4). This is especially helpful in large-scale screening experiments in which the organoids are seeded in 96-well plates. We believe this is a more accurate way to quantify the growth of the organoids.

3. The time point for organoid passage usually depends on the size of the organoids. As organoids grow, the central body

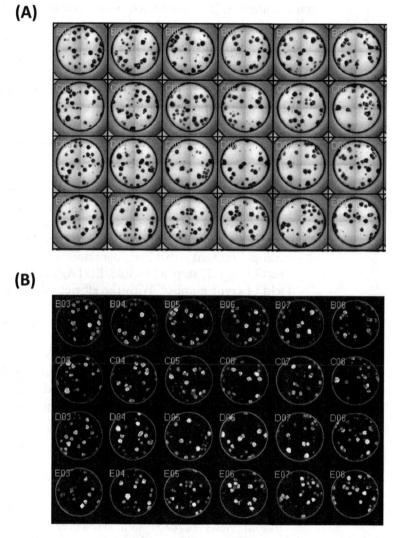

Fig. 4 Digital imaging analysis with ImageXpress Micro XLS Widefield High-Content Analysis System. The images are collected at 4× magnification from four sites per well with transmitted light (**a**) and TRITC fluorescence channel (**b**). By using a customized module in which TRITC autofluorescence intensity is set for organoid identification, the images are then stitched and processed through MetaXpress analysis software

becomes enlarged and filled with more and more nonviable cells over time. If not passaged, the organoids would burst and eventually go unhealthy and wither away (*see* Fig. 2c).

4. There is a recommended passage protocol from the manufacturer of Matrigel (BD Biosciences), which includes adding a "recovery solution" to depolarize Matrigel followed by a 1-h incubation on ice. In our experience, we found the recovery yield unsatisfied for a significant lower viability after passage. We therefore tried the method we described in Section 3.2 by adding the ice-cold PBS to homogenize the matrigel with

gentle pipetting. The subsequent wash steps with ice-cold PBS (Section 3.2, **steps 5–8**) are important to depolarize the gel.

5. During the passage procedure, the pipetting force should be as tender as possible. Try to keep the organoid structure intact during the wash steps. Only when ready to reseed the organoids, mix the organoids with crypt medium and dissociate the large organoids into smaller ones by pipetting (Section 3.2, **steps 10** and **11**). The dissociation step must be done before adding the matrigel (Section 3.2, **step 12**), since that it is difficult to break down the structure of organoids when coated by matrigel. The organoids after passage would grow and start to have buddings in 24 h after passage, which is a little faster than the ISCs first isolated from the tissue (*see* Fig. 2a, b; Supplementary **Videos 3** and **4**).

6. When harvesting the organoids for staining and confocal microscopic observation, make sure the remnant Matrigel around the organoids is removed by the wash step with ice-cold PBS. Repeat Section 3.2, **step 8** if necessary. The residual Matrigel may affect the subsequent staining and the effect of FocusClear. Furthermore, the gel can make the focus uneven under the confocal observation and imaging.

7. The time duration for permeabilization, antibody incubation, and FocusClear might be different depending on the size of the harvested organoids. For smaller round-shaped organoids (without the budding of crypt-like domain), the incubation time can be shortened; with the large and mature organoids with long crypt-like domains, the incubation time should be longer.

8. The preparation on the glass chambers to fix the organoids requires a few times of practice to leave no air within the chamber space. The number of reinforcement stickers for making the height of the chamber depends on the size of organoids. Make it high enough to keep the organoids in their original shapes without being compressed by the cover glass (Section 3.3, **steps 14–17**).

9. For the immunostaining in the organoids, we have tested certain marker proteins including Lgr5, Ki67, Muc2, chromogranin-A, lysozyme, villin, and apoptotic markers. The staining effect is highly variable among different antibodies as that modification of concentration and incubation time with antibody may refine the results. Choose the antibodies with proven application in immunohistochemistry on paraffin-fixed specimens.

10. During all the procedures for organoid transfer and wash, we prefer to handle with 1 mL micropipette. Use low binding

pipette tips can help prevent the organoids from retaining on the tip wall.

11. The viability of ISCs also is greatly dependent on the seeding density of crypts. We observed that higher seeding density resulted in a better growth rate of the organoids. However, the high density of cells could make the matrigel out of nutrient support and turn into yellow color within 5–7 days after seeding. The ideal density would allow abundant organoids to grow and keep the matrigel in red-to-orange color for 10–14 days.

12. According to our experience, the freshness of growth factors and medium is one of the critical points affecting the growth of ISCs. Make aliquots for stocks of all growth factors and avoid repeated freeze and thaw. Always prepare weekly amount of crypt medium and store the medium in the 4 °C refrigerator. Before use, aspirate out the needed volume of medium for prewarm; do not put the whole tube of medium in 37 °C water bath repeatedly.

Electronic Supplementary Material

Video 1 3-D reconstruction for the organoids. The organoids were stained with DAPI and the fluorescence signals were imaged by confocal microscopy under the 40× oil objective. The Z-stacked images were processed and reconstructed by Zen software (Carl Zeiss). This video covered part of the organoid with seven visible crypt-like domains (AVI 2061 kb)

Video 2 3-D reconstruction for the organoids. The organoids were stained with DAPI and the fluorescence signals were imaged by confocal microscopy under the 63× oil objective. The Z-stacked images were processed and reconstructed by Zen software (Carl Zeiss). This video covered only one crypt-like domain of an organoid (AVI 2440 kb)

Video 3 The growth of ISC into crypt organoids. The images were captured every 8 h on the first day and every 24 h afterward for 20 days. The images were captured and recorded by AZTEC CCM-1.4 II/M. The scale bar in the video represents 100 μm (AVI 31107 kb)

Video 4 The growth of organoids after passage. The images were captured every 24 h after passage for 9 days. The images were captured and recorded by AZTEC CCM-1.4 II/M. The scale bar in the video represents 100 μm (AVI 12728 kb)

References

1. Kuntz S, Rudloff S, Kunz C (2008) Oligosaccharides from human milk influence growth-related characteristics of intestinally transformed and non-transformed intestinal cells. Br J Nutr 99(3):462–471. doi:10.1017/S0007114507824068
2. Escaffit F, Perreault N, Jean D, Francoeur C, Herring E, Rancourt C, Rivard N, Vachon PH, Pare F, Boucher MP, Auclair J, Beaulieu JF (2005) Repressed E-cadherin expression in the lower crypt of human small intestine: a cell marker of functional relevance. Exp Cell Res 302(2):206–220. doi:10.1016/j.yexcr.2004.08.033
3. Francoeur C, Escaffit F, Vachon PH, Beaulieu JF (2004) Proinflammatory cytokines TNF-alpha and IFN-gamma alter laminin expression under an apoptosis-independent mechanism in human intestinal epithelial cells. Am J Physiol Gastrointest Liver Physiol 287(3):G592–G598. doi:10.1152/ajpgi.00535.2003
4. Ruemmele FM, Beaulieu JF, Dionne S, Levy E, Seidman EG, Cerf-Bensussan N, Lentze MJ (2002) Lipopolysaccharide modulation of normal enterocyte turnover by toll-like receptors is mediated by endogenously produced tumour necrosis factor alpha. Gut 51(6):842–848
5. Barker N, van Es JH, Kuipers J, Kujala P, van den Born M, Cozijnsen M, Haegebarth A, Korving J, Begthel H, Peters PJ, Clevers H (2007) Identification of stem cells in small intestine and colon by marker gene Lgr5. Nature 449(7165):1003–1007
6. Rouch JD, Scott A, Lei NY, Solorzano-Vargas RS, Wang J, Hanson EM, Kobayashi M, Lewis M, Stelzner MG, Dunn JC, Eckmann L, Martin MG (2016) Development of functional microfold (M) cells from intestinal stem cells in primary human enteroids. PLoS One 11(1):e0148216. doi:10.1371/journal.pone.0148216
7. Mohamed MS, Chen Y, Yao CL (2015) Intestinal stem cells and stem cell-based therapy for intestinal diseases. Cytotechnology 67(2):177–189. doi:10.1007/s10616-014-9753-9
8. Barthel ER, Speer AL, Levin DE, Sala FG, Hou X, Torashima Y, Wigfall CM, Grikscheit TC (2012) Tissue engineering of the intestine in a murine model. J Vis Exp 70:e4279. doi:10.3791/4279
9. Kuo WT, Lee TC, Yang HY, Chen CY, Au YC, Lu YZ, Wu LL, Wei SC, Ni YH, Lin BR, Chen Y, Tsai YH, Kung JT, Sheu F, Lin LW, Yu LC (2015) LPS receptor subunits have antagonistic roles in epithelial apoptosis and colonic carcinogenesis. Cell Death Differ 22(10):1590–1604. doi:10.1038/cdd.2014.240
10. Sato T, Vries RG, Snippert HJ, van de Wetering M, Barker N, Stange DE, van Es JH, Abo A, Kujala P, Peters PJ, Clevers H (2009) Single Lgr5 stem cells build crypt-villus structures in vitro without a mesenchymal niche. Nature 459(7244):262–265
11. Sato T, Clevers H (2013) Growing self-organizing mini-guts from a single intestinal stem cell: mechanism and applications. Science 340(6137):1190–1194. doi:10.1126/science.1234852
12. Chen Y, Lee SH, Tsai YH, Tseng SH (2014) Ischemic preconditioning increased the intestinal stem cell activities in the intestinal crypts in mice. J Surg Res 187(1):85–93. doi:10.1016/j.jss.2013.10.001
13. Chen Y, Tsai YH, Liu YA, Lee SH, Tseng SH, Tang SC (2013) Application of three-dimensional imaging to the intestinal crypt organoids and biopsied intestinal tissues. ScientificWorldJournal 2013:624342. doi:10.1155/2013/624342
14. Esposito G, Sarnelli G, Capoccia E, Cirillo C, Pesce M, Lu J, Cali G, Cuomo R, Steardo L (2016) Autologous transplantation of intestine-isolated glia cells improves neuropathology and restores cognitive deficits in beta amyloid-induced neurodegeneration. Sci Rep 6:22605. doi:10.1038/srep22605
15. Hanna J, Wernig M, Markoulaki S, Sun CW, Meissner A, Cassady JP, Beard C, Brambrink T, Wu LC, Townes TM, Jaenisch R (2007) Treatment of sickle cell anemia mouse model with iPS cells generated from autologous skin. Science 318(5858):1920–1923. doi:10.1126/science.1152092
16. Levin DE, Sala FG, Barthel ER, Speer AL, Hou X, Torashima Y, Grikscheit TC (2013) A "living bioreactor" for the production of tissue-engineered small intestine. Methods Mol Biol 1001:299–309. doi:10.1007/978-1-62703-363-3_25
17. Sala FG, Matthews JA, Speer AL, Torashima Y, Barthel ER, Grikscheit TC (2011) A multicellular approach forms a significant amount of tissue-engineered small intestine in the mouse. Tissue Eng Part A 17(13–14):1841–1850. doi:10.1089/ten.TEA.2010.0564
18. Levin DE, Barthel ER, Speer AL, Sala FG, Hou X, Torashima Y, Grikscheit TC (2013) Human tissue-engineered small intestine forms from postnatal progenitor cells. J Pediatr Surg 48(1):129–137. doi:10.1016/j.jpedsurg.2012.10.029
19. Choi RS, Riegler M, Pothoulakis C, Kim BS, Mooney D, Vacanti M, Vacanti JP (1998)

Studies of brush border enzymes, basement membrane components, and electrophysiology of tissue-engineered neointestine. J Pediatr Surg 33(7):991–996. discussion 996-997

20. Grant CN, Mojica SG, Sala FG, Hill JR, Levin DE, Speer AL, Barthel ER, Shimada H, Zachos NC, Grikscheit TC (2015) Human and mouse tissue-engineered small intestine both demonstrate digestive and absorptive function. Am J Physiol Gastrointest Liver Physiol 308 (8):G664–G677. doi:10.1152/ajpgi.00111. 2014

Human Intestinal Enteroids: New Models to Study Gastrointestinal Virus Infections

Winnie Y. Zou, Sarah E. Blutt, Sue E. Crawford, Khalil Ettayebi, Xi-Lei Zeng, Kapil Saxena, Sasirekha Ramani, Umesh C. Karandikar, Nicholas C. Zachos, and Mary K. Estes

Abstract

Human rotavirus (HRV) and human norovirus (HuNoV) infections are recognized as the most common causes of epidemic and sporadic cases of gastroenteritis worldwide. The study of these two human gastrointestinal viruses is important for understanding basic virus-host interactions and mechanisms of pathogenesis and to establish models to evaluate vaccines and treatments. Despite the introduction of live-attenuated vaccines to prevent life-threatening HRV-induced disease, the burden of HRV illness remains significant in low-income and less-industrialized countries, and small animal models or ex vivo models to study HRV infections efficiently are lacking. Similarly, HuNoVs remained non-cultivatable until recently. With the advent of non-transformed human intestinal enteroid (HIE) cultures, we are now able to culture and study both clinically relevant HRV and HuNoV in a biologically relevant human system. Methods described here will allow investigators to use these new culture techniques to grow HRV and HuNoV and analyze new aspects of virus replication and pathogenesis.

Keywords: Gastrointestinal viral infections, HIEs, Human intestinal enteroids, Human norovirus, Human rotavirus

1 Introduction

Diarrheal diseases are an important global health problem, causing an estimated 4 % of all deaths worldwide, including 1.3 million deaths per year for all ages with an estimated 499,000 childhood deaths in developing countries [1, 2]. Diarrheal disease can be caused by many pathogens including bacteria, parasites, and viruses, with the latter not being controllable through improvements in water, sanitation, and hygiene. Thus, viral gastroenteritis remains an important public health threat.

Human rotavirus (HRV) and human norovirus (HuNoV) are the most frequent causes of acute gastroenteritis worldwide [3–6]. Historically, HRV was the most common cause of severe diarrheal disease and the leading cause of diarrhea-related deaths in children [7]. The use of live-attenuated HRV vaccines has made a tremendous impact in reducing HRV disease in the developed world [8]. However, the efficacy of the HRV vaccines remains low in less-industrialized world [9]. Thus, many children continue to suffer from HRV-related disease today, and long-term consequences of diarrheal disease are increasingly being recognized [10]. With HRV disease decreasing markedly in developed countries, HuNoV infections have emerged as the most common cause of epidemic and sporadic cases of acute gastroenteritis worldwide in children [3, 11]. In addition, HuNoVs affect all age groups and are the leading cause of food-borne gastroenteritis. The major barrier to research and development of effective interventions for HuNoVs has been the lack of a robust and reproducible in vitro cultivation system. Such a system is critical to achieve a full mechanistic understanding of HuNoV replication, stability, evolution, pathogenesis, and vaccine development [12].

In 2007, Dr. Hans Clevers' group identified Lgr5 as a marker for stem cells in the intestinal epithelium in mice and showed multicellular intestinal epithelial cultures could be established ex vivo from isolated Lgr5+ cells [13]. With this knowledge, they subsequently developed human intestinal enteroids (HIEs), where three-dimensional (3D) in vitro cultures are derived from proliferating stem cells in crypts isolated from human gastrointestinal (GI) tract biopsies or surgical specimens [14, 15] (Fig. 1a). HIE cultures possess the multicellular complexity and organization of the intestinal epithelium and are physiologically active [16]. They contain all epithelial cell types of the normal GI tract and retain segment-specific properties [17]. These remarkable cultures are being developed for many basic, clinical, and translational applications, although infectious disease research was not highlighted initially [14]. With the advent of the HIEs, we evaluated and have shown the relevance of these novel cultures to model both HRV and HuNoV infections [12, 18, 19]. Importantly, we have shown for the first time that HIEs support replication of multiple strains of HuNoV as well as HRVs better than animal rotavirus strains, thus recapitulating the known biology of these two important human pathogens. Culturing each of these viruses requires both HIEs and the addition of intestinal factors to obtain optimal replication.

In this chapter, we describe our methods for the maintenance, expansion, and differentiation of 3D and monolayer HIEs as well as their utilization in HRV and HuNoV infections (Fig. 1). We do not describe the establishment of these cultures from patient tissues because this has been reported previously in several methods papers [12–15, 18–20]. We detail an HRV infection method using differentiated 3D HIEs in the presence of added pancreatin. Because of HRV's affinity toward differentiated cell types, proper and controlled differentiation of the 3D HIEs is crucial for successful

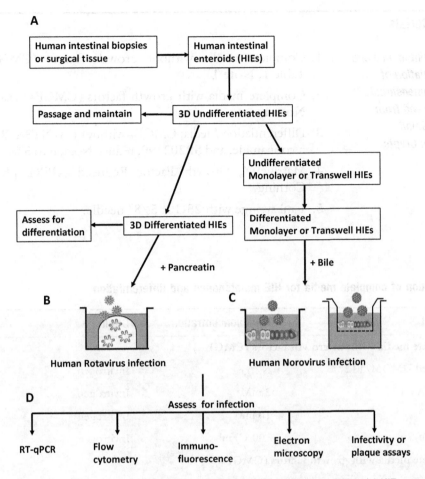

Fig. 1 Human rotavirus and norovirus infections in human intestinal enteroids (HIEs). (**a**) HIEs can be revived and passaged from frozen stocks and maintained in undifferentiated format. Differentiation is required for both rotavirus and norovirus infections. (**b**) Human rotavirus infection uses 3D HIEs in the presence of pancreatin. (**c**) Human norovirus infection uses monolayer or transwell HIEs in the presence of human, bovine, or porcine bile. (**d**) Infection can be monitored using various methods

infection. With HIE differentiation, HRV is able to infect without mechanical disruption, and the infected HIEs can be maintained in suspension until analysis [18]. The method for HuNoV infection uses monolayer or transwell cultures of HIEs and the addition of bile that affects the cells [12]. Different culture conditions also are needed to achieve replication of different HuNoV strains. Together, these protocols should facilitate the study of HRV and HuNoV in HIEs in many laboratories, and modifications of these basic procedures may be used for replication of other human enteric microbes.

2 Materials

2.1 Maintenance and Differentiation of Three-Dimensional HIEs Derived from Human Small Intestinal Crypts

1. Complete media without growth factors (CMGF−) (see Table 1, **Note 1**).
2. Complete media with growth factors (CMGF+) (see Table 1, **Note 1**).
3. Differentiation Media: CMGF+ without L-WNT3A, R-spondin, nicotinamide, and SB202190; reduce Noggin to 5 %.
4. Matrigel™, Growth Factor Reduced (GFR), phenol-free: Corning.
5. 1 ml syringe with 25G × 5/8″ needle.

Table 1
Composition of complete media for HIE maintenance and differentiation

Reagent	Final concentration	Origin
Complete media without growth factors (CMGF−)		
Advanced DMDM/F12	N/A	Invitrogen
GlutaMAX-1	2 mM	Invitrogen
HEPES	10 mM	Invitrogen
Penicillin/streptomycin	100 U/ml	Invitrogen
Complete media with growth factors (CMGF+)		
CMGF−	N/A	N/A
L-WNT3A-conditioned media	50 %	ATCC
R-Spondin-conditioned media	20 %	Trevigen
Noggin-conditioned media	10 %	A gift from Dr. Gijs van den Brink (University of Amsterdam)
B27	1×	Invitrogen
N2	1×	Invitrogen
N-acetylcysteine	1 mM	Sigma-Aldrich
Mouse recombinant EGF	50 ng/ml	Invitrogen
[Leu15]-Gastrin I	10 nM	Sigma-Aldrich
Nicotinamide	10 mM	Sigma-Aldrich
A-83-01	500 nM	Tocris
SB202190	10 µM	Sigma-Aldrich

6. 24-well Nunc cell culture-treated multiwell dishes: Thermo Scientific.
7. Accutase: BD Biosciences.
8. Fetal bovine serum: Corning.
9. Allophycocyanin (APC)-conjugated mouse anti-human CD44 antibody (Clone G44-26): BD Biosciences.
10. Round-bottom polystyrene tubes.
11. 300 µM DAPI.

2.2 Rotavirus Infection and Analysis of Differentiated 3D HIEs

1. CMGF−.
2. Differentiation Media.
3. Pancreatin from porcine pancreas: Sigma-Aldrich.
4. Rotavirus stock lysate and appropriate control lysate.
5. 0.22 µm syringe filter.
6. 3 ml syringe.
7. Trypsin: Worthington.
8. Matrigel™, GFR, phenol-free: Corning.
9. LoBind microcentrifuge tubes: Eppendorf.
10. Polyclonal rabbit anti-RV antibody: Laboratory generated [18].
11. Alexa Fluor 488 donkey anti-rabbit antibody: Invitrogen.
12. Cytofix/Cytoperm: BD Biosciences.
13. Perm/Wash buffer: BD Biosciences.

2.3 Establishment of Monolayer HIEs on Transwells and 96-Well Plates

1. CMGF−.
2. CMGF+.
3. Collagen type IV from human placenta: Sigma-Aldrich.
4. Glacial Acetic Acid: Thermo Scientific.
5. 1× DPBS, no calcium, no magnesium: Invitrogen.
6. 0.5 M EDTA.
7. 0.05 % Trypsin/0.5 mM EDTA: Invitrogen.
8. 5 mM ROCK inhibitor Y-27632: Sigma-Aldrich.
9. Fetal bovine serum: Corning.
10. Transwells: Costar.
11. 40 µm cell strainer.
12. 96-well tissue culture plates.

2.4 Norovirus Infection of Differentiated Monolayer HIEs

1. Human norovirus (HuNoV)-positive and human norovirus-negative stools.
2. Cup horn sonicator (Heat Systems Ultrasonics, Plainview, NY).
3. 5, 1.2, 0.8, 0.45, and 0.22 μm filters (PVDF membrane).
4. Round-bottom 96-well plate.
5. CMGF−.
6. Differentiation Media.
7. Bile (Human, bovine, or porcine): Sigma-Aldrich (*see* **Note 18**).
8. 0.05 % Trypsin-EDTA: Gibco.
9. DMEM: Invitrogen.
10. Fetal bovine serum.
11. Cytofix: BD Biosciences.
12. Stain Buffer containing 5 % BSA: BD Pharmingen.
13. Methanol.
14. PBS.
15. Guinea pig anti-GII.4/Sydney 2012 VP1 antibody: Laboratory generated [12, 21].
16. Alexa Fluor 488 goat anti-guinea pig antibody: Invitrogen.
17. 300 μM DAPI.

3 Methods

3.1 Maintenance and Differentiation of Three-Dimensional HIEs Derived from Human Small Intestinal Crypts

Three-dimensional HIEs were derived from human small intestinal biopsies or surgical specimens following protocols from the Clevers' laboratory and others [14, 15]. In our laboratory, HIEs have been derived from all segments of the small intestine, the colon, and the stomach.

3.1.1 Reviving Frozen Stocks of HIEs from Liquid Nitrogen

1. *Preparation:* Thaw appropriate amount of Matrigel™ overnight at 4 °C (30 μl/well).
2. Transfer a frozen vial containing HIEs from liquid nitrogen to dry ice.
3. Hold vial under room temperature (RT) tap water until ice detaches from the vial wall.
4. Transfer contents of the vial into a pre-chilled 15 ml conical tube containing 10 ml cold CMGF−.
5. Pellet cells in a swinging bucket rotor centrifuge at 80 × g, 4 °C for 5 min. Remove supernatant.

6. Leave 15 ml tube containing HIE pellet on ice. Resuspend pellet in the appropriate amount of Matrigel™ (30 μl/well) using cold P200 pipette tips. Plate HIEs as droplets in 24-well plates (*see* **Notes 2** and **3**).

7. Let gel solidify for 5–10 min at 37 °C.

8. Add 500 μl of RT CMGF+ to each well and culture in 37 °C incubator with 5 % CO_2.

9. Refresh culture with CMGF+ every other day until the HIEs are ready to be passaged (*see* **Note 4**).

3.1.2 HIE Passage and Differentiation

1. Remove old media from wells leaving the Matrigel™ plug intact.

2. Add 500 μl cold CMGF− to well and mechanically break up Matrigel™ by gently pipetting up and down with a P1000 pipet.

3. Using a 1 ml syringe with a 25G × 5/8″ needle, syringe up and down the contents of each well 2–3 times, then transfer the entire contents into a 15 ml conical tube (multiple wells of the same HIEs can be combined). Add an additional 2× volume of cold CMGF− to dissolve the Matrigel™ (*see* **Note 5**).

4. Pellet cells in a swinging bucket rotor centrifuge at 80 × g, 4 °C for 5 min. Remove supernatant.

5. Plate HIEs with Matrigel™ on 24-well plates as step 6–9 in Sect. 3.1.1.

6. If downstream analysis or infection is desired, switch CMGF+ to Differentiation Media after 4 days in culture for 3–5 days (Fig. 1a). Refresh culture with Differentiation Media every other day until use (*see* **Note 6**).

Table 2
RT-qPCR probes used to analyze differentiation status of HIEs

Marker	Changes during differentiation	Cat. No. (Life Technologies)
CD44	Down	Hs01075861_m1
LGR5	Down	Hs00969422_m1
MUC2	Up	Hs03005103_g1
PCNA	Down	Hs00427214_g1
Sucrase isomaltase	Up	Hs00356112_m1

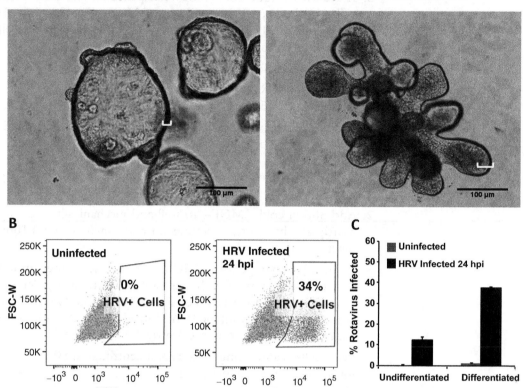

Fig. 2 Differentiation of human intestinal enteroids (HIEs). (**a**) Undifferentiated HIEs appear cystic and shining, whereas differentiated HIEs appear multi-lobular, with a thickened epithelium. White brackets denote epithelial lining. Note that differentiated HIEs can also often be circular depending on the intestinal segment and tissue origin. (**b**) Representative scheme for flow cytometry analysis of human rotavirus (HRV) infection. Single, live cells are gated using forward and side scatters. HRV+ cells can then be detected in the infected sample. (**c**) Differentiation of HIEs is required for HRV infection. Percentage of HRV-infected cells assessed by flow cytometry. $N = 3$ jejunal HIEs

3.1.3 Flow Cytometry Analysis of Undifferentiated and Differentiated HIEs

The differentiation status of HIEs is critical in HRV and HuNoV infections and can be monitored by many methods including RT-qPCR (some common probes used are listed in Table 2), light microscopy (Fig. 2a), immunofluorescent staining (some common antibodies used are listed in Table 3), and flow cytometry. Here, we describe an easy and quantitative method using CD44 staining in flow cytometry. CD44 has previously been shown to be a marker for crypt cells [22].

1. Remove old Differentiation Media from HIE wells, keeping Matrigel™ plug intact.
2. Add 500 μl ice-cold Accutase to each well. Vigorously pipette up and down with a P1000 to dissolve Matrigel™.
3. Incubate at 37 °C for 30 min. Pipette up and down 10 times every 10 min to further break up Matrigel™ and cell clumps.

Table 3
Antibodies used to analyze differentiation status of HIEs

Marker	Cell type marked	Company (concentration)
Sucrase isomaltase	Enterocytes	Santa Cruz (1:100)
Muc2	Goblet cells	Santa Cruz (1:500)
Chromagranin A	Enteroendocrine cells	Novus Biologicals (1:100)
UEA-1 lectin	Histo-blood group antigens	Sigma-Aldrich (1:500)
E-cadherin	Adherens junctions	BD Biosciences (1:100)
Lysozyme	Paneth cells	Dako (1:100)

4. Scrape and transfer all contents into a round-bottom polystyrene tube. Vortex for a final break up of cell clumps.

5. Add 3 ml CMGF− to dilute the Accutase enzyme reaction.

6. If desired, the cell solution can be passed through a 40 μm filter to exclude cell clumps.

7. Pellet cells in a swinging bucket rotor centrifuge at 400 × g, 4 °C for 5 min. Remove supernatant.

8. Wash cells with 3 ml CMGF− and pellet cells in a swinging bucket rotor centrifuge at 400 × g, 4 °C for 5 min. Remove supernatant.

9. Resuspend cells with 100 μl CMGF− containing 10 % FBS. Add APC-conjugated anti-human CD44 antibody at the manufacturer's recommended concentration. Vortex.

10. Incubate at RT for 30 min or 4 °C for 1 h.

11. Wash with 1 ml CMGF− and pellet cells in a swinging bucket rotor centrifuge at 400 × g, 4 °C for 5 min. Remove supernatant.

12. Resuspend cells in 500 μl CMGF− for flow cytometry analysis.

13. If desired, 5 μl of 300 μM DAPI can be added just before analysis. Dead, but not live, cells will incorporate DAPI thus can be excluded during flow analysis.

3.2 Rotavirus Infection and Analysis of Infected 3D HIEs

3.2.1 Rotavirus Infection of Differentiated 3D HIEs

1. *Preparation*: Make rotavirus incubation media by dissolving 0.25 mg/ml pancreatin in Differentiation Media. Allow the pancreatin to dissolve for about 30 min at RT. The undissolved particulates are then filtered out using a 0.22 μm syringe filter with a 3 ml syringe (*see* **Notes 7** and **8**).

2. Activate rotavirus and control stock lysates with 10 μg/ml trypsin at 37 °C for 30 min (*see* **Notes 9** and **10**).

3. Remove old Differentiation Media from wells. Keep Matrigel™ plug intact.

4. Add 500 μl ice-cold CMGF− to each well. Gently pipette up and down with P1000 to dissolve Matrigel™.

5. Scrape and transfer all contents into a 15 ml conical tube.

6. Add an additional 2× volume of cold CMGF− to dissolve Matrigel™.

7. Pellet in a swinging bucket rotor centrifuge at 80 × g, 4 °C for 5 min. Remove supernatant.

8. Remaining Matrigel™ residue can be seen as a clear clump above the cell clump after centrifugation. Washing steps 6–7 can be repeated until Matrigel™ residue has been completely dissolved.

9. Resuspend HIEs in 200 μl rotavirus incubation media containing the appropriate amount of HRV or control stock lysates. Transfer contents to a LoBind microcentrifuge tube.

10. Incubate at 37 °C, 5 % CO_2 for 1–2 h.

11. Wash unbound virus by adding 1 ml CMGF−.

12. Pellet in a swinging bucket rotor centrifuge at 80 × g, 4 °C for 5 min. Remove supernatant.

13. Repeat washing steps 11–12.

14. Add 500 μl of RT Differentiation Media to each tube and culture infected HIEs in suspension at 37 °C incubator with 5 % CO_2 until analysis.

15. If desired, 1 μg/ml trypsin can be added to differentiation media to allow for multiple rounds of infection. If longer infection duration is desired, 10 μM ROCK inhibitor Y27632 can be added to improve cell survival.

3.2.2 Flow Cytometry Analysis of Rotavirus-Infected 3D HIEs

Flow cytometry analysis measures percentage of rotavirus-infected cells (Fig. 2b, c). Dead cells due to HRV infection will be excluded in this analysis; therefore, this method is better used during early course of infection (up to 24 hpi). If analysis during late course of infection must be monitored, infectivity assays such as fluorescence focus assay and plaque assays can be used. These methods are discussed in Arnold et al. Curr Protoc Microbiology. 2009 [23].

1. Follow steps 1–8 in Sect. 3.1.3 to dissociate 3D HIEs into single cells.

2. Resuspend cells in 250 μl Cytofix/Cytoperm.

3. Incubate at 4 °C for 15–20 min.

4. Wash with 3 ml 1× Perm/Wash buffer. Pellet in a swinging bucket rotor centrifuge at 400 × g, 4 °C for 5 min. Remove supernatant (*see* **Note 11**).

5. Resuspend in 100 μl 1× Perm/Wash buffer. Add an appropriate amount of polyclonal rabbit anti-RV antibody. Vortex.
6. Incubate at RT for 30 min.
7. Wash with 1 ml 1× Perm/Wash buffer. Pellet in a swinging bucket rotor centrifuge at 400 × g, 4 °C for 5 min. Remove supernatant.
8. Resuspend in 100 μl 1× Perm/Wash buffer. Add Alexa Fluor 488 donkey anti-rabbit antibody at the manufacturer's recommended concentration. Vortex.
9. Incubate at RT for 30 min, protected from light.
10. Wash with 1 ml 1× Perm/Wash buffer. Pellet in a swinging bucket rotor centrifuge at 400 × g, 4 °C for 5 min. Remove supernatant.
11. Resuspend in 500 μl 1× Perm/Wash buffer for flow cytometry analysis.

3.3 Establishment of Monolayer HIEs on Transwell and 96-Well Plates

1. *Preparation*: Make 0.5 mM EDTA by diluting 0.5 M EDTA stock in cold 1× DPBS. Prepare 33 μg/ml Collagen IV by diluting the lyophilized powder in 0.6 % acetic acid (*see* **Note 12**).
2. Coat desired number of transwells or 96 wells using 100 μl of cold Collagen IV.
3. Allow Collagen IV to solidify by placing the coated plates at 37 °C for 2 h.
4. Take out the appropriate amount of undifferentiated, fully grown 3D HIEs (*see* **Note 13**).
5. Remove old media from wells leaving the Matrigel™ plug intact.
6. Collect HIEs by adding 500 μl per well of cold 0.5 mM EDTA in DPBS. Scrape and transfer all contents into a 15 ml conical tube.
7. Pellet in a swinging bucket rotor centrifuge at 200 × g, 4 °C for 5 min. Remove supernatant.
8. Dissociate HIEs into single cells by adding 500 μl 0.05 % trypsin/0.5 mM EDTA (*see* **Note 14**).
9. Incubate at 37 °C for 4 min.
10. Add 1 ml CMGF− containing 10 % FBS to inactivate trypsin.
11. Dissociate HIEs by vigorously pipetting up and down approximately 50 times using a P1000 (*see* **Note 15**).
12. Wet a 40 μm cell strainer with 1 ml CMGF− containing 10 % FBS. Allow the dissociated HIE solution to pass through the 40 μm cell strainer by gravity for 5 s into a 50 ml conical tube to exclude clumps of cells.

13. Pellet filtered cell solution in a swinging bucket rotor centrifuge at 400 × g, RT for 5 min. Remove supernatant.
14. Resuspend cells in the appropriate amount of CMGF+ containing 10 μM Y-27632 (see **Note 16**).
15. Take out Collagen IV-coated plates. Remove excess liquid from solidified Collagen IV-coated wells.
16. Plate cell solution on Collagen IV-coated plates, usually 100 μl per well in a transwell plate or 96-well plate (see **Note 17**).
17. If using transwells, add 500 μl RT CMGF+ to the bottom compartment and 100 μl to the upper compartment. If using 96-well tissue culture plates, add 100 μl CMGF+ to each well. Culture in 37 °C incubator with 5 % CO_2.
18. Switch CMGF+ to Differentiation Media the next day. Incubate for 3–5 days refreshing the culture with differentiation media every other day until use (see **Note 6**).

3.4 Human Norovirus Infection of Differentiated Monolayer HIEs

3.4.1 Preparation of 10 % HuNoV Stool Filtrates

1. Weigh 0.5 g of HuNoV-positive or HuNoV-negative stool in a 15 ml conical tube. Add 4.5 ml PBS to make a 10 % stool suspension.
2. Break up solids with a 1 ml pipette tip as necessary, then vortex for ~30 s.
3. Keep at RT for 5 min and vortex again.
4. Sonicate using a cup horn sonicator at a setting of 10, 3 times for 1 min, with a 1 min rest on ice between sonications.
5. Centrifuge at 1500 × g for 10 min to remove debris.
6. Collect supernatant in a new 15 ml centrifuge tube and repeat centrifugation one more time.
7. Serially pass the supernatant through 5 μm, then 1.2, 0.8, 0.45, and 0.22 μm filters.
8. Aliquot the resulting 10 % stool filtrate, 100 μl/tube and store at −80 °C.

3.4.2 Norovirus Infection of Monolayer HIEs on 96-Well Plate

Similar protocol can be used for HuNoV infection of transwell HIEs. In our laboratory, infection on monolayer HIEs usually gives a slightly higher yield when compared to transwell HIEs.

1. *Preparation*: Thaw HuNoV-positive and HuNoV-negative stool filtrates at RT. Make 1:10 and 1:100 dilutions of each stool filtrate.
2. In a round-bottom 96-well plate, add 228 μl of CMGF− medium with or without bile (see **Note 18**).
3. For each HuNoV-positive and HuNoV-negative stool filtrate, add 12 μl of each diluted sample into the 228 μl CMGF− medium. (This yields 240 μl inocula for infection. 100 μl

from each well will be used to inoculate two wells for the 1 hpi and 72 hpi timepoints, respectively.) Each sample should be prepared in triplicate.

4. Wash the HIE monolayers once with ice-cold CMGF−.
5. Gently aspirate the medium from each well.
6. Pipet the inocula up and down 5 times to mix. Transfer 100 μl inocula to the corresponding well containing HIEs. Pipet the inocula on the side of the well, careful not to disturb the HIE monolayer.
7. Incubate the infected plates for 1 h in a 37 °C incubator with 5 % CO_2.
8. Pipet the inocula off each well and gently wash each monolayer three times by applying 100 μl CMGF− medium to the wall of the well and aspirating the wash media off.
9. Add 100 μl of differentiation medium (with or without bile) to each well. The 1 hpi timepoint wells are ready for harvest. The 72 hpi timepoint wells are incubated in a 37 °C incubator with 5 % CO_2 until harvest.

3.4.3 Flow Cytometry Analysis of Norovirus-Infected Monolayer HIEs

Flow cytometry analysis measures percentage of HuNoV-infected cells. Dead cells due to HuNoV infection will not be included in this analysis; therefore, this method is better used during early course of infection (up to 24 hpi). If analysis during late course of infection must be monitored, infectivity assays such as RT-qPCR and fluorescence focus assay are more appropriate. Figure 3a shows

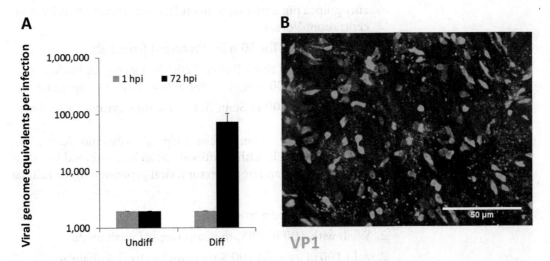

Fig. 3 Differentiation allows for human norovirus (HuNoV) infection. (**a**) Viral genome equivalents per infection in undifferentiated and differentiated jejunal HIE monolayers, assessed by RT-qPCR comparing 1 to 72 hpi. $N = 3$ jejunal HIE infections. (**b**) Expression of viral capsid protein VP1 in jejunal HIE monolayers detected using anti-VP1 antibody in immunofluorescent staining. Culture was innoculated with 9×10^7 HuNoV genome equivalents

sample analysis of HuNoV-infected HIE monolayers using RT-qPCR.

1. Remove old media from well.
2. Add 100 µl 0.05 % trypsin-EDTA.
3. Incubate at 37 °C for 5 min.
4. Add 1 ml DMEM containing 10 % FBS to stop the reaction.
5. Pellet in a swinging bucket rotor centrifuge at 400 × g, 4 °C for 5 min. Remove supernatant.
6. Resuspend in 500 µl Cytofix.
7. Incubate at RT for 10 min.
8. Add 1 ml of Stain Buffer and pellet cells in a swinging bucket rotor centrifuge at 400 × g, 4 °C for 5 min. Remove supernatant.
9. Resuspend in 900 µl −20 °C methanol.
10. Incubate at 4 °C for 30 min.
11. Wash with 3 ml PBS. Pellet in a swinging bucket rotor centrifuge at 400 × g, 4 °C for 5 min. Remove supernatant.
12. Repeat step 11.
13. Resuspend in 100 µl Stain Buffer. Add appropriate amount of guinea pig anti-GII.4/Sydney 2012 VP1 antibody. Vortex.
14. Incubate at RT for 30 min.
15. Wash with 1 ml Stain Buffer. Pellet in a swinging bucket rotor centrifuge at 400 × g, 4 °C for 5 min. Remove supernatant.
16. Resuspend in 100 µl Stain Buffer. Add Alexa Fluor 488 goat anti-guinea pig antibody at manufacturer's recommended concentration. Vortex.
17. Incubate at RT for 30 min. Protected from light.
18. Wash with 1 ml Stain Buffer. Pellet in a swinging bucket rotor centrifuge at 400 × g, 4 °C for 5 min. Remove supernatant.
19. Resuspend in 500 µl Stain Buffer for flow cytometry analysis.

3.4.4 Immunofluorescent Staining of Human Norovirus-Infected Monolayer HIEs

Immunofluorescent staining can help identify and locate the HuNoV-infected cells. Other antibodies can be combined to examine other structural and nonstructural viral proteins and/or cellular components.

1. Remove old media from well.
2. Wash with 100 µl PBS, then aspirate off wash media.
3. Add 100 µl ice-cold 100 % methanol to fix. Incubate at −20 °C for 20 min. Aspirate off fixative.

4. Wash with 100 μl PBS, then aspirate off wash media. Repeat twice.

5. Add 100 μl PBS containing 5 % FBS to block nonspecific binding of antibodies. Incubate at RT for 30 min. Aspirate off blocking buffer.

6. Add 100 μl PBS containing 5 % FBS and appropriate amount of guinea pig anti-GII.4/Sydney 2012 VP1 antibody.

7. Incubate at RT for 2 h or at 4 °C overnight.

8. Wash with 100 μl PBS, then aspirate off wash media. Repeat twice.

9. Add 100 μl PBS containing 5 % FBS and appropriate amount of Alexa Fluor 488 goat anti-guinea pig antibody at manufacturer's recommended concentration.

10. Incubate at RT for 1 h protected from light.

11. Wash with 100 μl PBS, then aspirate off wash media. Repeat twice.

12. Add 100 μl PBS containing 300 nM DAPI.

13. Incubate at RT for 1 min.

14. Wash with 100 μl PBS, then aspirate off wash media. Repeat twice.

15. Add 100 μl PBS and visualize under immunofluorescent microscope (Fig. 3b).

4 Notes

1. L-WNT3a cell line is commercially available from ATCC. R-Spondin 1 cell line is from Trevigen. Noggin cell line is a generous gift from Dr. Gijs van den Brink (University of Amsterdam). Conditioned media from L-WNT3a, R-Spondin 1, and Noggin can be aliquoted and stored at −20 °C for 1 month. Avoid freeze and thaw cycles. CMGF+ can be kept at 4 °C for up to 2 weeks. CMGF− and Differentiation Media can be kept at 4 °C for up to 1 month.

2. Matrigel™ is liquid at 4 °C and quickly solidifies at warmer temperature; therefore, all tubes and pipette tips must be pre-chilled at 4 °C before plating and kept on ice during plating.

3. Our frozen stock usually contains two wells of fully grown HIEs. Each frozen stock is plated onto four wells during revival.

4. Depending on plating density, HIEs are usually ready to be passaged after 6–7 days of growth. A 1:5 ratio can be used as a

starting point for passaging. It is important to passage HIEs a few times after revival before using for experiments.

5. The number of syringing must be experimentally determined. Depending on an individual's force and the number of days before the next passage, HIEs can be broken up into desired sizes.

6. The length of HIE differentiation must be experimentally determined for each HIE line and viral infection. Typical differentiation will change HIEs from a shining finish with thin epithelial lining to a multi-lobular structure with thickened epithelium (Fig. 2a), although variability is observed depending on the intestinal segment and tissue source.

7. Pancreatin is difficult to dissolve in aqueous solution. To ensure pancreatin is maximally dissolved, the solution should be vigorously vortexed and incubated at RT for at least 30 min.

8. Pancreatin is easily degraded. Make up fresh pancreatin solution every time.

9. For generating and quantifying rotavirus stocks, please refer to Arnold et al. *Curr Protoc Microbiology.* 2009 [23].

10. Rotavirus incubation with trypsin is important for the cleavage of the cell attachment protein VP4. This step is crucial for successful infection.

11. Fixed cells can be kept in $1 \times$ Perm/Wash buffer at 4 °C for up to 1 week.

12. Collagen IV solution can be aliquoted and stored at a stock concentration of 1 mg/ml at -20 °C for 1 year. Avoid freeze and thaw cycles. Dilute 1 mg/ml stock solution by 1:30 using cold H_2O before use. Collagen solution will solidify at RT. Keep cold on ice bath.

13. 3D HIEs should be in culture for 7 days before being used for monolayer plating. Each 3D HIE well should be examined under light microscope to ensure approximately 100 live, cystic HIEs. Dead HIEs will look small, dark, and rounded. In our laboratory, approximately one well of 3D HIEs can be used to plate 1 transwell or 2.5 wells on a 96-well tissue culture plate.

14. If pooling more than five wells of 3D HIEs in the same tube, the volume of 0.05 % trypsin/0.5 mM EDTA should be increased to 1 ml per well to ensure dissociation of all HIEs. We do not recommend pooling more than ten wells per tube for this dissociation process.

15. Avoid bubbles during dissociation by pipetting against the sides of a 15 ml conical tube. Vigorous pipetting is absolutely necessary to ensure single-cell dissociation from 3D HIEs. This

is also crucial to ensure all cells are passed through a 40 μm cell strainer at the later step.

16. Y-27632 inhibits Rho-associated, coiled-coil-containing protein kinase 1 and 2 (ROCK 1 and 2). Addition of Y-27632 enhances survival of dissociated single cells by preventing apoptosis.

17. HIE cell solutions should be added drop-wise onto the center of the monolayer or transwell. Cells will spread out and coat the transwell. Do not rotate or swirl the seeded plates.

18. Bile can be freshly isolated or lyophilized then reconstituted. Bovine and porcine bile is commercially available from Sigma-Aldrich. The percentage of bile used in HuNoV Incubation Media must be experimentally determined to maximize infection and minimize cell death for each virus strain and HIE line, respectively. In our hands, 5 % human bile and 0.03 % bovine bile gave the most optimal replication for GII.3 HuNoV. GII.4 HuNoV does not require bile for replication, but the presence of bile enhances replication [12].

Acknowledgments

The techniques described in this article were developed by research on rotaviruses and noroviruses that was supported by National Institutes of Health (NIH) Grants U19-AI116497, R01 AI080656, PO1 AI 057788 (to Mary K. Estes), U18-TR000552 (to Mark Donowitz), P30 DK-56338 (to Hashem El-Serag), F30 DK107173 (to Winnie Zou), Agriculture and Food Research Initiative competitive grant 2011-68003-30395 from the USDA National Institute of Food and Agriculture (to LeAnn Jaykus), and Howard Hughes Medical Institute Grant 570076890 (to Kapil Saxena).

References

1. Mortality and Causes of Death Collaborators (2016) Global, regional, national, and selected subnational levels of stillbirths, neonatal, infant, and under-5 mortality, 1980–2015: a systematic analysis for the Global Burden of Disease Study 2015. Lancet 388 (10053):1725–1774. doi:10.1016/s0140-6736(16)31575-6. Epub 2016/10/14. PubMed PMID: 27733285

2. Mortality and Causes of Death Collaborators (2016) Global, regional, and national life expectancy, all-cause mortality, and cause-specific mortality for 249 causes of death, 1980–2015: a systematic analysis for the Global Burden of Disease Study 2015. Lancet 388(10053):1459–1544. doi:10.1016/s0140-6736(16)31012-1. Epub 2016/10/14. PubMed PMID: 27733281

3. Glass RI, Parashar UD, Estes MK (2009) Norovirus gastroenteritis. New Engl J Med 361 (18):1776–1785. Epub 2009/10/30. doi: 10.1056/NEJMra0804575. PubMed PMID: 19864676; PMCID: PMC3880795

4. Parashar UD, Burton A, Lanata C, Boschi-Pinto C, Shibuya K, Steele D, Birmingham M, Glass RI (2009) Global mortality associated with rotavirus disease among children in 2004. J Infect Dis 200(Suppl 1):S9–S15. doi:10.

1086/605025. Epub 2009/10/13. PubMed PMID: 19817620

5. Estes MK, Greenberg HB (2013) Rotaviruses. In: Knipe DM, Howley PM (eds) Fields virology, 6th edn. Lippincott Williams & Wilkins, Philadelphia, PA, pp 1347–1401

6. Ahmed SM, Hall AJ, Robinson AE, Verhoef L, Premkumar P, Parashar UD, Koopmans M, Lopman BA (2014) Global prevalence of norovirus in cases of gastroenteritis: a systematic review and meta-analysis. Lancet Infect Dis 14(8):725–730. doi:10.1016/s1473-3099(14)70767-4. Epub 2014/07/02. PubMed PMID: 24981041

7. Bryce J, Boschi-Pinto C, Shibuya K, Black RE (2005) WHO estimates of the causes of death in children. Lancet 365(9465):1147–1152. doi:10.1016/s0140-6736(05)71877-8. Epub 2005/03/30

8. Vesikari T (2012) Rotavirus vaccination: a concise review. Clin Microbiol Infect 18(Suppl 5):57–63. doi:10.1111/j.1469-0691.2012.03981.x. Epub 2012/08/14. PubMed PMID: 22882248

9. Vesikari T (2016) Rotavirus vaccine and vaccination. In: Svensson L, Desselberger U, Greenberg HB, Estes MK (eds) Viral gastroenteritis: molecular epidemiology and pathogenesis. Academic Press, pp 301–328

10. McCormick BJ, Lang DR (2016) Diarrheal disease and enteric infections in LMIC communities: how big is the problem? Trop Dis Travel Med Vaccines 2(1):11

11. Payne DC, Vinje J, Szilagyi PG, Edwards KM, Staat MA, Weinberg GA, Hall CB, Chappell J, Bernstein DI, Curns AT, Wikswo M, Shirley SH, Hall AJ, Lopman B, Parashar UD (2013) Norovirus and medically attended gastroenteritis in U.S. children. New Engl J Med 368(12):1121–1130. doi: 10.1056/NEJMsa1206589. Epub 2013/03/22. PubMed PMID: 23514289; PMCID: PMC4618551

12. Ettayebi K, Crawford SE, Murakami K, Broughman JR, Karandikar U, Tenge VR, Neill FH, Blutt SE, Zeng XL, Qu L, Kou B, Opekun AR, Burrin D, Graham DY, Ramani S, Atmar RL, Estes MK (2016) Replication of human noroviruses in stem cell-derived human enteroids. Science. doi: 10.1126/science.aaf5211. PubMed PMID: 27562956. Epub 2016/08/27

13. Sato T, Vries RG, Snippert HJ, van de Wetering M, Barker N, Stange DE, van Es JH, Abo A, Kujala P, Peters PJ, Clevers H (2009) Single Lgr5 stem cells build crypt-villus structures in vitro without a mesenchymal niche. Nature 459(7244):262–265. doi:10.1038/nature07935. Epub 2009/03/31. PubMed PMID: 19329995

14. Sato T, Clevers H (2013) Growing self-organizing mini-guts from a single intestinal stem cell: mechanism and applications. Science 340(6137):1190–1194. doi:10.1126/science.1234852. Epub 2013/06/08. PubMed PMID: 23744940

15. Sato T, Stange DE, Ferrante M, Vries RG, van Es JH, van den Brink S, van Houdt WJ, Pronk A, van Gorp J, Siersema PD (2011) Long-term expansion of epithelial organoids from human colon, adenoma, adenocarcinoma, and Barrett's epithelium. Gastroenterology 141(5):1762–1772

16. In JG, Foulke-Abel J, Estes MK, Zachos NC, Kovbasnjuk O, Donowitz M (2016) Human mini-guts: new insights into intestinal physiology and host-pathogen interactions. Nat Rev Gastroenterol Hepatol 13(11):633–642

17. Middendorp S, Schneeberger K, Wiegerinck CL, Mokry M, Akkerman RD, Wijngaarden S, Clevers H, Nieuwenhuis EE (2014) Adult stem cells in the small intestine are intrinsically programmed with their location-specific function. Stem Cells 32(5):1083–1091

18. Saxena K, Blutt SE, Ettayebi K, Zeng XL, Broughman JR, Crawford SE, Karandikar UC, Sastri NP, Conner ME, Opekun AR, Graham DY, Qureshi W, Sherman V, Foulke-Abel J, In J, Kovbasnjuk O, Zachos NC, Donowitz M, Estes MK (2016) Human intestinal enteroids: a new model to study human rotavirus infection, host restriction, and pathophysiology. J Virol 90(1):43–56. doi:10.1128/jvi.01930-15. Epub 2015/10/09. PubMed PMID: 26446608; PMCID: PMC4702582

19. Saxena K, Simon LM, Zeng XL, Blutt SE, Crawford SE, Sastri NP, Karandikar UC, Ajami NJ, Zachos NC, Kovbasnjuk O, Donowitz M, Conner ME, Shaw CA, Estes MK (2017) A paradox of transcriptional and functional innate interferon responses of human intestinal enteroids to enteric virus infection. Proc Natl Acad Sci USA 114(4):E570-E579. doi:10.1073/pnas.1615422114. Pubmed PMID: 28069942, PMCID: PMC5278484

20. Mahe MM, Sundaram N, Watson CL, Shroyer NF, Helmrath MA (2015) Establishment of human epithelial enteroids and colonoids from whole tissue and biopsy. J Vis Exp. (97). doi:10.3791/52483. PubMedPMID: 25866936, PMCID: PMC4401205

21. Jiang X, Wang M, Graham DY, Estes MK (1992) Expression, self-assembly, and antigenicity of the Norwalk virus capsid protein. J Virol 66(11):6527–6532
22. Wang F, Scoville D, He XC, Mahe MM, Box A, Perry JM, Smith NR, Lei NY, Davies PS, Fuller MK, Haug JS, McClain M, Gracz AD, Ding S, Stelzner M, Dunn JC, Magness ST, Wong MH, Martin MG, Helmrath M, Li L (2013) Isolation and characterization of intestinal stem cells based on surface marker combinations and colony-formation assay. Gastroenterology. 145(2):383–395.e1-21. doi: 10.1053/j.gastro.2013.04.050. Epub 2013/05/07. PubMed PMID: 23644405; PMCID: PMC3781924
23. Arnold M, Patton JT, McDonald SM (2009) Culturing, storage, and quantification of rotaviruses. Current protocols in microbiology. Chapter 15:Unit 15C.3. Epub 2009/11/04. doi: 10.1002/9780471729259.mc15c03s15. PubMed PMID: 19885940; PMCID: PMC3403738

Study Bacteria–Host Interactions Using Intestinal Organoids

Yong-guo Zhang and Jun Sun

Abstract

The intestinal epithelial cells function to gain nutrients, retain water and electrolytes, and form an efficient barrier against foreign microbes and antigens. Researchers employed cell culture lines derived from human or animal cancer cells as experimental models in vitro for understanding of intestinal infections. However, most in vitro models used to investigate interactions between bacteria and intestinal epithelial cells fail to recreate the differentiated tissue components and structure observed in the normal intestine. The in vitro analysis of host–bacteria interactions in the intestine has been hampered by a lack of suitable intestinal epithelium culture systems. Here, we present a new experimental model using an organoid culture system to study bacterial infection.

Keywords: Host–bacteria interactions, Inflammation, Intestine, Infection, Organoid, *Salmonella*, Stem cells, Tight junctions

1 Introduction

The gastrointestinal (GI) tract is lined by a single layer of epithelial cells that serve to facilitate digestion and absorption of nutrients [1, 2]. Intestinal epithelial cells (IECs) are consistently exposed to pathogenic microorganisms and foreign antigens, which play a key role in normal intestinal development and innate immunity [3–5]. Researchers have employed several in vitro models used to investigate interactions between bacteria and intestinal epithelial cells, including culture lines derived from human or animal cancer cells, suspension culture technology using a rotating wall vessel bioreactor that allows cells to remain in suspension with bubble-free aeration [6, 7]. Studies by Clevers and colleagues established the isolation and culture of primary small intestinal epithelial stem cells [8–11]. Isolated crypts form "organoid structures" contain multiple cell types, including enterocytes, goblet cells, enteroendocrine cells, and Paneth cells. Furthermore, some functional and physiological properties of the organoids have been demonstrated by the presence of brush borders on enterocytes, production of mucin by goblet cells. This culture system is particularly useful for studying the regulation of intestinal stem cell self-renewal and differentiation, and of host–pathogen interactions. We sought to establish a bacteria-infected organoid culture system using crypt-derived intestinal organoids [12]. In the infected organoids, we were able to visualize the invasiveness of *Salmonella* and the morphologic changes of the organoids.

2 Materials

2.1 Reagents

- Sterile PBS (Ca^{2+}/Mg^{2+} free).
- Buffer #1: 2 mM EDTA in PBS.
- Buffer #2: 54.9 mM D-sorbitol and 43.4 mM sucrose in PBS.
- Growth factor reduced, Phenol red free Matrigel.
- Minigut Medium:
  - Advanced DMEM/F12.
  - Penicillin (100 U/ml)/Streptomycin (100 μg/ml).
  - 10 mM HEPES.
  - 1:100 N2 supplement.
  - 1:50 B27 supplement.
- Mouse Recombinant EGF (50 μg/ml) (**Note 1**).
- Mouse Recombinant Noggin (100 μg/ml) (**Note 1**).
- Rho Kinase Inhibitor Y27632 (10 mM).
- R-spondin solution (**Note 2**).

3 Methods

3.1 Organoids Isolation

1. Murine small intestine (mostly jejunum and ileum) is removed immediately after cervical dislocation, using scissors to remove fat/mesentery. Dissect out 10 cm of jejunum and 10 cm of ileum.

2. The stool is flushed out with ice-cold PBS (Penicillin, 100 U/ml/Streptomycin, 100 μg/ml) in small intestine. Cut intestines longitudinally with special intestine scissors (Made in Germany FST 14080-11) and the small intestines are cut into small (~1 cm) pieces.

3. Use the forceps to transfer all the pieces to a 15 ml conical tube containing 5 ml of ice-cold PBS, rock for 10 min in 4 °C.

4. Aspirate PBS from the top of the tissue and replace with cold Buffer #1, rock for 30 min in 4 °C.

5. Aspirate PBS from the top of the tissue and replace with cold Buffer #2, and then shake for 2 min manually (~80 shakes/min).

Note: Jejunum will dissociate easier than ileum, which may require extra shaking.

6. Take 20 μl droplet of Buffer #2 after shaking and check under the microscope. Villi and debris are large chunks. Many crypts with granular paneth cells are at the bottom. Do another 1 min of shaking in the same buffer if many crypts not observed.

7. Equilibrate 70 μm sterile cell strainer with Buffer #2. Rinse all filters with 5 ml of Buffer #2 to ensure that enough crypts have passed through the filters. Only crypts and small fragments should pass through this filter (i.e., large chunks of tissue will be removed in this step.)

8. Count number of crypts: Mix 10 μl of crypts with 10 μl of trypan blue. Apply all 20 μl to a microscope slide and count all the crypts in the entire drop using a quality inverted microscope.

9. Transfer enough medium/crypts into one tube so that there are 1500 crypts in a round bottom 5 ml polypropylene tube, centrifuge tubes for 2 min at $150 \times g$ at 4 °C.

10. While centrifuging mixture, prepare Matrigel™ on ice (Matrigel™ will polymerize at room temperature).

    50 μl Matrigel/well:
    (a) 0.25 μl of 100 μg/ml EGF (final concentration 50 ng/ml)
    (b) 1 μl of 50 μg/ml Noggin (final concentration 100 ng/ml)
    (c) 1 μl of Y27632 (ROCK inhibitor, final concentration 10 μM)
    (d) 5 μl of R-spondin

11. After the spin, remove supernatant with a pasteur pipet attached to a vacuum, being careful not to touch the pellet. Leave ~50 μl of medium in the tube.

12. Resuspend crypt pellet in the 50 μl that remained in the tube. Add an additional 100 μl of Matrigel and mix well with the pipette, pipette up and down being careful to avoid bubbles.

    Note: Remember to place the tube with the Matrigel back on ice immediately after using it.

13. Draw up the entire 50 μl volume with Matrigel/crypt into a pipette, being careful to avoid any bubbles. Put 50 μl droplet of mix in center of a well of a 12 well plate. Incubate 24-well plate at 37 °C for 30 min to polymerize Matrigel. After 30 min of polymerization, 500 μl of minigut medium was overlain [10].

    Note: Not supplemented with growth factors at this point.

14. Change medium every 3–4 days with minigut medium containing growth factors (Noggin: 100 ng/ml; EGF: 50 ng/ml; Y27632:10 μM; R-spondin: 1/10).

3.2 Observe Organoids Under Microscope

See Fig. 1.

3.3 Passage of Organoids

1. The passage was performed every 7–10 days with a 1:4 split ratio.
2. Leave 12-wells cell culture plate on ice for 30 min. Aspirate medium and add 1 ml of cold PBS to each well.
3. Mix twice through a 1 ml syringe with a 27G 1/2 needle and transfer into a 5 ml tube. Spin down for 5 min at 200 × g at 4 °C.
4. Remove supernatant, and leave ~50 μl of medium in the tube for resuspension.
5. Add an additional 150 μl of Matrigel and mix well with the pipette avoiding bubbles.
6. Add 50 μl droplet of mix to the center of a well of a 12 well plate. Incubate cell plate at 37 °C for 30 min. After 30 min of polymerization, 500 μl of minigut medium containing growth factors was overlain.
7. Change medium every 3–4 days with minigut medium containing growth factors.

3.4 Bacterial Colonization of Organoids

1. Organoids cells colonized with bacteria: Organoid cells (6 days after passage) were colonized with the indicated bacterial strain for 30 min, washed with HBSS, and incubate in minigut medium containing gentamicin (500 mg/ml) for 1 h.

   Note: After extensive HBSS washing, the extracellular bacteria were washed away. Incubation with gentamicin inhibited the growth of extracellular bacteria [13].

2. Organoid cells Immunoblotting: The organoid cells were rinsed three times in ice-cold HBSS and then suspended in ice-cold HBSS. The organoid cells were then spun down at

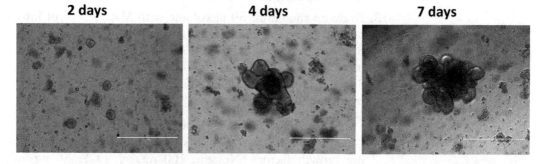

Fig. 1 Observing organoids under the microscope. At 2 days post-prep, the organoid begins to take shape. At 3–4 days post-prep, the organoid begins to bud. At 5–7 days post-prep, the organoid has many buds with lots of luminal debris, which is a point for passage

200 × g for 10 min at 4 °C. Next, using a pipette to aspirate the PBS at the top, the organoid cells were lysed in lysis buffer (1 % Triton X-100, 150 mM NaCl, 10 mM Tris-HCl pH 7.4, 1 mM EDTA, 1 mM EGTA pH 8.0, 0.2 mM sodium orthovanadate, protease inhibitor cocktail) and then sonicated. The protein concentration was then measured. Next, equal amounts of protein (20 μg/well) were separated by SDS–polyacrylamide gel electrophoresis, transferred to nitrocellulose, and immunoblotted with primary antibodies. Following the primary antibody step, the nitrocellulose membranes were incubated with secondary antibodies and visualized by ECL.

3. Organoid cells embedded in a paraffin block: The organoid cells were rinsed three times in ice-cold HBSS and then suspended in cold HBSS. The organoid cells were spun down at 200 × g for 10 min at 4 °C. To fix organoid cells, the following steps are used: 10 % formalin 30 min; 75 % alcohol 5 min; 100 % alcohol 10 min; xylene 5 min; xylene 10 min; and paraffin (65 °C) 60 min. The paraffin sections were processed with standard techniques [14, 15].

4. Immunofluorescent staining: The immunofluorescence measurements were performed on paraffin-embedded sections (4 μm) of organoid cells. After preparation of the slides, as described previously [16], the slides were permeabilized for 20 min with 0.2 % Triton X-100, followed by three rinses with HBSS, and incubation for 1 h in 3 % BSA + 1 % goat serum in HBSS to reduce nonspecific background. The permeabilized organoid cell samples were incubated with primary antibodies overnight in 37 °C. The samples were then incubated with goat anti-rabbit Alexa Fluor 488 or goat anti-mouse Alexa Fluor 488 (Molecular Probes, CA; 1:200) and DAPI (Molecular Probes 1:10,000) for 1 h at room temperature. The organoid cells were mounted with SlowFade (SlowFade® AntiFade Kit, Molecular Probes) followed by a coverslip, and the edges were sealed to prevent drying. The specimens were examined with a Zeiss 710 Laser Scanning confocal microscope.

4 Notes

1. To make stock solution of mouse recombinant EGF, mouse recombinant noggin, reconstitute powder in sterile PBS then followed by sterilization through 0.22 μm filter. Aliquot in 100 μl volumes and store at −20 °C.

2. To make solution of R-spondin Day1: check the confluence of HEK293-RSPO2 (~90 %), aspirate normal medium, wash the cells gently with several milliliters of OPTI-MEM, then replace

them with ~30 ml of OPTI for every T150 flask. Day4: collect the first wave conditional medium, add fresh OPTI again. Day7: collect the second wave conditional medium. Mix the first wave and second wave medium together and aliquot them. Add 1 ml of conditional medium to 9 ml of Minigut medium which is good for organoid growth.

Acknowledgements

We would like to acknowledge the NIDDK grant R01 DK105118 (J.S.).

References

1. Turner JR (2009) Intestinal mucosal barrier function in health and disease. Nat Rev Immunol 9(11):799–809
2. Robinson A, Keely S, Karhausen J, Gerich ME, Furuta GT, Colgan SP (2008) Mucosal protection by hypoxia-inducible factor prolyl hydroxylase inhibition. Gastroenterology 134(1):145–155
3. Hooper LV (2009) Do symbiotic bacteria subvert host immunity? Nat Rev Microbiol 7(5):367–374
4. Camp JG, Kanther M, Semova I, Rawls JF (2009) Patterns and scales in gastrointestinal microbial ecology. Gastroenterology 136(6):1989–2002
5. Neish AS (2009) Microbes in gastrointestinal health and disease. Gastroenterology 136(1):65–80
6. Nickerson CA, Ott CM (2004) A new dimension in modeling infectious disease. ASM News 70(4):169–175
7. Unsworth BR, Lelkes PI (1998) Growing tissues in microgravity. Nat Med 4(8):901–907
8. Sato T, Stange DE, Ferrante M, Vries RG, Van Es JH, Van den Brink S et al (2011) Long-term expansion of epithelial organoids from human colon, adenoma, adenocarcinoma, and Barrett's epithelium. Gastroenterology 141(5):1762–1772
9. Sato T, van Es JH, Snippert HJ, Stange DE, Vries RG, van den Born M et al (2011) Paneth cells constitute the niche for Lgr5 stem cells in intestinal crypts. Nature 469(7330):415–418
10. Wang N, Zhang H, Zhang BQ, Liu W, Zhang Z, Qiao M et al (2014) Adenovirus-mediated efficient gene transfer into cultured three-dimensional organoids. PLoS One 9(4), e93608
11. Sato T, Clevers H (2013) Primary mouse small intestinal epithelial cell cultures. Methods Mol Biol 945:319–328
12. Zhang YG, Wu S, Xia Y, Sun J (2014) Salmonella-infected crypt-derived intestinal organoid culture system for host-bacterial interactions. Physiol Rep 2(9)
13. Sun J, Hobert ME, Rao AS, Neish AS, Madara JL (2004) Bacterial activation of beta-catenin signaling in human epithelia. Am J Physiol Gastrointest Liver Physiol 287(1):G220–G227
14. Wu S, Ye Z, Liu X, Zhao Y, Xia Y, Steiner A et al (2010) Salmonella typhimurium infection increases p53 acetylation in intestinal epithelial cells. Am J Physiol Gastrointest Liver Physiol 298(5):G784–G794
15. Ye Z, Petrof EO, Boone D, Claud EC, Sun J (2007) Salmonella effector AvrA regulation of colonic epithelial cell inflammation by deubiquitination. Am J Pathol 171(3):882–892
16. Wu S, Liao AP, Xia Y, Li YC, Li JD, Sartor RB et al (2010) Vitamin D receptor negatively regulates bacterial-stimulated NF-kappaB activity in intestine. Am J Pathol 177(2):686–697

Disaggregation and Reaggregation of Zebrafish Retinal Cells for the Analysis of Neuronal Layering

Megan K. Eldred, Leila Muresan, and William A. Harris

Abstract

The reaggregation of dissociated cells to form organotypic structures provides an in vitro system for the analysis of the cellular interactions and molecular mechanisms involved in the formation of tissue architecture. The retina, an outgrowth of the forebrain, is a precisely layered neural tissue, yet the mechanisms underlying layer formation are largely unexplored. Here we describe the protocol to dissociate, re-aggregate, and culture zebrafish retinal cells from a transgenic, Spectrum of Fates, line where all main cell types are labelled with a combination of fluorescent proteins driven by fate-specific promoters. These cells re-aggregate and self-organize in just 48 h in minimal culture conditions. We also describe how the patterning in these aggregates can be analyzed using isocontour profiling to compare whether different conditions affect their self-organization.

Keywords: 3D Petri dish, Aggregation, Isocontour profiling, Müller Glia, Organoid, Retina, Self-organization, SoFa, Zebrafish

1 Introduction

The brain is a well-structured tissue, comprised of multiple neuronal cell types and glia, organized into distinct layers. But little is understood about the cellular and molecular mechanisms behind the development of these layers. The retina, an outgrowth of the forebrain, is a good model for investigating this due to it being simple, easier to access, and easy to manipulate genetically and pharmacologically.

The retina is made up of just five main neuronal cell types and one glial type. At the apical side, photoreceptors receive light input which is transduced to an electrical signal and passed through bipolar cells to the retinal ganglion cells which project into the brain. These signals are modulated by the inhibitory horizontal and amacrine cells. The sole glial cell type, the Müller glia, span the width of the retina and form connections with all cell types. All these cell types layer in a specific order which helps them form the appropriate connections for the tissue to function correctly. However, it is unclear how these layers form. The simplest explanation would be that cells are layering according to order of birth, but they are in fact born during overlapping periods [1] and experience significant interdigitation before sorting into the correct layers [2, 3]. There are some suggestions that cells of different types communicate with each other to find their appropriate positions [3, 4], but it is difficult to investigate these interactions in vivo. Knockout studies in vivo where individual cell types or combinations of them were removed from the retina show that the remaining cell types are still able to layer correctly [5]. The retina is also supported by other surrounding tissues such as the retinal pigment epithelium (RPE) of which the relationship between them is yet unexplored. To decipher the cellular and molecular mechanisms of retinal lamination, in the absence of extrinsic cues, we must isolate the cells themselves.

A common technique used in studying tissue organization is re-aggregation. This consists of breaking the tissue down into its simplest components, the cells, and allowing them to re-aggregate in a simple environment, void of surrounding tissues. This allows the investigator to observe how these cells interact to form the tissue and to manipulate the components to find out which are important in this process. Re-aggregation studies have been used since the early twentieth century, first using simple organisms such as sea urchin and sponge [6, 7], and later using more complex, multi-layered tissues such as limb buds and retina from chick [8, 9]. These studies have revealed the innate ability of these multi-layered tissues to self-organize in vitro in the absence of many extrinsic cues and scaffolds.

Previous retinal re-aggregation studies in chick show that retinal cells are capable of self-organizing into fully stratified retinospheroids [10–12]. These studies used a simple rotary culture whereby cells in suspension were continuously agitated in a swirling motion on a gyratory shaker to encourage close proximity between cells. These cells re-aggregated quickly, but needed to be cultured for 12 days, and in the presence of an RPE monolayer co-culture to achieve laminar organization. The aggregates were then harvested, fixed, and sectioned for immunohistochemistry staining to identify the cell types. These studies revealed the ability of retinal cells to self-organize in a simple culture environment and began to look at some of the mechanisms governing this, but the culture protocols were complex and time intensive. They were also limited by the lack of genetic tools available in chick. To overcome these challenges and further develop these studies, the zebrafish retina is the ideal model to use due to the variety of genetic and pharmacological tools available with which to label, manipulate, and image cells and molecular components.

Here we describe a re-aggregation culture protocol for zebrafish retinal cells in non-adhesive agarose micro-well dishes in a minimal culture medium. In just 48 h of culture these cells re-aggregate, differentiate, and self-organize. Using the Spectrum of Fates (SoFa) transgenic line [2] whereby cells are labelled with a combination of fluorescent proteins tagged to fate-specific promoters to allow visualization of the main neuronal cell fates, the patterning of these cells can be analyzed. These aggregates can also be stained with immunohistochemistry for further analysis of expression of various markers. Finally we describe the use of isocontour pattern profiling to gain a quantitative measure of organization for investigating the cellular and molecular mechanisms of retinal self-organization. Using this protocol we have revealed the role of Müller glia in the self-organization of zebrafish retinal cells, as shown in our previous publication [13].

2 Materials

2.1 General Comments

1. All procedures should be carried out using sterile technique.
2. All equipment should be sterilized with 70% Ethanol.
3. Once prepared, all solutions used should be sterilized by filtering through 0.22 μm filters and kept on ice or at 4 °C until use.
4. Culture medium should be kept on ice until use and then allowed to warm to room temperature before adding to cells.
5. Culture medium and dissociation mediums should be made fresh on the day of dissection and culture.
6. The quantities of dissociation medium and culture medium described in this protocol are sufficient for carrying out the dissection of roughly 20 retinas, which will provide enough material to seed three dishes of 15 wells each (45 wells in total).
7. The protocol described here is for retinas from 24hpf zebrafish embryos. To dissect and dissociate retinas from other stages will require alterations to the dissection technique and length of trypsin incubation. The cells must be harvested from 24hpf embryos to achieve the best self-organization after culture.
8. Solutions are prepared with double-distilled sterile water, unless otherwise stated.
9. Before use, thaw all cell culture reagents and supplements on ice.

2.2 Zebrafish Embryo Maintenance

1. Adult zebrafish are maintained and bred at 26.5 °C. Embryos are collected on the morning of the day before you wish to start the culture. Embryos are maintained in embryo medium at 28.5 °C and staged in hours post fertilization (hpf) according to morphological features, as previously described [14].
2. Embryos are treated with 0.003% N-Phenylthiourea (PTU) from 8hpf onwards to prevent pigment formation.
3. Embryos used in this protocol are from the Spectrum of Fates (SoFa1) transgenic line [2], GFAP:GFP transgenic line [15] or AB or TL wild type (WT).

2.3 Solutions and Cell Culture Reagents

1. L-15 (Leibovitz's L-15 Medium with L-glutamine).
2. PSF (Antibiotic-Antimycotic (100×). Aliquot and store at −20 °C.
3. PTU (*N*-Phenylthiourea, Sigma) (warning: toxic): In a fume hood, dissolve 1.2 g in 1 l water to make a 40× concentrated stock and store at 4 °C. This may take a few hours to dissolve with stirring.
4. FBS (Fetal Bovine Serum, Thermo Fisher Scientific): Aliquot (100 µl) and store at −20 °C.
5. MS-222 (Tricaine methanesulfonate, Sigma): Dissolve 4 g in 100 ml water to make a 100× concentrated stock, aliquot (1 ml) and store at −20 °C.
6. Trypsin-EDTA 0.25% (Sigma). Thaw on ice, aliquot (1 ml) and store at −20 °C.
7. N2 supplement (Thermo Fisher Scientific). Thaw on ice, aliquot (100 µl) and store at −20 °C.

2.3.1 Dissection and Dissociation of Zebrafish Retinas

1. Calcium-free medium: The final working concentrations (1×) are: 116.6 mM NaCl, 0.67 mM KCl, 4.62 mM Tris, 0.4 mM EDTA. Dissolve in water to make a 10× concentrated stock, adjust the pH to 7.8 and store at 4 °C for up to 6 months.
2. Heparin solution (Heparin sodium salt, Sigma): The final working concentration (1×) is 100 µg/ml. Dissolve in water to make a 10× concentrated stock and store at 4 °C for up to 6 months.
3. Dissociation medium: From the stocks previously prepared, add 2 ml calcium-free medium, 2 ml heparin solution, 200 µl MS-222, 200 µl PSF and 500 µl PTU. Fill up to 20 ml with water (*see* **Note 1**).
4. Cell washing medium: L-15 supplemented with 3% FBS.

2.3.2 Culture of Zebrafish Retinal Cells

1. Calcium-free Ringer's solution: Working concentrations are: 116 mM NaCl, 2.9 mM KCl, 5.0 mM HEPEs. Dissolve in water to make a 1× working stock concentration, adjust pH to 7.2, and store at 4 °C for up to 6 months.
2. Embryo extract: Grow wild type zebrafish embryos until 72hpf and ensure they are all dechorionated before proceeding. Chill the embryos on ice and transfer to a Dounce homgonizer. Remove most of the embryo medium and rinse with 0.5% bleach for 2 min (re-suspending the embryos a few times), then with calcium-free Ringer's solution for 2 min. Remove as much liquid as possible and homogenize thoroughly. For every 200 embryos, add 1 ml L-15 supplemented with 1% PSF. Aliquot and store at −20 °C for up to 1 month, or at −80 °C for up to 6 months. Recipe adapted from ZFin: (https://zfin.org/zf_info/zfbook/chapt6.html).

3. Culture medium: L-15 supplemented with 10% embryo extract, 3% FBS, 2% N2, 1% PTU, 1% PSF. Make enough for 750 µl per culture dish (*see* **Note 2**).

2.4 Equipment for Isolation of Zebrafish Retinal Cells

1. Dissecting stereo microscope with transmitted light base.
2. Sylgard dissecting dish: Prepare dishes prior to starting the protocol using the Sylgard 184 Silicone Elastomer Kit (Dow Corning). Mix the base and curing agents together in a 10:1 ratio. Pour into the bottoms of 3 cm culture dishes and bake at 65 °C overnight to set. Many dissecting dishes can be made and kept for later use.
3. Two dissecting pin holders with 0.125 mm tungsten needles (Fine Science Tools).
4. Two glass pasteur pipettes, narrowed using a Bunsen burner: one wide enough for transferring embryos and one narrower, for dissociating with. Prepare new pipettes for each dissociation.
5. 10 ml petri dishes.
6. Three-well glass dish (Pyrex, Corning).
7. 1.5 ml tubes.
8. Table top centrifuge, accurate at low speeds (around $300 \times g$).
9. Cell counter (Improved Neubauer).
10. Large bucket of ice.
11. Waste container with bleach.

2.5 Agarose Microwell Culture Dishes

1. 35-Well PDMS moulds from the 3D Petri Dish range (Microtissues): use the 24–35 moulds (these fit into a 24-well plate and have pegs for 35 wells). Alternatively, modify this mould by cutting the outer rows of pegs with a sharp scalpel to create a mould for just 15 wells.
2. Saline solution: Dissolve NaCl in water to a concentration of 0.9% (w/v). Filter and store at room temperature.
3. UltraPure LMP Agarose (Invitrogen): Prepare a 2% agarose stock by dissolving 2 g agarose powder in 100 ml saline solution whilst heating and stirring. This may take around an hour to dissolve. Aliquot into 1 ml aliquots and store at room temperature.
4. Equilibration medium: Supplement L-15 with 1% PSF, filter and keep at room temperature.
5. 4-Well culture dish (Nunclon, Thermo Fisher Scientific).

2.6 Sample Fixation and Mounting

1. 4% PFA (in PBS).
2. 1× PBS.
3. Microscope slides (Menzel Gläser, Thermo Fisher Scientific).
4. 13 mm round coverslips (Thickness No.1, Cat: 631–0149, VWR).

5. Whatman lens cleaning tissues (GE Healthcare, Life Sciences).
6. Reinforcement ring labels (transparent, standard office supplier).
7. VectaShield hard-set mounting medium with DAPI (Vector Laboratories).
8. Microscope slide holder.
9. Forceps.

2.7 Immunostaining

1. 1× PBS.
2. PBS-T: Dissolve Triton X-100 (Sigma) in PBS to make the following stock concentrations: 0.5, 0.1, and 0.05%.
3. HIGS (Heat-inactivated goat serum, Thermo Fisher Scientific). Aliquot and store at −20 °C.
4. BSA (Bovine serum albumin, Sigma): Dissolve in 1× PBS with stirring to make a stock concentration of 10%. Aliquot and store at −20 °C.
5. Blocking buffer: 10% HIGS, 1% BSA, 0.5% Triton, in PBS.
6. Antibodies used in this protocol are: mouse anti-Zn5 (1:100 ZIRC) and mouse anti-GFAP (1:100 zrf1, ZIRC).

3 Methods

Carry out all procedures at room temperature, using sterile technique, and prepare all culture mediums and culture plates in a sterile culture hood.

3.1 Preparation of Equipment and Materials

1. Dechorionate embryos using fine no.5 forceps, wash with filtered embryo medium then transfer to a new 10 ml petri dish with fresh embryo medium, supplemented with 1% PSF.
2. Prepare dissociation medium (as described in Sect. 2.3.1) and keep on ice.
3. Prepare culture medium (as described in Sect. 2.3.2) and keep on ice.
4. Sterilize all dissection and dissociation equipment (as described in Sect. 2.4) with 70% ethanol and leave to dry. Place all tools on a sterile tray and keep next to the microscope. When the 3-well glass dish is dry, place it on ice.
5. Fill a 10 ml petri dish with some dissociation medium to use as a washing dish.
6. When the dissecting dish is dry, add a little dissociation medium, just enough to cover the bottom.

3.1.1 Preparation of Agarose Micro-well Culture Dishes

Agarose micro-well dishes are prepared as previously described in [16]:

1. Sterilize 3D Petri Dish moulds in 70% culture dish grade ethanol for 20 min then allow to air dry in the culture hood.
2. Heat an aliquot of agarose in a heat block at 65 °C for at least 20 min (*see* **Note 3**).
3. In the culture hood, pipette 330 μl agarose into the 35-well mould, or 350 μl if the mould has been modified for 15 wells (*see* **Note 4**).
4. Allow to set for 20 min and then push out the agarose dishes from the moulds into the 4-well culture plate (*see* **Note 5**).
5. Prepare equilibration medium and add 750 μl to each dish. Leave for at least 20 min to allow the dishes to equilibrate.
6. Just before seeding, replace the equilibration medium once more.

3.2 Isolation of Zebrafish Retinal Cells

3.2.1 Dissection of Retinas from 24hpf Zebrafish Embryos

1. Transfer a maximum of 5 embryos at a time using the wide bored glass pipette to the washing dish in a minimal amount of embryo medium. Swirl the dish to wash, then transfer embryos to the dissecting dish.
2. Dissect out retinas using the two dissecting pins by gently teasing apart the tissues surrounding the retinas. At this stage the tissues are very soft, so be careful not to cut into the retinas in order to keep them intact (*see* **Note 6**).
3. Remove and discard all debris from the dish in the waste container before transferring the retinas in dissecting medium to the glass dish on ice (*see* **Note 7**).
4. Repeat the above steps as necessary until you have collected the desired number of retinas (*see* **Note 8**).

3.2.2 Dissociation of Retinas into a Single Cell Suspension

1. Remove the glass dish containing retinas and the trypsin aliquot from the ice and allow to come to room temperature.
2. Gently remove as much dissociation medium as possible from the dish.
3. Incubate with 200 μl Trypsin-EDTA per 20 eyes for 12 min at room temperature.
4. Remove Trypsin-EDTA and replace with 200 μl dissociation medium per 20 eyes.
5. Dissociate the retinas first by gentle trituration using the narrow-bore glass pipette until mostly dissociated then triturate more vigorously using a P-200 pipette until it is a single cell suspension (*see* **Notes 9** and **10**).

6. Transfer the cell suspension to a 1.5 ml tube and top up to 500 μl with cell washing medium and mix well.

7. Balance the centrifuge and spin at $300 \times g$ for 7 min.

8. Gently remove all but 200 μl of the supernatant and top up to 500 μl again with cell washing medium (*see* **Note 11**).

9. Spin again at $300 \times g$ for 7 min and concentrate cells appropriately by removing the desired amount of supernatant (*see* **Notes 12 and 13**).

10. Leave tube on ice until ready to seed, but don't leave longer than 20–30 min as the cells will begin to re-aggregate.

3.3 Culture of Zebrafish Retinal Cells

1. Remove the equilibration medium from the seeding chambers of the agarose micro-well dishes.

2. Re-suspend the cells and seed 75 μl into the seeding chamber of each dish if using the 35-well dish, or 45 μl if using the modified 15-well dishes (*see* **Note 14**).

3. Allow the cells to settle for 15–20 min (*see* **Note 15**).

4. Meanwhile, allow the culture medium to come to room temperature.

5. Add 750 μl culture medium to each dish via the medium exchange port (*see* **Note 16**).

6. Gently transfer dishes to a 28 °C incubator and allow to culture for 48 h. Cells will begin to aggregate immediately and be fully aggregated by 15 h in culture (Fig. 1).

3.4 Fixation, Mounting and Imaging

Aggregate fixation and staining can all be carried out within the agarose micro-well dishes. You can exchange the media via the medium exchange ports (the indents at the sides of the dish) gently to avoid flushing out the aggregates. You can also remove media from the seeding chamber, but do so while viewing under a microscope to ensure you don't disturb the aggregates. The aggregates can be viewed using an upright microscope with a water dipping lens, but for best image quality they should be mounted and imaged using confocal microscopy. To fix and mount the aggregates, carry out the following steps:

1. Rinse aggregates with PBS.

2. Remove PBS and fix aggregates with 4% PFA at room temperature for 15 min. From this point, keep fixed aggregates away from light to avoid photo-bleaching of fluorophores.

3. Wash for 5 min with PBS. Repeat twice more.

4. While washing, prepare the microscope slides:

5. Place reinforcement rings onto the microscope slides (max 2 per slide) and secure them, by pushing out any air bubbles,

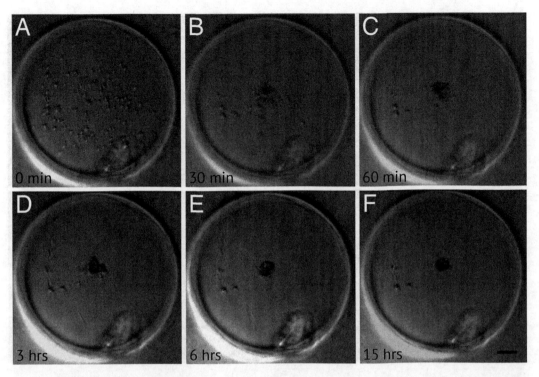

Fig. 1 Time lapse of cells aggregating in one of the wells of a micro-well plate. A time-point movie was set up to take images every 2 min for 15 h. (A 3-h movie can be viewed at Ref. [13]). Within 30 min of seeding, cells settled to the bottom of the well and began aggregating. Cells are fully aggregated by 3 h and have undergone compaction by 15 h. Time in minutes and hours after seeding. Scale bar = 100 μm. Figure adapted from Ref. [13] with permission from development

using the blunt end of a pair of forceps, or other blunt, straight edged object.

6. Transfer aggregates using a P20 pipette: set the pipette to 5 μl and fill the tip with PBS from the dish. Flush downwards into each well to release the aggregate and collect them whilst they are floating. Transfer up to 5 aggregates to the area on the microscope slide within the reinforcement ring label.

7. To this, add 5 μl VectaShield, then using the broken off tip of a 20 μl microloader tip (or similarly soft/hair-like object), gently mix the mounting medium with the PBS, trying not to disturb the aggregates (*see* **Note 17**).

8. Clean a 13 mm round coverslip with microscope lens paper to remove any debris or smudges, and gently place on top of the spacer using forceps (*see* **Note 18**).

9. Leave to set at room temperature for 20 min before imaging or leave at 4 °C overnight. Once mounted, samples can be stored at 4 °C away from light for up to 2 months, depending on the stability of the fluorophores.

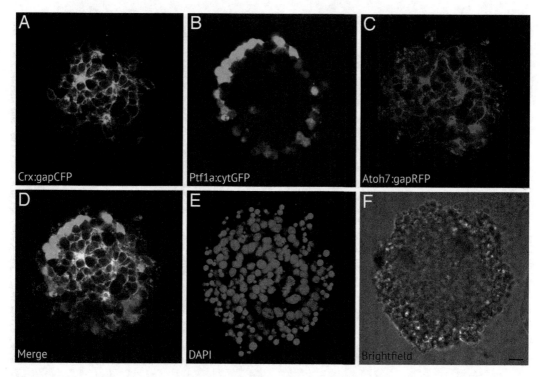

Fig. 2 A central sagittal slice of an aggregate cultured using the Spectrum of Fates (SoFa) transgenic line. The SoFa transgenic line labels each of the five main retinal cell types using a combination of three fluorophores driven by cell type specific promoters [2]. Visualization of the central slice of an aggregate cultured using this line allows us to see that zebrafish retinal cells are capable of self-organizing. Crx:gapCFP is expressed in photoreceptors and bipolar cells; Ptf1a:cytGFP is expressed in amacrine and horizontal cells; Atoh7:gapRFP is expressed in retinal ganglion cells, photoreceptors, amacrine and horizontal cells. It can be seen clearly that Ptf1a:cytGFP expressing cells organize in a ring around the Crx:gapCFP expressing cells. Scale bar = 5 μm. Figure adapted from Ref. [13] with permission from development

10. Samples can be imaged on any fluorescence microscope, using the appropriate filters for the fluorophores. See example of an aggregate cultured from the SoFa line using a confocal microscope (Fig. 2).

3.5 Immunostaining

Aggregates can also be stained whole-mount with antibodies whilst in the agarose micro-well dishes. First fix them as described in step 1 and 2 above, but replace the washing steps with those outlined below, keeping dishes away from light between each step:

1. Rinse once quickly and then incubate for 10 min with 0.1% PBS-T.
2. Wash for 10 min with PBS.
3. Prepare antibodies diluted in PBS only.
4. Blocking step: Some antibodies require blocking (such as for Zn5 described in Fig. 3): Incubate with blocking buffer for

20 min at room temperature then wash off with PBS. For antibodies that don't require blocking, skip to step 5.

5. Remove as much PBS as possible from the dishes.
6. Gently add 200 μl antibody mix to the seeding chamber only (*see* **Note 19**).
7. Seal the culture plates with parafilm to prevent the dishes drying out.
8. Leave at 4 °C overnight.
9. The next day, dilute secondary antibodies 1:500 in PBS (plus DAPI 1:1,000 if required).
10. Gently remove the primary antibody.
11. Wash for 10 min with 0.05% PBT.
12. Add secondary antibodies and place on a rocker at a very slow speed for 2 h at room temperature.
13. Wash for 10 min with PBS and place on the rocker.
14. Do one final quick wash with PBS.
15. To mount and image the aggregates follow the same protocol as from step 4 in the section above. Examples of aggregates stained for Zn5 and GFAP are shown in (Fig. 3).

3.6 Analysis of Patterning by Isocontour Profiling

You can find the Matlab scripts necessary for the next steps at: http://caic.bio.cam.ac.uk/dataanalysis/projects/aggregateimages/scripts (*see* **Notes 20** and **21**).

1. To analyse the patterning of these aggregates, we use two ways of computing isocontour profiles: integrated fluorescence along concentric bands up to a maximal radius (Center to Perhiphery: CtoP) or starting from the perimeter towards the center, along shrinking bands (Periphery to Center: PtoC). The script "ComputePatterns" computes both versions. The width of the band can be set as parameter w in line 13 (by default $w = 5$ pixels).
2. The image files must be in multipage Tiff stack format. In our case, the order of the images in the stack is: CFP channel, GFP channel, RFP channel, brightfield, DAPI. If DAPI is missing, it will be replaced by the brightfield image for the subsequent computations (*see* **Note 22**).
3. Open "ComputePatterns" Script.
4. Select first Tiff stack file. The first part of the script will run and prompt you to draw the outline of the aggregate you wish to analyse.
5. The center point and outline of the aggregate will be shown superimposed to the DAPI image. If the figure looks incorrect, run the programme again and correct the outline.

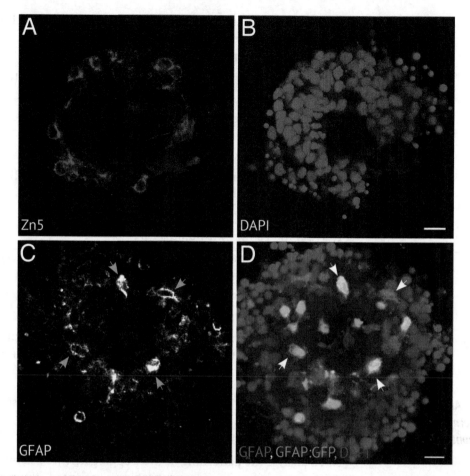

Fig. 3 Central sagittal slices of aggregates after whole-mount staining. (**a, b**) Zn5 staining showing position of retinal ganglion cells. (**c, d**) GFAP staining showing position of Müller glia. Scale bars = 10 μm. Figure adapted from Ref. [13] with permission from development

6. The fluorescence profiles are computed and saved in csv files in the same folder as the input image stack, along with the mask generated from the outline.

7. Repeat steps 4–6 until you have analyzed all samples of that condition.

8. Next, run "MakeFigures" script.

9. Select the source folder, where all your csv files are saved for that condition.

10. This script will produce the following graphs (the suffix PtoC in the figure names indicates profiles on isocontours measured from periphery to center and the suffix CtoP indicates isocontours measured from center to periphery):

    - [filename] Showing the fluorescence profiles for each individual sample. By default, this part of code is commented out (remove comments from lines 79–98 if you wish to generate the figures).

- [FigAll_samples_profiles] Showing all fluorescence profiles on the same graph.

- [Figfilename_ecdf]. This shows the fluorescence profiles plotted as an empirical cumulative distribution function (ecdf) for each individual sample. By default, this part of code is commented out (remove comments from lines 110–126 if you wish to generate the figures).

- [FigAll samples_profiles_with error]. This shows the fluorescence profiles for all samples in that condition, with sample error. Shown only if at least two profiles available.

- [FigAll samples_ecdf_with error]. This shows the ecdf plots for all samples in that condition, with sample error. Shown only if at least two profiles available.

- The parameter CorP in line 8 of MakeFigures script indicates if the figures are generated based on CtoP isocontours (CorP = 1), or PtoC isocontours (CorP = 2). The data shown in Fig. 4 was generated from PtoC isocontours (*see* **Note 23**).

11. This script will then produce a csv file called [Area_measurements.csv] containing the areas measured under the ecdf curves for the RFP, CFP and GFP channels for each aggregate and the area between the CFP and GFP ecdf curves. This is the measure of organization of the aggregates.

12. Use the data in the final column (AUC CFP–AUC GFP) to compare this condition to another. An example analysis of the difference in organization between aggregates in control conditions and those where the birth of Müller Glia has been prevented, is shown in (Fig. 4).

4 Notes

1. Depending on how many eyes you need to dissect, and how clean your dissection is, make between 10 and 50 ml dissociation medium.

2. You need to make enough for 750 µl per dish, plus an extra 100 µl due to some medium being lost during the filtering step.

3. Ensure the agarose is well mixed by inverting the tube a few times before use. If not well mixed, this will cause the microwell dishes to crack and contents to leak out once seeded.

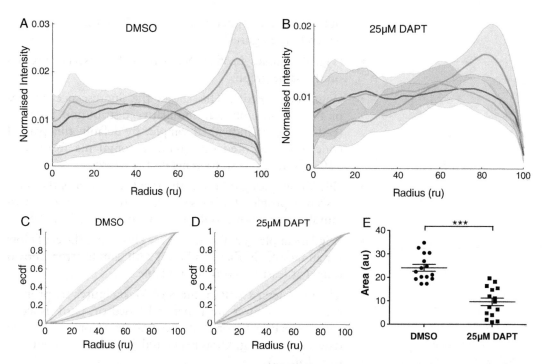

Fig. 4 Isocontour profiling and analysis of aggregates cultured with DMSO control or with 25 μM DAPT, from the equivalent of 45hpf onwards, to prevent Müller glia differentiation. (**a, b**) Average fluorescence profiles with sample error for the CFP, GFP and RFP channels of aggregates cultured with DMSO (**a**) ($n = 15$) and 25 μM DAPT (**b**) ($n = 15$). (**c, d**) Average empirical cumulative distribution frequency (ecdf) plots with sample error for the CFP and GFP channel of aggregates cultured with DMSO (**c**) and 25 μM DAPT (**d**). (**e**) Quantitation of the area (au) between the CFP and GFP ecdf plots for the DMSO and 25 μM DAPT conditions. The areas between the CFP and GFP ecdf plots for the 25 μM DAPT condition are significantly smaller than those for the DMSO control, indicating the aggregates lacking Müller glia are less organized. Figure adapted from Ref. [13] with permission from development

4. Pipette slowly and try not to introduce air bubbles. Also use a new tip for each pipette as the agarose may get stuck to the inside of the tip and the amounts will vary.

5. Release the dish from the edges of the mould first, then push it out from the middle at the back whilst also applying pressure from slightly off-center. This may take some practice to release the dish without cracking it.

6. Dissection tips:

    (a) First sever the heads using a scissor movement with the pins (from the heart to the hindbrain) and discard the bodies before continuing with the dissection of the retinas. This will make it easier to avoid perforating the yolk sac which would leak its contents and make a mess of the dissecting dish.

    (b) Once the heads are severed, to make it easier to access the retinas, break the tissue covering the retinas by making an incision through the forebrain (between the retinas). The retinas will then face upwards, with the covering tissue facing downwards. You can then gently roll the retinas away from this tissue.

(c) If you find the tissues are getting very sticky, it may be that you have a lot of debris in the dish. Replace the dissociation medium with fresh medium every few rounds of dissection to avoid this.

7. Before using the glass pipette, rinse it once with dissociation medium to avoid retinas sticking to the inside. Do the same for the pipette used for dissociation.

8. It's important to replace the medium every few rounds of dissection, otherwise it will become more difficult to dissect as tissues will get stuck to the dissecting pins.

9. Try not to introduce air bubbles during trituration otherwise cells will burst.

10. Check status of dissociation momentarily at increased magnification. Since the cells are transparent, you can also alter the mirror angle on the microscope to see them better.

11. As the cells have only been spun gently, the pellet may be loose, so check the supernatant under the microscope to check you haven't removed too many cells.

12. Leave at least 20 µl extra to allow for air bubbles when seeding.

13. This amount of material from 20 retinas is enough for seeding three dishes of the 15 well variety, but if you require a higher or lower concentration, count the cells after the first wash (step 8) by re-suspending the cells and counting a 10 µl sample using the Neubauer Chamber. If count number is too low, it will be inaccurate. If so, count after the next concentration steps. Try to count 100 cells for a <20% error rate.

14. When seeding, add cells in a dropwise manner from directly above the dish to allow equal distribution of cells and prevent them washing to one side.

15. Do not disturb the dishes at this time otherwise cells will not be equally distributed amongst the wells.

16. The medium exchange ports are the indents of the agarose dish on two of the sides. They allow you to place your pipette tip down the outside of the agarose dish, inside the well of the 4-well plate. Add the culture medium slowly, to prevent washing out the cells from the dish. The dish itself may also float, so prevent this by pipetting at an angle so the pipette tip holds the dish down at the same time.

17. It's better to keep the aggregates in PBS and mix with the mounting medium, rather than first removing the PBS, as the

aggregates are still quite delicate. The mounting medium will still set hard as it usually does.

18. The best technique to avoid introducing air bubbles is to start by placing the coverslip down onto the spacer on one side, then slowly lower the other side down. You will notice that the mounting medium is attracted to the coverslip. Once this happens, you can let go. You can also slide, or apply gentle downward pressure to the coverslip on one side if the aggregates are moving too close to the edge.

19. Do this under a microscope and pipette slowly down the inside edge of the seeding chamber to ensure you don't disturb the aggregates.

20. All provided scripts have to be in Matlab path.

21. The authors of the script imgpolarcoordMod.m are Juan Carlos Gutierrez and Javier Montoya and of the script shadedErrorBar.m is Rob Campbell. Both scripts can be downloaded from Matlab File Exchange (http://www.mathworks.com/matlabcentral/fileexchange).

22. Stacks can be created with software such as Volocity or Fiji and saved into separate folders for each condition.

23. We recommend generating data from periphery to center (PtoC) isocontours as these account for the slight differences in aggregate shape, whereas CtoP isocontours are based on perfectly round shapes.

Acknowledgements

We are grateful to Alexandra D. Almeida for useful discussions during the preparation of this manuscript. We also thank Mark Charlton-Perkins, Ryan MacDonald, and Afnan Azizi for their discussions and contributions to the original research manuscript where the protocols were developed.

This work was funded by a Wellcome Trust Senior Investigator Award to W.A.H. (100329/Z/12/Z) and a Biotechnology and Biological Sciences Research Council.

Studentship Award to M.K.E. (BB/J014540/1).

References

1. Holt CE, Bertsch TW, Ellis HM, Harris WA (1988) Cellular determination in the xenopus retina is independent of lineage and birth date. Neuron 1:15–26. doi:10.1016/0896-6273(88)90205-X
2. Almeida AD, Boije H, Chow RW et al (2014) Spectrum of fates: a new approach to the study of the developing zebrafish retina. Development 141:1971–1980. doi:10.1242/dev.104760
3. Chow RW-Y, Almeida AD, Randlett O et al (2015) Inhibitory neuron migration and IPL formation in the developing zebrafish retina. Development. doi:10.1242/dev.122473

4. Icha J, Kunath C, Rocha-Martins M, Norden C (2016) Independent modes of ganglion cell translocation ensure correct lamination of the zebrafish retina. J Cell Biol 215:259–275. doi:10.1083/jcb.201604095
5. Randlett O, MacDonald RB, Yoshimatsu T et al (2013) Cellular requirements for building a retinal neuropil. Cell Rep 3:282–290. doi:10.1016/j.celrep.2013.01.020
6. Herbst C (1900) über das Auseinandergehen von Furchungs- und Gewebezellen in kalkfreiem Medium. Arch für Entwicklungsmechanik der Org 9:424–463. doi:10.1007/BF02156626
7. Wilson HV (1907) On some phenomena of coalescence and regeneration in sponges. J Exp Zool 5:245–258. doi:10.1002/jez.1400050204
8. Moscona A, Moscona H (1952) The dissociation and aggregation of cells from organ rudiments of the early chick embryo. J Anat 86:287–301
9. Moscona A (1961) Rotation-mediated histogenetic aggregation of dissociated cells. Exp Cell Res 22:455–475. doi:10.1016/0014-4827(61)90122-7
10. Layer PG, Willbold E (1993) Histogenesis of the avian retina in reaggregation culture: from dissociated cells to laminar neuronal networks. Int Rev Cytol 146:1–47
11. Layer PG, Willbold E (1994) Regeneration of the avian retina by retinospheroid technology. Prog Retin Eye Res 13:197–230. doi:10.1016/1350-9462(94)90010-8
12. Rothermel A, Willbold E, Degrip WJ, Layer PG (1997) Pigmented epithelium induces complete retinal reconstitution from dispersed embryonic chick retinae in reaggregation culture. Proc Biol Sci 264:1293–1302. doi:10.1098/rspb.1997.0179
13. Eldred MK, Charlton-Perkins M, Muresan L, Harris WA (2017) Self-organising aggregates of zebrafish retinal cells for investigating mechanisms of neural lamination. Development 144:1097–1106. doi:10.1242/dev.142760
14. Kimmel CB, Ballard WW, Kimmel SR et al (1995) Stages of embryonic development of the zebrafish. Dev Dyn 203:253–310. doi:10.1002/aja.1002030302
15. Bernardos RL, Raymond PA (2006) GFAP transgenic zebrafish. Gene Expr Patterns 6:1007–1013. doi:10.1016/j.modgep.2006.04.006
16. Napolitano AP, Dean DM, Man AJ et al (2007) Scaffold-free three-dimensional cell culture utilizing micromolded nonadhesive hydrogels. Biotechniques 43(494):496–500

Antibody Uptake Assay in the Embryonic Zebrafish Forebrain to Study Notch Signaling Dynamics in Neural Progenitor Cells In Vivo

Kai Tong, Mahendra Wagle, and Su Guo

Abstract

Stem cells can generate cell fate heterogeneity through asymmetric cell division (ACD). ACD derives from the asymmetric segregation of fate-determining molecules and/or organelles in the dividing cell. Radial glia in the embryonic zebrafish forebrain are an excellent model for studying the molecular mechanisms regulating ACD of stem cells in vertebrates, especially for live imaging concerning in vivo molecular and cellular dynamics. Due to the current difficulty in expressing fluorescent reporter-tagged proteins at physiological levels in zebrafish for live imaging, we have developed an antibody uptake assay to label proteins in live embryonic zebrafish forebrain with high specificity. DeltaD is a transmembrane ligand in Notch signaling pathway in the context of ACD of radial glia in zebrafish. By using this assay, we have successfully observed the in vivo dynamics of DeltaD for studying ACD of radial glia in the embryonic zebrafish forebrain.

Keywords: Antibody uptake assay, Asymmetric cell division, Live imaging, Notch signaling, Radial glia, Stem cell, Zebrafish

1 Introduction

Stem cells have the amazing ability to maintain a stem cell population while producing differentiating progeny, which critically underscores the generation of cell fate diversity during development. One important approach to achieve this is through asymmetric cell division (ACD), which produces one self-renewing daughter cell and the other committed to differentiation. How ACD of stem/progenitor cells is regulated has been extensively studied in invertebrates. Typically, cell fate asymmetry can be mediated by the asymmetric segregation of fate-determining molecules/organelles between the two daughter cells [1–4]. In contrast, the molecular details underlying ACD of vertebrate stem cells are only beginning to be characterized. Radial glia, multi-potent neural progenitor cells in the developing vertebrate central nervous system, predominantly undergo ACD during the peak phase of neurogenesis in mammals [5–8] and zebrafish [9, 10]. This feature, combined with the accessibility and transparency of zebrafish embryos, makes radial glia in zebrafish an excellent model for addressing questions regarding ACD of vertebrate stem cells, especially when concerning in vivo molecular or cellular dynamics [11].

To observe the in vivo dynamics of molecular and cellular processes in zebrafish, exogenous fluorescent fusion proteins are often transiently (e.g., by microinjecting the mRNA) or stably (e.g., by establishing the transgenic line) introduced. However, dosage sensitivity of certain proteins (in particular, those with important signaling activity) precludes such approach, as introducing exogenous copies of DNA or RNA molecules can lead to overexpression and disruption of normal phenotypes. Knock-in of the fluorescent protein sequences into the endogenous genomic loci is ideal, but remains challenging for zebrafish despite a few published successes [12–15]. Given the requirement of observing protein dynamics at its physiological level in zebrafish, antibody uptake assay can be a suitable approach, under the circumstances that the protein is transmembrane with an extracellular region and frequently undergoes endocytosis. Briefly, the complexes of primary antibodies and fluorescently labeled secondary antibodies are formed in vitro and then applied into the extracellular space. The antibody complex specifically binds to the extracellular region of the protein and is uptaken into the cell upon endocytosis of the labeled protein, which allows the tracking of in vivo dynamics of this protein with fluorescent confocal imaging. Previously, the antibody uptake assay was used to study the intracellular trafficking and asymmetric segregation of Delta and/or Notch, in the context of ACD of sensory organ precursor (SOP) cells in *Drosophila* [16–21] and neural precursor cells in zebrafish spinal cord [22]. Here, we describe how the antibody uptake assay is performed to study the in vivo dynamics of DeltaD in radial glia of embryonic zebrafish forebrain, as well as demonstrate its broad applicability to various developmental stages and brain regions.

2 Materials

1. Collect and raise zebrafish embryos: mating cages, dividers, plastic Petri dishes, 28.5 °C incubator.

2. Embryonic medium: 0.12 g $CaSO_4$, 0.2 g "Instant Ocean" Sea Salts, and 0.3 g Methylene Blue added to 1 l distilled water.

3. 0.3% Phenylthiourea (PTU).

4. Glass pipette.

5. Fine forceps (Inox 5, Dumont Electronic, Switzerland).

6. Stereo dissection microscope (Zeiss Stemi 2000 with a maximum magnification of 50×).

7. Tricaine.

8. Monoclonal mouse IgG1 anti-DeltaD primary antibody (zdd2, Abcam ab73331, 1 µg/µl) directed against the extracellular region of DeltaD.
9. Zenon® Alexa Fluor® Mouse IgG1 labeling reagent (Molecular Probes, Z-25008 for Alexa Fluor 647, 0.2 µg/µl).
10. PBS 1×.
11. Low-melting point agarose (Shelton Scientific, Inc. catalog #IB70050).
12. Heat block.
13. Glass capillaries (1.2 mm O.D., 0.94 mm I.D., with filament).
14. Flaming-Brown P897 puller (Sutter Instruments, Novato, CA, USA).
15. Eppendorf Microloader Tips for filling Femtotips and other glass microcapillaries (0.5–20 µl, Catalog No. 930001007).
16. Micromanipulator (WPI M3301R, World Precision Instruments, Sarasota, FL, USA).
17. Air pressure microinjector (Narishige IM 300 microinjector) with a nitrogen cylinder.
18. Halocarbon oil.
19. Stage micrometer.
20. Microsurgical knife.
21. 35 mm glass-bottom culture dish (MatTek corporation, Part# P35GC-1.5-10-C, www.glass-bottom-dishes.com).
22. Confocal microscope with a temperature controller.

3 Methods

All steps are performed at room temperature unless otherwise specified.

3.1 Preparation of Zebrafish Embryos

1. Two days before the antibody uptake assay, set up mating cages of zebrafish (*see* **Note 1**) with dividers to separate males and females.
2. One day before the antibody uptake assay, pull the dividers in mating cages.
3. Collect zebrafish embryos and incubate 50 fertilized embryos in a plastic Petri dish with 30 ml of embryonic medium containing 0.003% phenylthiourea (PTU) in an incubator at 28.5 °C.
4. On the day of the antibody uptake assay, when the embryos reach the developmental stage of 29 hpf (*see* **Note 2**), dechorionate the embryos manually with fine forceps under a stereo dissection microscope.
5. Add 1.2 ml tricaine stock (4 mg/ml, 25×) to the embryonic medium to anesthetize the embryos.

3.2 Preparation for Antibody Injection

1. Mix 0.5 μl of mouse IgG1 anti-DeltaD primary antibody (1 μg/μl) and 2.5 μl of Alexa-labeled Zenon® mouse IgG1 labeling reagent (0.2 μg/μl). This labeling reagent is Alexa-labeled goat Fab fragments targeted against the Fc tail of mouse IgG1 primary antibodies. Incubate the mixture at room temperature for 30 min, protected from light, and then dilute with PBS 1× to 5 μl. The antibody complex of primary and secondary antibodies is thus ready for the antibody uptake assay (*see* **Note 3**).

2. Prepare 1% (w/v) low-melting point agarose in embryonic medium. Aliquot the agarose solution in 2 ml microcentrifuge tubes and add 80 μl of tricaine stock (4 mg/ml, 25×) into each tube after the agarose solution cools down a bit. Keep the aliquots in a heat block at ~37 °C.

3. Glass injection needles are pulled from capillaries on a puller. The needle tip is broken off with fine forceps to get a sharp end, whose diameter should be around 10 μm (can be checked with a stage micrometer) and whose angle should be around 30°. The appropriate diameter and angle of the tip are very important for inserting the needle into the forebrain ventricle and injecting a suitable volume of antibody complex solution.

3.3 Antibody Uptake Assay

1. To mount the embryos, transfer two dechorionated embryos into 1% low-melting point agarose with a glass pipette. After a brief immersion, immediately remove the embryos along with small amount of agarose. Under a stereo dissection microscope, place each embryo on an inverted plastic Petri dish lid in an individual drop of agarose.

2. Orientate the embryos gently with a fiber probe (can be cut from an Eppendorf microloader tip) to a dorsal-up position before the agarose solidifies, so that the forebrain is accessible to the microinjection needle (*see* **Note 4**).

3. To prevent the Alexa Fluor® fluorescent tag in the antibody complex from photobleaching, from the next step, turn off the light in the room and turn down the light source of the stereo dissection microscope.

4. Backload the microinjection needle with 2 μl of antibody complex solution using an Eppendorf microloader tip. Make sure no air bubble is in the solution.

5. In order to adjust the volume of one single injection to 4 nl, manipulate the position of the microinjection needle with a micromanipulator to inject one drop of solution with an air pressure microinjector into the halocarbon oil (or other similar oil) placed on a stage micrometer. A drop with a diameter of 0.2 mm is about 4 nl in volume. Manipulate the injection parameters (pressure and/or time) with the air pressure microinjector to achieve the expected injection volume.

6. Insert the microinjection needle into the forebrain ventricle of the embryo (Fig. 1) (*see* **Note 5**) and inject 8 nl of antibody complex solution by injecting 4 nl twice (*see* **Note 6**), and the swelling of the forebrain ventricle can be easily observed (*see* **Note 7**).

7. Add 5 ml of embryonic medium containing 0.16 mg/ml tricaine to cover the agarose drops and carefully peel the agarose with a microsurgical knife to release the embryos.

8. To mount the embryos for imaging, transfer one embryo into 1% low-melting point agarose with a glass pipette. After a brief immersion, immediately remove the embryo from the agarose. Place the embryo on the center of a small glass-bottom culture dish in a single agarose drop (*see* **Note 8**).

9. Orientate the embryo gently with a fiber probe to a dorsal-down position (or a dorsal-up position if an up-right objective is used) before the agarose solidifies.

10. Place the dish properly on the temperature controlled stage of the confocal microscope with an inverted microscope. Adjust the temperature to 28.5 °C.

11. Add 3 ml 28.5 °C preheated embryonic medium containing 0.003% PTU and 0.16 mg/ml tricaine to cover the embryo.

12. Perform in vivo time-lapse confocal imaging with a fixed interval, and the embryo should be ~30 hpf now (*see* **Note 9**) (Fig. 2).

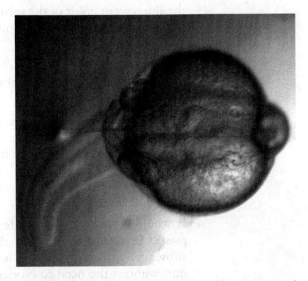

Fig. 1 One zebrafish embryo at 29.5 hpf is mounted in an agarose drop, with a dorsal-up orientation and its forebrain ventricle readily approachable by the microinjection needle. The microinjection needle backloaded with the antibody complex solution is inserted into the forebrain ventricle, followed by the injection of 8 nl of the antibody complex solution using the air pressure microinjector

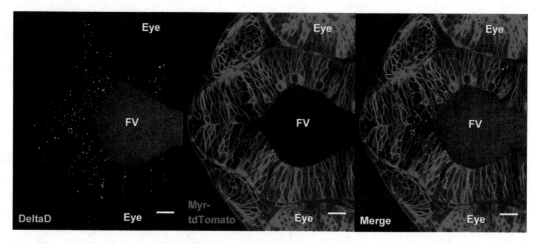

Fig. 2 After injecting the anti-DeltaD antibody complex solution (the secondary antibody in antibody complex is conjugated with Alexa Fluor 647) into the forebrain ventricle of the zebrafish embryo of EF1α:myr-tdTomato transgenic line (myr-tdTomato is a red cell membrane marker), the far-red fluorescent signals can be observed both inside the radial glia cells of the forebrain and on their cell membranes (FV, forebrain ventricle; Scale bar, 20 μm), indicating that anti-DeltaD antibody complex binds to DeltaD and undergoes uptake into the cell. Time-lapse confocal imaging can thus be performed to observe the dynamics of DeltaD

4 Notes

1. In order to better visualize the intracellular dynamics of DeltaD, a transgenic zebrafish line with a cell membrane marker (e.g., EF1α:myr-tdTomato) is often used. Under such circumstance, the fluorescent color of the secondary antibody for forming the antibody complex should not overlap with that of the cell membrane marker.

2. Antibody uptake assay can also be applied to other developmental stages for the embryonic zebrafish forebrain, at least from 24 to 32 hpf.

3. The antibody complex solution can be stored at 4 °C, protected from light, for at least a month. In addition to labeling the primary antibody to form the in vitro primary–secondary antibody complexes by the Zenon® labeling technology, the directly labeled antibody is presumed to be also workable in such antibody uptake assay.

4. It would be better to orientate the brains of all embryos on a plastic Petri dish lid towards the same direction, so that the subsequent microinjection can be more conveniently carried out, without the need to reorient the Petri dish lid every time before injecting the next embryo. Moreover, the process from mounting the embryos to releasing the embryos from the agarose should be done within half an hour; otherwise a large extent

of the water in the agarose may evaporate, causing the agarose to shrink and the embryos to be crushed.

5. Antibody uptake assay can also be applied to other brain regions, like midbrain and hindbrain, by microinjecting into the midbrain ventricle and hindbrain ventricle, respectively (the injection volume can be adjusted according to the injection position). At this developmental stage (~29.5 hpf), the ventricles of forebrain, midbrain, and hindbrain are interconnected, and the antibody complexes microinjected into the forebrain ventricle can diffuse through the ventricles and thus can be uptaken into cells not only in the forebrain but also in the midbrain and hindbrain, if the protein endocytosis does happen there.

6. Sometimes inserting the needle into the forebrain ventricle can block the needle to some extent and thus affect the injection volume. Under such concern, it would be better to check the injection volume of each injection. A simple way to do this is to, after determining the injection volume by the stage micrometer, touch the tip of the needle onto the surface of an agarose drop used to mount the embryo, inject once and keep in mind the size of the solution drop formed on the surface of the agarose drop. Then after each injection into an embryo, inject on the surface of the agarose drop to ensure a solution drop is formed with a similar size. A significantly smaller size of the drop would indicate that the last injection volume is likely smaller than it should be and the needle is likely blocked. When this happens, the embryo just injected should be discarded, and a larger injection pressure or a "Clear" function of the microinjector should be used to clear the needle tip while a certain amount of solution remains in the needle, and then adjust the injection volume to 4 nl again by the stage micrometer.

7. If the ventricle swelling is not observed but the injection of the solution is indeed observed, then the solution may not be injected into the forebrain ventricle but probably some deeper tissues or even the yolk. The embryo should then be discarded.

8. Sometimes in a later step when covering the solidified agarose drop with the embryonic medium, the agarose can float up. To prevent this, one can apply the agarose across the whole concave glass area of the glass-bottom culture dish.

9. Clear intracellular fluorescent signals inside the forebrain should be observed within 15 min after injecting the antibody complex (15 min is roughly the minimal time required for releasing and remounting embryos followed by setting the microscope), and if the microscope settings are well controlled to minimize photobleaching (and phototoxicity), the fluorescent signals can be observed for at least 2 h. Clear fluorescent signals should also be observed in the forebrain ventricle, and their absence indicates the failure in the injection of the antibody complex solution, its

preparation/storage, or the prevention of photobleaching during the experiment. In order to ensure the intracellular fluorescent signals are not the result of any nonspecific uptake of the antibody complex, necessary controls like injecting only the fluorescently labeled secondary antibody should be performed.

Acknowledgments

This work was supported by NIH (R01 NS095734) and Fudan Bio-elite program. We thank Xiang Zhao for discussions on the antibody uptake assay and comments on the manuscript, as well as Maximilian Fürthauer for sharing experience on the antibody uptake assay in the zebrafish spinal cord.

References

1. Gonczy P (2008) Mechanisms of asymmetric cell division: flies and worms pave the way. Nat Rev Mol Cell Biol 9(5):355–366. doi:10.1038/nrm2388
2. Neumuller RA, Knoblich JA (2009) Dividing cellular asymmetry: asymmetric cell division and its implications for stem cells and cancer. Genes Dev 23(23):2675–2699. doi:10.1101/gad.1850809
3. Knoblich JA (2010) Asymmetric cell division: recent developments and their implications for tumour biology. Nat Rev Mol Cell Biol 11(12):849–860. doi:10.1038/nrm3010
4. Schweisguth F (2015) Asymmetric cell division in the Drosophila bristle lineage: from the polarization of sensory organ precursor cells to Notch-mediated binary fate decision. Wiley Interdiscip Rev Dev Biol 4(3):299–309. doi:10.1002/wdev.175
5. Chenn A, McConnell SK (1995) Cleavage orientation and the asymmetric inheritance of Notch1 immunoreactivity in mammalian neurogenesis. Cell 82(4):631–641
6. Miyata T, Kawaguchi A, Saito K, Kawano M, Muto T, Ogawa M (2004) Asymmetric production of surface-dividing and non-surface-dividing cortical progenitor cells. Development 131(13):3133–3145. doi:10.1242/dev.01173
7. Noctor SC, Martinez-Cerdeno V, Ivic L, Kriegstein AR (2004) Cortical neurons arise in symmetric and asymmetric division zones and migrate through specific phases. Nat Neurosci 7(2):136–144. doi:10.1038/nn1172
8. Noctor SC, Martinez-Cerdeno V, Kriegstein AR (2008) Distinct behaviors of neural stem and progenitor cells underlie cortical neurogenesis. J Comp Neurol 508(1):28–44. doi:10.1002/cne.21669
9. Alexandre P, Reugels AM, Barker D, Blanc E, Clarke JD (2010) Neurons derive from the more apical daughter in asymmetric divisions in the zebrafish neural tube. Nat Neurosci 13(6):673–679. doi:10.1038/nn.2547
10. Dong Z, Yang N, Yeo SY, Chitnis A, Guo S (2012) Intralineage directional Notch signaling regulates self-renewal and differentiation of asymmetrically dividing radial glia. Neuron 74(1):65–78. doi:10.1016/j.neuron.2012.01.031
11. Dong Z, Wagle M, Guo S (2011) Time-lapse live imaging of clonally related neural progenitor cells in the developing zebrafish forebrain. J Vis Exp (50). doi:10.3791/2594
12. Kimura Y, Hisano Y, Kawahara A, Higashijima S (2014) Efficient generation of knock-in transgenic zebrafish carrying reporter/driver genes by CRISPR/Cas9-mediated genome engineering. Sci Rep 4:6545. doi:10.1038/srep06545
13. Hisano Y, Sakuma T, Nakade S, Ohga R, Ota S, Okamoto H, Yamamoto T, Kawahara A (2015) Precise in-frame integration of exogenous DNA mediated by CRISPR/Cas9 system in zebrafish. Sci Rep 5:8841. doi:10.1038/srep08841
14. Li J, Zhang BB, Ren YG, Gu SY, Xiang YH, Du JL (2015) Intron targeting-mediated and endogenous gene integrity-maintaining knockin in zebrafish using the CRISPR/Cas9 system. Cell Res 25(5):634–637. doi:10.1038/cr.2015.43
15. Hoshijima K, Jurynec MJ, Grunwald DJ (2016) Precise editing of the Zebrafish genome made

simple and efficient. Dev Cell 36(6):654–667. doi:10.1016/j.devcel.2016.02.015

16. Le Borgne R, Schweisguth F (2003) Unequal segregation of Neuralized biases Notch activation during asymmetric cell division. Dev Cell 5 (1):139–148

17. Le Borgne R, Remaud S, Hamel S, Schweisguth F (2005) Two distinct E3 ubiquitin ligases have complementary functions in the regulation of delta and serrate signaling in Drosophila. PLoS Biol 3(4):e96. doi:10.1371/journal.pbio.0030096

18. Daskalaki A, Shalaby NA, Kux K, Tsoumpekos G, Tsibidis GD, Muskavitch MA, Delidakis C (2011) Distinct intracellular motifs of Delta mediate its ubiquitylation and activation by Mindbomb1 and Neuralized. J Cell Biol 195 (6):1017–1031. doi:10.1083/jcb.201105166

19. Couturier L, Vodovar N, Schweisguth F (2012) Endocytosis by Numb breaks Notch symmetry at cytokinesis. Nat Cell Biol 14 (2):131–139. doi:10.1038/ncb2419

20. Giagtzoglou N, Yamamoto S, Zitserman D, Graves HK, Schulze KL, Wang H, Klein H, Roegiers F, Bellen HJ (2012) dEHBP1 controls exocytosis and recycling of Delta during asymmetric divisions. J Cell Biol 196 (1):65–83. doi:10.1083/jcb.201106088

21. Couturier L, Schweisguth F (2014) Antibody uptake assay and in vivo imaging to study intracellular trafficking of Notch and Delta in Drosophila. Methods Mol Biol 1187:79–86. doi:10.1007/978-1-4939-1139-4_6

22. Kressmann S, Campos C, Castanon I, Furthauer M, Gonzalez-Gaitan M (2015) Directional Notch trafficking in Sara endosomes during asymmetric cell division in the spinal cord. Nat Cell Biol 17(3):333–339. doi:10.1038/ncb3119

Scaffold-Based and Scaffold-Free Testicular Organoids from Primary Human Testicular Cells

Yoni Baert, Charlotte Rombaut, and Ellen Goossens

Abstract

Organoid systems take advantage of the self-organizing capabilities of cells to create diverse multi-cellular tissue surrogates that constitute a powerful novel class of biological models. Clearly, the formation of a testicular organoid (TO) in which human spermatogenesis can proceed from a single-cell suspension would exert a tremendous impact on research and development, clinical treatment of infertility, and screening of potential drugs and toxic agents. Recently, we showed that primary adult and pubertal human testicular cells auto-assembled in TOs either with or without the support of a natural testis scaffold. These mini-tissues harboured both the spermatogonial stem cells and their important niche cells, which retained certain specific functions during long-term culture. As such, human TOs might advance the development of a system allowing human in vitro spermatogenesis. Here we describe the methodology to make scaffold-based and scaffold-free TOs.

Keywords Testis, Organoid, Primary cells, Extracellular matrix, Scaffold, In vitro spermatogenesis

1 Introduction

Organoid systems leverage the self-organization ability of cells to form tissue-specific multi-cellular structures. They are capable of recapitulating many important properties of a stem cell niche, thereby offering a promising new class of biological models. As organoids provide a more advanced in vitro tool which is amendable to extended cultivation and manipulation, they enable physiologically relevant experiments that cannot be conducted in animals or humans. De novo formation of functional testicular tissue from isolated mammalian testicular somatic and germ cells has been demonstrated in vivo and in vitro [1].

On the one hand, when immature rodent and pig testicular cells, obtained by enzymatic digestion, were transplanted under the dorsal skin of mice, the cells were able to rearrange into a functional endocrine and spermatogenic unit, supporting complete maturation and development of functional haploid male gametes [2, 3]. Other reports confirmed this morphogenic ability of isolated testicular cells in vivo in ovine and bovine models [4, 5]. On the other hand, various approaches aimed at in vitro sperm production using reassembled testicular somatic cells and germ cells either with or without scaffold support. In a mouse study, single-cell suspensions were turned into aggregates in suspension culture and afterwards cultured at the gas–liquid interphase. In this scaffold-free approach, and thus only by relying on the auto-assembly properties of the testicular cell suspension, the seminiferous tubular structure was reorganized. However, spermatogenesis was arrested at the meiotic phase [6]. More promising results were obtained when artificial 3D scaffolds were employed. Cultivation in a 3D collagen gel matrix system succeeded in generating spermatids from recombined immature rat testicular cells [7]. Taking it further, the combination of a primary pre-meiotic testicular cell suspension and a 3D agarose or methylcellulose matrix resulted in morphologically normal mouse spermatozoa [8].

Recently, we published the first human in vitro study and reported the generation of scaffold-based and scaffold-free testicular organoids (TOs) with biomimetic activities using primary testicular cells from either adult or pubertal donors (Fig. 1). Although, the testicular cells were not able to reorganize into the typical testis cytoarchitecture during long-term culture, the niche cells maintained testis-specific functions, including de novo matrix production by elongating peritubular myoid cells, testosterone production by Leydig cells, inhibin b secretion and tight junction formation by Sertoli cells, and germ cell renewal. Cytokine secretion profiling confirmed recapitulation of testicular processes in vitro [9]. The necessity of the scaffold in TO formation is debatable, given that the spatial–temporal behaviour and hormone and cytokine secretion profiles of testicular cells in scaffold-free TOs were comparable. Nevertheless, the testis scaffold might still be of value for future experiments as tissue-derived scaffolds generally contain important regulatory cues [10, 11].

Human TOs take the development of a human in vitro spermatogenesis culture system a step forward. Such a system would be of great value in basic studies revealing the detailed mechanisms behind spermatogenesis and its disorders, and has an enormous clinical value in preventing and curing male infertility [12]. Also, it has potential to become an industrial tool to find male contraceptives and to screen for reprotoxic compounds [13, 14]. The methodology to prepare scaffold-based and scaffold-free TOs is summarized in Fig. 2 and described below.

2 Materials

2.1 Digestion of Testicular Tissue

1. Shaking water bath.
2. Petri dishes with a diameter of 100 mm.
3. Sterile surgical tweezer and scissor or scalpel.

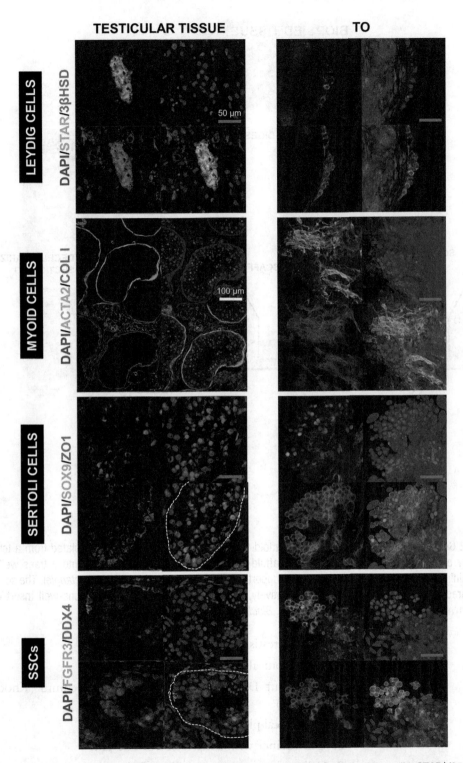

Fig. 1 Long-term cultured TOs contain spermatogonial stem cells (SSCs) and their niche cells. STAR$^+$/3β-HSD$^+$ cells represent steroidogenic Leydig cells (first row), SOX9$^+$/ZO-1$^+$ indicates tight junction-forming Sertoli cells (second row), and ACTA2$^+$/COL I$^+$ cells are ECM-producing peritubular myoid cells (third row). Spermatogonial stem cells (SSCs) are stained by combining the spermatogonial marker FGFR3$^+$ with the germ cell lineage marker DDX4$^+$ (fourth row). The *dotted lines* delineate seminiferous tubules. *Purple scale bars*: 50 μm; *white scale bars*: 100 μm

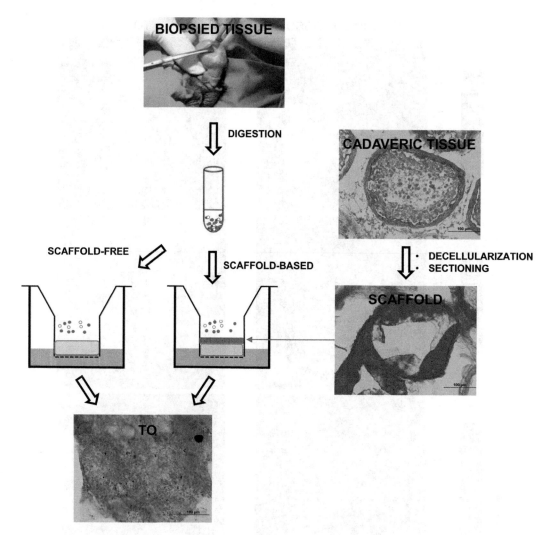

Fig. 2 Generation of scaffold-based and scaffold-free TOs. Testicular cells can be isolated from a testicular biopsy by enzymatic digestion. To form scaffold-based TOs, the cells are seeded into a trans-well insert containing a testicular scaffold (*red arrow*) supported by an agarose hydrogel (*grey rectangle*). The scaffold is prepared as described before [15]. Alternatively, addition of the scaffold to the trans-well insert can be circumvented to generate scaffold-free TOs. *Black scale bars*: 100 μm

4. Human testis biopsy.
5. 50 ml conical tubes.
6. KnockOut Dulbecco's Modified Eagle Medium (KnockOut DMEM).
7. Serological pipettes.
8. Cell strainer with a mesh size of 100 μm.
9. Cell strainer with a mesh size of 40 μm.
10. Cell counter.

11. Parafilm.
12. Analytical balance.
13. Centrifuge.
14. Sterile enzyme mix 1: DMEM-F12 containing collagenase IA (1 mg/ml).
15. Sterile enzyme mix 2: DMEM-F12 containing collagenase (1 mg/ml), DNAse type I (0.5 mg/ml), and hyaluronidase type I-S (0.5 mg/ml).

2.2 Generation of Organoids

1. Sterile 0.7% agarose in PBS (m/v).
2. Testicular scaffold discs [15].
3. Microwave.
4. 24-Well plate.
5. Hanging culture inserts for 24-well plate.
6. Culture medium: 10% (v/v) CTS KnockOut SR XenoFree medium, 1× GlutaMAX and 1% (v/v) penicillin–streptomycin diluted in KnockOut DMEM [9].
7. Incubator at 35 °C with 5% CO_2 and humidified atmosphere.

3 Methods

3.1 Digestion of Testicular Tissue

1. Preheat the water bath at 37 °C.
2. Weigh the testicular tissue and prepare sterile enzyme mix 1 and 2 according to the following rule: 10 ml enzyme mix/1 g of tissue.
3. Add 10 ml of enzyme mix 1 to a petri dish.
4. Place the testicular tissue piece into the enzyme mix and cut the tissue into fragments of 2 mm^3 with a sterile surgical scissor or scalpel.
5. Transfer the tissue pieces and the enzyme mix 1 with a serological pipette to a 50 ml conical tube containing the remaining volume of enzyme mix 1 (*see* **Note 1**). Aspirate several times.
6. Seal the tube with parafilm and place in the preheat water bath (37 °C) for 10 min at 120 RPM.
7. Take the tube out of the water bath and let the remaining fragments (tubules) sediment for 10 min.
8. Collect the supernatant (interstitial cells) in a new 50 ml Falcon tube and centrifuge at 300 × *g* for 10 min. Resuspend the pellet in KnockOut DMEM with DNAse type I (0.5 mg/ml) and store until step 13 is reached.
9. Add enzyme mix 2 to the sedimented tissue pieces and aspirate several times (*see* **Note 1**).

10. Seal the 50 ml tube with parafilm. Place it in the shaking water bath (37 °C) for 10 min at 120 RPM.
11. Take the tube out of the water bath and aspirate several times to mechanically help the digestion.
12. After re-sealing the tubes, shake for another 10 min in the water bath. Aspirate several times. If tubular fragments are still visible, extend the incubation time.
13. Run the cell suspensions from steps 8 and 12 through a 100 μm-mesh sized cell strainer to remove large non-digested tubular fragments.
14. Run the obtained cell suspension over a cell strainer with a mesh size of 40 μm to obtain a homogenous single-cell suspension.
15. Centrifuge the tube at $300 \times g$ for 10 min and resuspend the cell pellet in KnockOut DMEM.
16. Count the cells.

3.2 Generation of Organoids

1. Heat the 0.7% agarose in PBS (m/v) in the microwave for 1 min (500 W) until it is completely liquid.
2. Dilute the liquid agarose 1/2 with culture medium to obtain a 0.35% agarose solution.
3. Place the trans-well inserts in a 24-well plate.
4. Pipette 75 μl of the 0.35% agarose mix in each trans-well insert (*see* **Note 2**).
5. Let the agarose solution jellify for 5 min at room temperature.
6. Place a testicular scaffold disc in each trans-well insert if scaffold-based TOs are needed (*see* **Note 3**). Skip this step to generate scaffold-free TOs (Fig. 2).
7. Spin the cells down at $300 \times g$ for 10 min and adjust the cell concentration to 10^6 cells/10 μl culture medium.
8. Seed 10 μl of cell suspension on top of the agarose gel (scaffold-free) or on top of the thin scaffold disc (scaffold-based) in each trans-well insert (Fig. 2).
9. Pipette 600 μl of culture medium in the wells of the 24-well plate by pipetting next to the inserts (*see* **Note 4**).
10. Place the 24-well plate with the inserts in the incubator at 35 °C with 5% CO_2 and a humidified atmosphere.
11. Refresh the medium every 7 days.

4 Notes

1. It is recommended to not completely fill the tubes with enzyme mix. Leave the solution with tissue pieces some room for shaking.
2. A height of approximately 2 mm is advised for the agarose support. The presence of the agarose support is crucial in the formation of TOs for two reasons. On the one hand, it prevents cell attachment to the membrane of the trans-well insert. On the other hand, the surface of the agarose hydrogel is non-adhesive and, consequently, promotes cell–cell attachment and 3D growth.
3. The protocol to produce thin testicular scaffold discs was described before [15].
4. Culture medium may be adapted according to the experimental setup. Check regularly whether the level of culture medium reaches the cells in the trans-well insert to avoid dehydration.

Acknowledgements

Financial support was obtained from the Agency for Innovation by Science and Technology (IWT) and the Vrije Universiteit Brussel, Y.B. is a postdoctoral fellow of the Scientific Research Foundation Flanders (FWO).

References

1. Yin X et al (2016) Engineering stem cell organoids. Cell Stem Cell 18(1):25–38
2. Honaramooz A, Megee SO, Rathi R, Dobrinski I (2007) Building a testis: formation of functional testis tissue after transplantation of isolated porcine (Sus scrofa) testis cells. Biol Reprod 76(1):43–47
3. Kita K et al (2007) Production of functional spermatids from mouse germline stem cells in ectopically reconstituted seminiferous tubules. Biol Reprod 76(2):211–217
4. Arregui L et al (2008) Xenografting of sheep testis tissue and isolated cells as a model for preservation of genetic material from endangered ungulates. Reproduction 136(1):85–93
5. Zhang Z, Hill J, Holland M, Kurihara Y, Loveland KL (2008) Bovine sertoli cells colonize and form tubules in murine hosts following transplantation and grafting procedures. J Androl 29(4):418–430
6. Yokonishi T et al (2013) In vitro reconstruction of mouse seminiferous tubules supporting germ cell differentiation. Biol Reprod 89:1–6
7. Lee JH, Kim HJ, Kim H, Lee SJ, Gye MC (2006) In vitro spermatogenesis by three-dimensional culture of rat testicular cells in collagen gel matrix. Biomaterials 27(14):2845–2853
8. Stukenborg J-B et al (2009) New horizons for in vitro spermatogenesis? An update on novel three-dimensional culture systems as tools for meiotic and post-meiotic differentiation of testicular germ cells. Mol Hum Reprod 15(9):521–529
9. Baert Y et al (2017) Primary human testicular cells self-organize into organoids with testicular properties. Stem Cell Reports 8(1):30–38
10. Brown BN, Badylak SF (2014) Extracellular matrix as an inductive scaffold for functional tissue reconstruction. Transl Res 163(4):268–285

11. Baert Y et al (2015) Derivation and characterization of a cytocompatible scaffold from human testis. Hum Reprod 30(2):256–267
12. Valli H, Gassei K, Orwig KE (2015) Stem cell therapies for male infertility: where are we now and where are we going? Bienn Rev Infertil 4 (3):1017–1039
13. Kanakis G, Goulis D (2015) Male contraception: a clinically-oriented review. Hormones 14 (4):598–614
14. Chapin RE et al (2013) Assuring safety without animal testing: the case for the human testis in vitro. Reprod Toxicol 39:63–68
15. Baert Y, Goossens E (2017) Preparation of scaffolds from decellularized testicular matrix. Methods Mol Biol

Use of a Super-hydrophobic Microbioreactor to Generate and Boost Pancreatic Mini-organoids

Tiziana A.L. Brevini, Elena F.M. Manzoni, Sergio Ledda, and Fulvio Gandolfi

Abstract

Cell remarkable ability to self-organize and rearrange in functional organoids has been greatly boosted by the recent advances in 3-D culture technologies and materials. This approach can be presently applied to model human organ development and function "in a dish" and to predict drug response in a patient specific fashion.

Here we describe a protocol that allows for the derivation of functional pancreatic mini-organoids from skin biopsies. Cells are suspended in a drop of medium and encapsulated with hydrophobic polytetrafluoroethylene (PTFE) powder particles, to form microbioreactors defined as "Liquid Marbles," that stimulate cell coalescence and 3-D aggregation. The PTFE shell ensures an optimal gas exchange between the interior liquid and the surrounding environment. It also makes it possible to scale down experiments and work in smaller volumes and is therefore amenable for higher throughput applications.

Keywords: Fibroblast, Epigenetic, Microbioreactor, PTFE, Pancreatic organoid

1 Introduction

The last decade has seen an impressive increase in the number of publications where single or groups of cells are grown in 3-D conditions and lead to the formation of structures referred to as organoids [1]. Today's methods are the product of what started in 1906 with the hanging drop tissue culture approach, described by Ross Harrison in an attempt to recapitulate organogenesis in culture. Presently, organoids have been derived from different types of stem cells: pluripotent embryonic stem (ES) cells, induced pluripotent stem (iPS) cells, and organ-restricted adult stem cells (aSC) cells, using growth factor cocktails that mimic the embryonic organ differentiation process and organ stem cell niches [2]. Various culture conditions, both 2-D and 3-D based, have been successfully applied to the purpose and presently, organoids are produced to model human organ development and various human pathologies "in a dish."

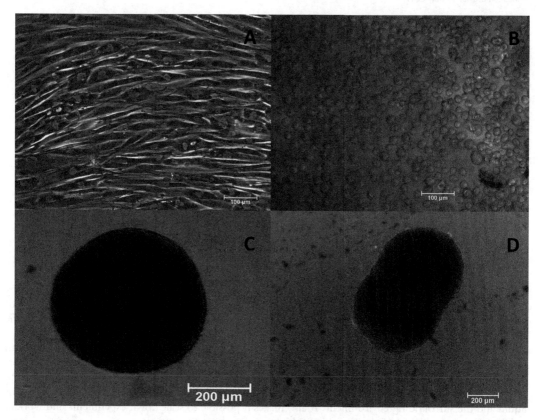

Fig. 1 Morphological changes in epigenetically erased fibroblasts encapsulated in liquid marbles. (**a**) Fibroblasts plated on plastic dishes. (**b**) PTFE is non-adhesive and allows cells to freely interact with each other. (**c, d**) Cells encapsulated in PTFE formed three dimensional spheroids

Here we describe a protocol that allows for the derivation of functional pancreatic organoids from skin biopsies. Human fibroblasts are encapsulated in super-hydrophobic microbioreactor that have been previously shown to support the growth of living microorganisms [3], tumor spheroids [4], fibroblasts [5], red blood cells [6], embryonic stem cells [7], and oocytes [8]. Structures defined as Liquid Marbles (LM) are formed by enveloping cells suspended in a drop of medium, with hydrophobic Polytetrafluoroethylene (PTFE) powder with particle size of 1 μm, to form an elastic shell with fine pores [9]. PTFE particles adhere to the surface of the medium drop, isolating the liquid core from the supporting surface, while allowing an optimal gas exchange between the interior liquid and the surrounding environment. The coating material acts as a confined space, which is non-adhesive and allows cells to freely interact with each other [10]. This method makes it possible to scale down experiments and is therefore amenable for higher throughput applications. Furthermore, work in smaller volumes allows us to study the effect of paracrine/autocrine signalling of the rich environment established within the microbioreactor.

In the protocol described, cells are encapsulated, in order to stimulate them to coalesce and form spheroids, erased using an epigenetic modifier, that increases their plasticity, and then subjected to a three step-differentiation protocol that drives them toward the pancreatic lineage [11–13]. At the end of the 36-days differentiation protocol, cells are arranged in three dimensional structures that exhibit self-organization and ability to produce insulin in response to glucose challenge (*see* Fig. 1).

2 Materials

2.1 Fibroblast Isolation and Culture

1. Dulbecco's Phosphate Buffered Saline (PBS): 2.7 mM KCl, 1.5 mM KH_2PO_4, 136.9 mM NaCl, 8.9 mM Na_2HPO_4 (anhydrous).
2. Fibroblast culture medium: 77% (v/v) Dulbecco's Modified Eagle Medium (DMEM) high glucose, 20% (v/v) Fetal Bovine Serum (FBS), 1% (v/v) L-Glutamine solution, 2% (v/v) antibiotic antimycotic solution.
3. 100 mm Petri dish.
4. Porcine Gelatin 0.1%: dissolve 0.1 g of porcine gelatin in 100 ml of water. Sterilize solution with autoclave.
5. Sterile-filtered water.
6. 35 mm Petri dishes.
7. Sterile scalpels.
8. Trypsin-EDTA solution (1×): 0.5 g/l porcine trypsin, 0.2 g/l EDTA·4Na in Hank's Balanced Salt Solution with phenol red.

2.2 Microbioreactor Preparation

1. Polytetrafluoroethylene (PTFE) powder (*see* **Note 1**).
2. 96-Well plates.
3. 1000 μl pipette tips cut at the edge.
4. Counting chamber.

2.3 5-aza-CR Treatment

1. 5-Azacytidine (5-aza-CR) stock solution: dissolve 2.44 mg of 5-aza-CR in 10 ml of DMEM high glucose medium. Sterilize by filtration (*see* **Note 2**).
2. 10 ml syringes.
3. 0.22 μm pore size hydrophilic Polyethersulfone (PES) filters.

2.4 Pancreatic Differentiation

1. Replenish medium (RP): 40% (v/v) Nutrient Mixture F-10 Ham, 40% (v/v) Dulbecco's Modified Eagle Medium (DMEM) low glucose pyruvate, 10% (v/v) KnockOut Serum Replacement, 5% (v/v) Fetal Bovine Serum (FBS), 1% (v/v) Antibiotic Antimycotic Solution (100×), 1% (v/v) L-Glutamine

solution 200 mM, 1% (v/v) Minimum Essential Medium (MEM) Non-Essential Amino Acids Solution (100×), 1% (v/v) 2-Mercaptoethanol stock (*see* **Note 3**), 1% (v/v) Nucleoside mix stock (*see* **Note 4**), 0.1% (v/v) Leukemia Inhibitory Factor (LIF), 0.1% (v/v) Recombinant Human FGF basic (bFGF) stock (*see* **Note 5**).

2. Pancreatic Basal Medium: 93% (v/v) Dulbecco's Modified Eagle Medium: Nutrient Mixture F-12 (DMEM/F-12), 2% (v/v) B-27 Supplement (50×) Minus Vitamin A, 1% (v/v) N-2 Supplement, 1% (v/v) Antibiotic Antimycotic Solution (100×), 1% (v/v) L-Glutamine solution 200 mM, 1% (v/v) Minimum Essential Medium (MEM) Non-Essential Amino Acids Solution (100×), 1% (v/v) 2-Mercaptoethanol stock (*see* **Note 3**), 1% (v/v) Bovine Serum Albumin (BSA) stock (*see* **Note 6**).

3. Activin A Recombinant Human Protein stock: dissolve 5 μg of activin A Recombinant Human Protein in 166.6 μl of sterile water.

4. Retinoic Acid stock: add 16.6 ml of DMSO to 50 mg of Retinoic Acid.

5. Insulin-Transferrin-Selenium (ITS): 1000.0 mg/l Insulin, 550.0 mg/l Transferrin, 0.67 mg/l Sodium Selenite.

3 Methods

All the procedures described below must be performed under laminar flow hood in sterile conditions. Make sure that all culture procedures are carried out on thermostatically controlled stages and cells are maintained at 37 °C throughout their handling.

3.1 Skin Fibroblast Isolation

1. Prepare fresh Culture Dish adding 1.5 ml of sterile 0.1% porcine gelatin to 35 mm Petri dishes. Wait 2 h to coat, maintaining them at room temperature (*see* **Note 7**).

2. Wash biopsies with new PBS supplemented with 2% antibiotic antimycotic solution.

3. Place biopsies in a 100 mm Petri dish and cut into approximately 2 mm^3 fragments with sterile scalpels.

4. Remove the porcine gelatin excess immediately prior to plating fragments.

5. Place 5–6 skin fragments into the pre-coated 35 mm Petri dish.

6. Prepare fibroblast culture medium (*see* Sect. 2.1, item 2).

7. Add a droplet of fibroblast medium over each fragment (usually 100 μl per fragment) and culture them at 37 °C in 5% CO_2.

8. After 24 h, add 500 µl of fibroblast culture medium over the fragments to keep them wet at 37 °C in 5% CO_2.
9. Change the medium with a pipette at least once every 48 h.
10. After 6 days of incubation, remove tissue fragments carefully and discard them (*see* **Note 8**).
11. Refresh medium, add 2 ml of fibroblast culture medium and continue cell monolayer culture at 37 °C in 5% CO_2 incubator.

3.2 Fibroblast Culture

1. Culture fibroblasts at 37 °C in 5% CO_2 until 80% confluence.
2. For passaging, aspirate fibroblast culture medium from tissue culture dishes. Wash cells three times with 4 ml of PBS supplemented with 1% antibiotic antimycotic solution.
3. Add a thin layer (10% of the culture medium volume) of trypsin-EDTA solution and incubate at 37 °C until cell monolayer begins to detach from the bottom of the tissue culture dish and cells dissociate.
4. Dilute cell suspension with nine parts of fibroblast culture medium to neutralize trypsin action. Centrifugation is not necessary.
5. Plate cells in new culture dishes (without gelatine) and culture at 37 °C in 5% CO_2 incubator. Keep the passage ratio between 1:2 and 1:4, depending on growth rate.
6. When cells reach around 80% confluence, passage them (usually twice per week).

3.3 Preparation of Liquid Marbles Microbioreactors and Exposure to 5-aza-CR

1. Day 0: prepare fresh 5-aza-CR stock solution (*see* Sect. 2.3, item 1).
2. Dilute 1 µl of 5-aza-CR stock solution in 1 ml of fibroblast culture medium (final concentration 1 µM).
3. Remove fibroblast culture medium from culture dishes. Wash cells three times with PBS supplemented with 1% antibiotic antimycotic solution.
4. Add a thin layer (10% of the culture medium volume) of trypsin-EDTA solution and incubate at 37 °C until cell monolayer begins to detach from the bottom of the tissue culture dish and cells dissociate.
5. Dilute cell suspension with nine parts of fibroblast culture medium to neutralize trypsin action.
6. Count cells using a counting chamber under a microscope at room temperature. Calculate the required volume of 1 µM 5-aza-CR stock solution to re-suspend cells in order to obtain 4×10^4 cells in 30 µl of 1 µM 5-aza-CR (*see* **Note 9**).

7. Centrifuge cell suspension at 150 × *g* for 5 min at room temperature. Remove supernatant and re-suspend cells with the previously calculated volume of 1 μM 5-aza-CR stock solution.
8. Prepare a Petri dish containing a polytetrafluoroethylene (PTFE) powder bed-average particle size of 1 μm (*see* Fig. 2).
9. Dispense 30 μl single droplet containing 40,000 cells onto the powder bed.
10. Gently rotate the plate in a circular motion to ensure that the powder particles completely cover the surface of the liquid drop and form a liquid marble.

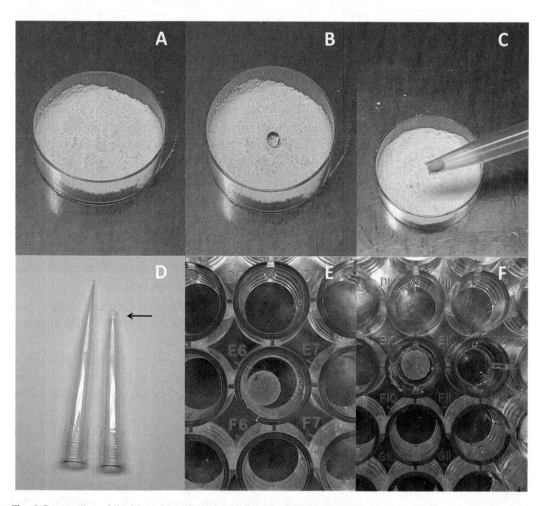

Fig. 2 Preparation of liquid marble containing cells. (**a**) A PTFE powder bed is prepared in a Petri dish. (**b**) A droplet of cells is deposited on top of the bed and then coated with PTFE by rotating the marble with circular movements. (**c**) The marble is collected and (**d**) transferred, using a 1000 μl pipette tip (*black arrow* indicates the cut end of the tip), (**e**) to a 96-well plate and (**f**) floated on the media

11. Pick up the marbles using a 1000 μl pipette tip, cut at the edge, to accommodate the diameter of the marble (see **Note 10**).

12. Place the marbles into a 96-well plate (one marble/well).

13. To float the marbles, add 100 μl of media from the margin of the well to slowly bathe the marble (see **Notes 11** and **12**).

14. Incubate marbles for 18 h at 37 °C in 5% CO_2 incubator (see **Note 13**).

15. Day 1: prepare fresh Replenish Medium (RP) medium (see Sect. 2.4, item 1).

16. After incubation with 1 μM 5-aza-CR, incubate 5-aza-CR treated fibroblasts with RP medium for 3 h (recovery period) at 37 °C in 5% CO_2.

3.4 Pancreatic Differentiation Protocol

1. Days 1–6: Prepare fresh Pancreatic Basal Medium (see Sect. 2.4, item 2).

2. Break the marbles by puncturing with a needle and recover forming organoids using a 200 μl pipette tip, cut at the edge (see **Note 14**).

3. Prepare a Petri dish containing a polytetrafluoroethylene (PTFE) powder bed (average particle size of 1 μm).

4. Dispense the organoid in a droplet of 30 μl Pancreatic Basal Medium, supplemented with 1 μl/ml activin A stock solution, onto the powder bed.

5. Gently rotate the plate in a circular motion to ensure that the powder particles completely cover the surface of the liquid drop and form a new liquid marble.

6. Culture marbles for 6 days at 37 °C in 5% CO_2 incubator. Change medium daily following the procedure described in Sects. 3.4, items 3–5.

7. Days 7–8: Culture marbles in Pancreatic Basal Medium supplemented with 1 μl/ml activin A stock solution and 1 μl/ml retinoic acid stock solution for 2 days at 37 °C in 5% CO_2. Change medium daily following the procedure described in Sects. 3.4, items 3–5

8. Days 9–36: Culture marbles in Pancreatic Basal Medium supplemented with 1% (v/v) Insulin-Transferrin-Selenium (ITS), 2% (v/v) B27, and 0.1% (v/v) Recombinant Human FGF basic (bFGF) stock solution at 37 °C in 5% CO_2 incubator. Change the medium daily for the first 15 days following the procedure described in Sects. 3.4, items 3–5

9. From day 16 onward, refresh medium every other day.

4 Notes

1. It is crucial to use clean PTFE powder because previously used PTFE powder tends to cause aggregation of particles.
2. Prepare 5-aza-CR stock immediately prior to use. 5-aza-CR solutions are very unstable.
3. 2-Mercaptoethanol stock preparation: dilute 3.5 μl of 2-Mercaptoethanol in 5 ml of sterile PBS. Store in dark at +4 °C. Use within 2 weeks.
4. Nucleoside mix stock preparation: dissolve 0.042 g Guanosine, 0.040 g Adenosine, 0.036 g Cytidine, 0.036 g Uridine, and 0.012 g Thymidine in 50 ml of sterile water. Melt at 50 °C to dissolve. Sterilize by filtration and store at +4 °C.
5. Recombinant Human FGF basic (bFGF) stock preparation: add 5 ml of 0.1% Bovine Serum Albumin (BSA) in PBS at 25 μg of (bFGF).
6. Bovine Serum Albumin (BSA) stock preparation: dissolve 250 mg of BSA in 50 ml of water. Sterilize by filtration and store at +4 °C.
7. Human skin biopsies are collected by excision under local anesthesia from an avascular area of the anterior aspect of the forearm and stored in Dulbecco's Phosphate Buffered Saline (PBS) supplemented with 2% antibiotic antimycotic solution.
8. After 6 days, fibroblasts start to grow out of the tissue fragments and begin to form a cell monolayer.
9. It is important to optimize the cell seeding density inside the LM and determine the ideal volume of LM for organoid production, based on the characteristics of the cell used during the experiment.
10. The approximate diameter of the opening pipette tip is slightly less than the marble diameter which creates a friction fit to grip the marble inside the tip.
11. Direct liquid contact disrupts the hydrophobicity of the coated PTFE powders and breaks the marbles.
12. In alternative, liquid marbles can be cultured in a 35-mm Petri dish placed within a bigger Petri dish containing sterile water to prevent evaporation. Ensure to cap both Petri dishes.
13. The fine coating of hydrophobic powders allows to monitor Liquid Marble content using optical-fluorescent microscopy. At the same time, it ensures optimal gas exchange between the interior liquid and the surrounding environment, thanks to the powder particle size of 1 μm.

14. This method makes it possible to scale down experiments and is therefore amenable for higher throughput applications. Furthermore, work in smaller volumes allows to study the effect of paracrine/autocrine signalling of the rich environment established within the microbioreactor. For this reason, it is possible to recover the medium contained in the marbles for further analysis.

Acknowledgement

This work was supported by Carraresi Foundation. Authors are members of the COST Actions CA16119. TALB participates to COST Action CM1406.

References

1. Simian M, Bissell MJ (2017) Organoids: a historical perspective of thinking in three dimensions. J Cell Biol 216(1):31–40. doi:10.1083/jcb.201610056
2. Clevers H (2016) Modeling development and disease with organoids. Cell 165(7):1586–1597. doi:10.1016/j.cell.2016.05.082
3. Tian J, Fu N, Chen XD, Shen W (2013) Respirable liquid marble for the cultivation of microorganisms. Colloids Surf B Biointerfaces 106:187–190. doi:10.1016/j.colsurfb.2013.01.016
4. Arbatan T, Al-Abboodi A, Sarvi F, Chan PP, Shen W (2012) Tumor inside a pearl drop. Adv Healthc Mater 1(4):467–469. doi:10.1002/adhm.201200050
5. Serrano MC, Nardecchia S, Gutierrez MC, Ferrer ML, del Monte F (2015) Mammalian cell cryopreservation by using liquid marbles. ACS Appl Mater Interfaces 7(6):3854–3860. doi:10.1021/acsami.5b00072
6. Arbatan T, Li LZ, Tian JF, Shen W (2012) Liquid marbles as micro-bioreactors for rapid blood typing. Adv Healthc Mater 1(1):80–83. doi:10.1002/adhm.201100016
7. Sarvi F, Jain K, Arbatan T, Verma PJ, Hourigan K, Thompson MC, Shen W, Chan PPY (2015) Cardiogenesis of embryonic stem cells with liquid marble micro-bioreactor. Adv Healthc Mater 4(1). doi:10.1002/adhm.201400138
8. Ledda S, Idda A, Kelly J, Ariu F, Bogliolo L, Bebbere D (2016) A novel technique for in vitro maturation of sheep oocytes in a liquid marble microbioreactor. J Assist Reprod Genet 33(4):513–518. doi:10.1007/s10815-016-0666-8
9. Vadivelu RK, Ooi CH, Yao RQ, Tello Velasquez J, Pastrana E, Diaz-Nido J, Lim F, Ekberg JA, Nguyen NT, St John JA (2015) Generation of three-dimensional multiple spheroid model of olfactory ensheathing cells using floating liquid marbles. Sci Rep 5:15083. doi:10.1038/srep15083
10. Sarvi F, Arbatan T, Chan PPY, Shen W (2013) A novel technique for the formation of embryoid bodies inside liquid marbles. RSC Adv 3(34):14501–14508. doi:10.1039/c3ra40364e
11. Pennarossa G, Maffei S, Campagnol M, Tarantini L, Gandolfi F, Brevini TA (2013) Brief demethylation step allows the conversion of adult human skin fibroblasts into insulin-secreting cells. Proc Natl Acad Sci U S A 110(22):8948–8953. doi:10.1073/pnas.1220637110
12. Brevini TA, Pennarossa G, Rahman MM, Paffoni A, Antonini S, Ragni G, deEguileor M, Tettamanti G, Gandolfi F (2014) Morphological and molecular changes of human granulosa cells exposed to 5-azacytidine and addressed toward muscular differentiation. Stem Cell Rev. doi:10.1007/s12015-014-9521-4
13. Manzoni EF, Pennarossa G, deEguileor M, Tettamanti G, Gandolfi F, Brevini TA (2016) 5-Azacytidine affects TET2 and histone transcription and reshapes morphology of human skin fibroblasts. Sci Rep 6:37017. doi:10.1038/srep37017

Tissue Engineering of 3D Organotypic Microtissues by Acoustic Assembly

Yuqing Zhu, Vahid Serpooshan, Sean Wu, Utkan Demirci, Pu Chen, and Sinan Güven

Abstract

There is a rapidly growing interest in generation of 3D organotypic microtissues with human physiologically relevant structure, function, and cell population in a wide range of applications including drug screening, in vitro physiological/pathological models, and regenerative medicine. Here, we provide a detailed procedure to generate structurally defined 3D organotypic microtissues from cells or cell spheroids using acoustic waves as a biocompatible and scaffold-free tissue engineering tool.

Keywords 3D organotypic microtissues, Acoustic assembly, Acoustic waves, Cardiac scaffold, Cardiomyocyte, Cell assembly, Cell spheroids, In vitro tissue model, Regenerative medicine, Spheroid assembly, Tissue engineering

1 Introduction

In vitro 3D human organotypic models are being increasingly investigated in drug screening and basic biomedical research applications as more physiologically relevant platforms in comparison to the conventional cell culture models [1–3]. The ability to generate 3D organotypic microtissues representing basic cellular structure, such as the hepatic acinus in the liver and the nephron in the kidney, will enable diverse applications in basic and transitional medical research. 3D bioprinting has recently drawn great attention in biofabrication of 3D human tissue/organ mimics due to its unsurpassed flexibility in defining the spatial organization of heterogeneous cell and biomaterial constructs [4, 5]. However, a number of challenges still constrain the clinical applications of bioprinting modalities including (1) long processing time in large-scale biofabrication technologies, (2) low cytocompatibility particularly during high-speed printing, and (3) difficulty to achieve physiologically relevant cell packing density and close cell proximity resembling those in native tissues, due to the limitations in the mixing ratio of cells in the bioink (biomaterial). Herein, we demonstrate a unique tissue engineering approach for generating 3D organotypic microtissues directly from cells or cell spheroids without the need of scaffold/bioink during structure formation. This approach utilizes Faraday wave, a type of surface acoustic wave existing at the air-liquid interface, to drive cells in a fluidic environment and pack them into 3D architecture at the bottom of an assembly chamber [6–8]. The geometry of the generated 3D architecture can be pre-designed by a mathematical-physical model using numerical simulation and can be flexibly tuned by acoustic frequency and amplitude. This acoustic assembly process only takes a few seconds regardless of the number of cells in the chamber, thus representing a more efficient method to generate 3D tissue constructs with high cell packing density, in comparison to 3D bioprinting and other bioengineering approaches [9]. In this chapter, we describe two experimental procedures including (1) assembly of fibroblast spheroids into organotypic microtissues [6] and (2) assembly of human induced pluripotent stem cell derived cardiomyocytes (hiPSC-CMs) into organotypic microtissues with physiologically relevant structure and function (Fig. 1) [8]. We envision that this tissue engineering tool will find broad applications in designing and developing 3D tissue models for drug screening and personalized medicine.

2 Materials

Prepare solutions with Milli-Q or equivalently purified water. Diligently follow all waste disposal regulations when disposing waste materials. All reagents, unless specified, are purely analytical chemicals from Sigma.

2.1 Devices for Acoustic Assembly

1. Arbitrary function generator (33510B, Agilent, CA, USA) (Fig. 2).

2. Audio power amplifier (Dayton Audio DTA-120 Class T Mini Amplifier 60 WPC, USA) (*see* **Note 1**).

3. Vibration exciter (U56001, 3B Scientific, Tucker, GA, USA).

4. Assembly chamber (20 mm × 20 mm × 1.55 mm) (*see* **Note 2**).

5. Metric tilt platform (Edmund Optics, NJ, USA) (*see* **Note 3**).

6. Vibration isolation pad (Grainger, CA, USA) (*see* **Note 4**).

7. Accelerometer (MMA7341L, Freescale Semiconductor, TX, USA).

8. Bubble level (Spirit Level, Hoefer, MA).

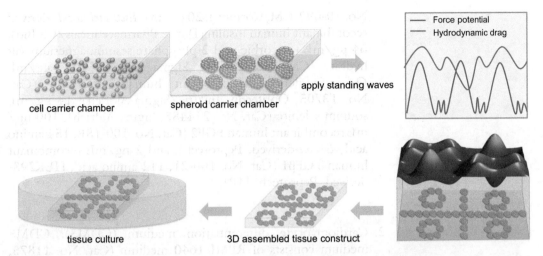

Fig. 1 Illustration of acoustic node assembly of organotypic microtissues. Cells/spheroids are first loaded in the assembly chamber. Then standing waves are applied to assemble cells/spheroids into pre-defined patterns based on node and antinode zones on standing waves in a fluidic environment. Further assembled cells/spheroids are immobilized in fibrin hydrogel and maturated in culture media

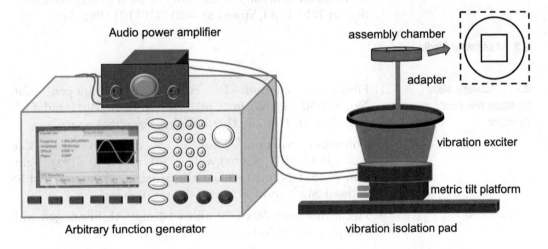

Fig. 2 Schematic of experimental setup for acoustic assembly

2.2 Culture Medium

2.2.1 Mouse Fibroblast Cell Culture Medium

Mouse fibroblast cells culture medium: Dulbecco's Modified Eagle Medium (DMEM)–High Glucose (HG) (Cat. No. 11965092, ThermoFisher Scientific) supplemented with 10% Fetal Bovine Serum (FBS) (Cat. No. F2442, Sigma-Aldrich), and 1% penicillin-streptomycin (Cat. No. 15140122, ThermoFisher Scientific).

2.2.2 hiPSC Culture and Differentiation Media

1. hiPSC culture medium (E8): E8 medium consists of DMEM-F12 (50:50 mixture of DMEM and Ham's F12 medium; Cat.

No. 10-092-CM, Corning), 20 μg/mL *Escherichia coli*-derived recombinant human insulin (Dance Pharmaceuticals/CS Bio), 64 μg/mL L-ascorbic acid 2-phosphate sesquimagnesium salt hydrate (Cat. No. A8960, Sigma-Aldrich), 10.7 μg/mL *O. sativa*-derived recombinant human transferrin (Cat. No. T3705, Optiferrin, Invitria/Sigma-Aldrich), 14 ng/mL sodium selenite (Cat. No. 214485, Sigma-Aldrich), 100 ng/mL recombinant human FGF2 (Cat. No. 100-18B, 154 amino acid, *E. coli*-derived, Peprotech), and 2 ng/mL recombinant human TGFβ1 (Cat. No. 100-21, 112 amino acid, HEK293-derived, Peprotech) [10].

2. Cardiomyocyte differentiation medium (CDM3): CDM3 medium consists of RPMI 1640 medium (Cat. No. 11875, ThermoFisher Scientific), 500 μg/mL *O. sativa*–derived recombinant human albumin (Cat. No. A0237, Sigma-Aldrich, 75 mg/mL stock solution in water for injection (WFI) quality H_2O, stored at −20 °C), and 213 μg/mL L-ascorbic acid 2-phosphate (A8960, Sigma-Aldrich, 64 mg/mL stock solution in WFI H_2O, stored at −20 °C) [10] (Fig. 3).

2.3 Assembly Media

2.3.1 Assembly Media for Mouse Fibroblast Spheroids

1. Fibrinogen solution: Dissolve 20 mg fibrinogen (Cat. No. F8630, Sigma) in 2 mL PBS without Mg^{2+} and Ca^{2+} (Cat. No. 10010023, ThermoFisher Scientific) on ice.

2. Thrombin solution: Dissolve 5 mg thrombin (Cat. No. T4648-1KU, Sigma) with 35 mL dH_2O to storage solution (125 UN/mL); dilute the storage solution with PBS without Mg^{2+} and Ca^{2+} in a 1–50 ratio on ice.

3. Fibrin hydrogel: Mix 500 μL of 10 mg/mL fibrinogen with 150 μL 2.5 UN/mL thrombin on ice (*see* **Note 5**).

2.3.2 Assembly Media for hiPSC-Derived Cardiomyocytes

1. Hank's Balanced Salt Solution with calcium and magnesium HBSS (Cat. No. 55037C, Sigma-Aldrich).

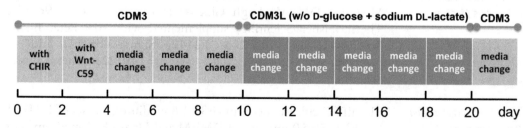

Fig. 3 Schematic of optimized chemically defined cardiac differentiation protocol [10]

2. Fibrinogen solution: Dissolve 10 mg fibrinogen in 2 mL HBSS with calcium and magnesium on ice.

3. Thrombin solution: Dilute the storage solution (125 UN/mL) with HBSS with calcium and magnesium in a 1–50 ratio on ice.

4. Fibrin hydrogel: Mix 550 μL of 5 mg/mL fibrinogen with 150 μL 2.5 UN/mL thrombin on ice.

2.4 Reagents for the Immunohisto chemistry

2.4.1 Solutions

1. Phosphate-buffered saline (PBS), Cat. No. 10010023 Gibco, ThermoFisher Scientific.

2. Live/Dead viability/cytotoxicity kit (L-3224, Molecular Probes, ThermoFisher Scientific).

3. The Live/Dead staining solution: Using Live/Dead viability/cytotoxicity kit mix 20 μL ethidium homodimer and 5 μL Calcein AM in 10 mL PBS (*see* **Note 6**).

4. 4% Paraformaldehyde (Cat. No. 15710, Electron Microscopy Sciences): Dilute 16% paraformaldehyde to 4% with H_2O.

5. Blocking Buffer (PBS/5% normal goat serum (Cat. No. 5425, Cell Signaling Technology)/0.1% Saponin, Cat. No. 47036, Sigma-Aldrich): Add 2.5 mL serum from the same species as the secondary antibody (normal goat serum) to 47.5 mL 1× PBS and mix well. While stirring, add Saponin to final concentration of 0.1% (*see* **Note 7**).

6. Antibody Dilution Buffer (1% BSA/0.1% Saponin): Add 0.4 g BSA to 40 mL 1× PBS and mix well. While stirring, add Saponin to final concentration of 0.1% (*see* **Note 8**).

2.4.2 Antibodies

1. Primary antibodies for mouse fibroblast cells: Mouse Anti-collagen type 1 (Cat. No. ab6308, Abcam), Rabbit Anti-Ki67 (Cat. No. ab15580, Abcam).

2. Secondary antibodies for mouse fibroblast cells: Donkey anti-mouse Alexa Fluor 488 (Cat. No. 715-545-150), and Rhodamine Red™-X (RRX) AffiniPure Donkey Anti-Rabbit IgG (H + L) (Cat. No. 711-295-152, Jackson Immuno Research).

3. Primary antibodies for hiPSC-CMs: mouse anti-sarcomeric α-actinin (Cat. No. A7811, Sigma–Aldrich), mouse anti-α-smooth muscle actin (Cat. No. A2547, Sigma-Aldrich), rabbit anti-vimentin (Cat. No. ab92547, Abcam), rabbit anti-nanog (Cat. No. ab21624, Abcam), rabbit anti-CD31 (Cat. No. ab28364, Abcam), rabbit anti-Connexin 43 (Cat. No. ab11370, Abcam), goat anti-AFP (Cat. No. sc-8108,

Santa Cruz Biotechnology), and mouse anti-NeuN (Cat. No. MAB377, Sigma-Aldrich).

4. Secondary antibodies for iPSC-CMs: Alexa Fluor 488 goat anti-rabbit and anti-mouse, Alexa Fluor 568 goat anti-mouse, Alexa Fluor 546 goat anti-mouse, Alexa Fluor 647 goat anti-rabbit, and Alexa Fluor 594 donkey anti-goat (all Thermo-Fisher Scientific) (*see* **Note 9**).

3 Methods

3.1 Mouse Fibroblast Cells

3.1.1 Spheroid Formation and Assembly

1. Culture NIH 3T3 mouse fibroblast cells in DMEM supplemented with 10% (v/v) FBS, 1% penicillin, and streptomycin.
2. Harvest and plate cells in 60 mm Petri dishes (pure virgin polystyrene) (ES3510, Medsupply Partners, USA) at 1×10^6 cells/dish for 2 days (*see* **Note 10**).
3. Rinse the dish with PBS and collect formed cell spheroids.
4. Centrifuge for 3 min at a speed of 800 rpm.
5. Discard the solution. Resuspend spheroids in 500 μL 10 mg/mL human fibrinogen (Cat. No. F8630, Sigma) prepared in PBS.
6. Add 150 μL of thrombin (Cat. No. T6884, Sigma) at 0.5 IU/mL final concentration and mix homogenously.
7. Transfer the solution to the assembly chamber (20 mm × 20 mm × 1.5 mm) with a 1 mL pipette (*see* **Note 11**).
8. Apply acoustic surface standing waves at 20–200 Hz frequency and 1.2–1.8 g amplitude ranges, resulting in formation of various assembly patterns (*see* **Note 12**).
9. Terminate acoustic surface standing waves at $t = 15$ s.
10. Immobilize (cross-linking of the fibrin) the assembled structures in formed fibrin hydrogel for 20 min at room temperature or 10 min at 37 °C. Room temperature control is critical to sustain consistent cross-linking rate.
11. Recover constructs on cover slip and culture in corresponding cell culture media [DMEM supplemented with 10% FBS, 1% penicillin-streptomycin (Cat. No. 15140-122, ThermoFisher Scientific), and 0.1 TIU/mL aprotinin (Cat. No. A1153, Sigma-Aldrich)] at 5% CO_2 and 37 °C at a humidified atmosphere.

3.1.2 Cell Viability and Proliferation Assays

1. Cut the hydrogel into nine aliquots with a blade and place each hydrogel construct in a 24-well plate (*see* **Note 13**).
2. Add 0.5 mL the Live/Dead staining solution; make sure the construct is submerged by the staining solution.
3. Incubate constructs at 37 °C in humidified incubator for 15 min in Live/Dead staining solution (*see* **Note 14**).
4. Rinse once with PBS.
5. Image with fluorescent microscope immediately.

3.1.3 Immunocytochemistry

1. Fix fibrin hydrogels containing microtissues with 4% paraformaldehyde (*see* **Note 15**).
2. Allow cells to fix for 20 min at room temperature.
3. Aspirate fixative; wash construct with excessive PBS three times for 5 min each.
4. Add 0.3% Triton-X 100 (Cat. No. T9284, Sigma-Aldrich) and block with 1% BSA (Cat. No. A2153, Sigma-Aldrich).
5. Place the plate in the shaker for 60 min.
6. Aspirate blocking solution; apply diluted primary antibody.
7. Stain spheroids overnight at 4 °C for primary antibodies.
8. Rinse three times with PBS for 5 min each in the shaker.
9. Stain 2 h for secondary antibodies at room temperature in the shaker in the dark.
10. Rinse in PBS as in step 8.
11. Stain cytoskeleton with phalloidin Alexa Fluor 647 (Cat. No. A22287, ThermoFisher Scientific) and use DAPI (Cat. No. 62247, ThermoFisher Scientific) as nuclear staining.
12. Keep immunostained constructs at dark and analyze with confocal microscopy as soon as possible (*see* **Note 16**).

3.2 hiPSCs

3.2.1 Culture and Assembly of Cells

1. Isolate the obtained hiPSC clones and culture (using E8 medium) on six-well tissue culture plates (Greiner) coated with 1:200 growth factor-reduced Matrigel (9 μg/cm^2) (Cat. No. 354230, Corning).
2. Medium is changed every other day (48 h). Cells are passaged every 4 days (split at 1:10 or 1:12 ratios) at ~85% confluence using 0.5 mM EDTA. 2 μM Thiazovivin (Cat. No. S1459, Selleck Chemicals) is added for the first 24 h after each passage.
3. Treat hiPSCs with a small molecule inhibitor of GSK3B signaling, 6 μM CHIR99021 (Cat. No. C6556, LC Laboratories), in CDM3 media for 2 days (*see* Fig. 3).

4. Treat cells subsequently with CDM3 media supplemented with a Wnt signaling inhibitor, 2 μM Wnt-C59 (Cat. No. S7037, Selleck Chemicals), for another 2 days.

5. At days 4–8 of differentiation, use CDM3 media without any factors. Change the media every other day (*see* **Note 17**).

6. Culture (starve) cardiomyocytes (CMs) in CDM3 media without glucose, supplemented with 5 mM sodium DL-lactate (Cat. No. 71720, Sigma-Aldrich), for 4 days. This step is necessary to purify the culture from the non-CM cells.

7. Use TrypLE Express (Cat. No. 12605010, ThermoFisher Scientific) for 3–5 min to dissociate CMs into single cells.

8. Count CMs with cell count plate; ensure cells are at a density of 1.0×10^6 cells/mL (or desired cell density).

9. Once dissociated, transfer the hiPSC-CMs cells into a 15 mL centrifuge tube on ice and pipette a few times to break them into single cell suspension.

10. Centrifuge cell suspension at a speed of 1200 rpm for 4 min.

11. Discard the supernatant culture medium and resuspend hiPSC-CMs in fibrin prepolymer solution on ice at a density of 1.7×10^6 cells/mL.

12. Place an 18 mm × 18 mm coverslip at the bottom chamber in advance (*see* **Note 18**).

13. Transfer the solution to the assembly chamber (20 mm × 20 mm × 1.5 mm) with a 1 mL pipette (*see* **Note 19**).

14. Wait until the solution gravitationally sediments onto the glass coverslip at the bottom of the chamber (~1 min). Apply the Faraday waves (10 s for 5 times with intervals of 10 s). Randomly distributed cells at the substrate are driven by the hydrodynamic drag field and patterned into the pre-defined 3D pattern. For the selected circle-square hybrid patterning of hiPSC-CMs, the waveform generator should be excited at 127 Hz and 110 mV (*see* **Note 20**).

15. Terminate the acoustic assembly device.

16. Add an additional 25 μL of thrombin (2.5 UN/mL) to the cast solution to accelerate the cross-linking of fibrinogen and wait for ~10 min.

17. Transfer the assembly chamber to a tissue culture incubator and incubate at 37 °C for another 15 min.

18. Detach cell-encapsulating fibrin scaffold from the assembly chamber using a blade and transfer the scaffold into a well of six-well tissue culture plate.

19. Put the six-well plate into cell incubator and culture for further development and maturation of the 3D tissue (*see* **Note 21**).

3.2.2 Characterization of Developed Tissue

Immunohistochemical Analysis of Patterned Constructs

1. Rinse 3D fibrin scaffolds (encapsulating hiPSC-CMs) three times for 20 min each with DPBS (Cat. No. 14190144, ThermoFisher Scientific).
2. Fix with 4% paraformaldehyde for 60 min on rocker at room temperature.
3. Rinse with DPBS three times for 20 min each.
4. Permeabilize with 0.1% Saponin for 60 min at room temperature with gentle agitation.
5. Block in blocking solution (10% goat serum, 0.1% Saponin, and 1% BSA solution) for 3 h on rocker at room temperature.
6. Drain and stain samples with 1:200 dilution of primary antibodies.
7. Incubate overnight at 4 °C in the blocking solution.
8. Wash with DPBS, four times for 30 min each on the rocker.
9. Incubate for 3 h at room temperature in the dark with 1:200 secondary antibodies in the blocking solution.
10. Wash with DPBS, three times for 30 min each.
11. Stain cell nuclei with NucBlue Fixed Cell Stain (Cat. No. R37606, ThermoFisher Scientific) in DPBS.
12. Perform the confocal imaging (*see* **Note 22**).
13. Count (NucBlue stained) cell nuclei for total cell number and α-actinin + NucBlue stained cells for the CM number.

hiPSC-CM Metabolic Activity Assay

1. Add AlamarBlue reagent (Cat. No. DAL1025, ThermoFisher Scientific) to each well at 10% of the culture volume (CDM3) at different time points in culture.
2. Incubate cellular constructs at 37 °C for 4 h.
3. Use acellular fibrin gels as control, for background reference subtraction.
4. Read the absorbance of 100 mL of medium at 550 and 600 nm using a microplate reader (Cytation 5 Cell Imaging Multi-Mode Reader, BioTek Instruments) and calculate the percentage of reduced AlamarBlue according to the manufacturer's instructions.

Motion Analysis of Beating Cardiomyocytes

1. At (desired) serial time points during culture, use an optical microscope with 10× objective at a speed of 12.5 frames per second.

2. Transform the raw videos to 8 bit gray scale and then analyze using an image velocimetry script in MATLAB (MathWorks) [11] (see **Note 23**).

4 Notes

1. The audio power amplifier is used to amplify signals from an arbitrary function generator to electrically drive vibration exciter.
2. The assembly chamber is fabricated from poly (methyl methacrylate) (PMMA) plates, double-sided adhesive (DSA) (iTapestore, Scotch Plains, NJ, USA) using a laser cutter (VersaLaser, Scottsdale, AZ, USA) [6]. The size and shape of the chamber can be customized according to the experimental requirements.
3. The metric tilt platform is used to adjust the horizontal level of liquid surface in the chamber together with a bubble level, which is placed on the chamber.
4. The vibration isolation pad is installed at the bottom of metric tilt platform to isolate mechanical perturbation.
5. The formulation of the solution should be carried out on ice; if not, the solution would link floc, which would result in Flocculent precipitate appear.
6. Live cells stained with Calcein AM (green) and dead cells stained for EthD-1 (red).
7. Dispense the blocking buffer with 1 mL each centrifuge tube and preserve at 4 °C.
8. When using any primary or fluorochrome-conjugated secondary antibody for the first time, titrate the antibody to determine which dilution allows for the strongest specific signal with the least background for your sample.
9. Using secondary antibodies that have been pre-absorbed with the serum of the other species (the mouse specific IgG antibody pre-absorbed with rat serum and vice versa) to avoid cross-reaction between these highly related animal species.
10. Spheroids are formed after 2 days of incubation. Spheroids sediment and distribute randomly on the substrate.
11. Wait until spheroids have landed onto cover slip (18 × 18 mm) placed at the bottom of the assembly chamber.
12. The force potential, U, that affects CMs with a radius R at the substrate exposed to a hydrodynamic drag field can be described as follows:

$$U = \left[-\frac{4}{3}(\rho_{cell} - \rho_{liq})\omega g R^3 - \frac{24\nu\rho_{liq}R\omega e^{-\left(\frac{\lambda}{2}-H\right)/\delta} \sin\left(\frac{\frac{\lambda}{2}-H}{\delta}\right)}{k \sin h(kH)} \zeta_{sh} \right.$$

$$+ \frac{3\pi^2 \nu \rho_{liq} R h \omega}{kL(\sin h(kH))^2} \left\{ -3 + e^{-2\left(\frac{\lambda}{2}-H\right)/\delta} \right.$$

$$+ 8e^{-\left(\frac{\lambda}{2}-H\right)/\delta} \sin\left(\frac{\frac{\lambda}{2}-H}{\delta}\right) + 2e^{-\left(\frac{\lambda}{2}-H\right)/\delta} \cos\left(\frac{\frac{\lambda}{2}-H}{\delta}\right)$$

$$\left. \left. -2\sqrt{2}\left(\frac{\frac{\lambda}{2}-H}{\delta}\right) e^{-\left(\frac{\lambda}{2}-H\right)/\delta} \cos\left(\frac{\frac{\lambda}{2}-H}{\delta}+\frac{\pi}{4}\right) \right\} \zeta_h \right]$$

Here, ρ_{cell} represents the density of the cell, ρ_{liq} shows the density of liquid, ω is the Faraday wave frequency, g is the gravitational constant, ν is the kinematic viscosity of the liquid, H is the thickness of the liquid, k is the Faraday wavenumber, h is the height of Faraday wave, λ is the Faraday wavelength, δ is the Stokes characteristic length, and ζ_{sh} is the sub-harmonic component of deformation of liquid surface. ζ_h and ζ_{sh} show wave functions and are adjusted by ω [8].

13. Each construct can be cultured in a well of a 24-well culture plate and can be used for different staining assays.
14. The staining time is related to the volume of the hydrogel, generally no more than 20 min.
15. Paraformaldehyde is toxic; use only in fume hood.
16. For long-term storage, store slides flat at 4 °C, protected from light.
17. Contracting cells were noted from day 7.
18. Make sure there is no bubble between the chamber and the coverslip.
19. Add the solution along the edge of the chamber and ensure that the coverslip is completely submerged.
20. Adjust the amplitude from low to high until the pattern appears.
21. Add 1 TIU/mL aprotinin (Cat. No. A1153, Sigma-Aldrich) into the culture media to inhibit cell-induced fibrin degradation.
22. 3D reconstruction and panoramic videos of the z-stacks are generated by the confocal microscope software (ZEN 2012,

Carl Zeiss Microscopy GMBH, MA). Cell density measurements using optical images are conducted by counting the number of cells in each focal plane (in focus).

23. Calculate contractile stress values from video recordings, based on the stiffness and depth of material, and the cell displacement, which is relative to the reference frame at resting states between contractions.

Acknowledgments

U.D. acknowledges that this material is based in part upon work supported by the NSF CAREER Award Number 1150733. S.G. would like to thank the Turkish Scientific Research and Technology Council (TÜBİTAK) award number 115C125, the Turkish Academy of Sciences GEBIP Award, and the Academy of Sciences BAGEP Award. P.C. and Y. Z. would like to acknowledge Wuhan University Program 2042017kf0230. V.S. acknowledges NIH Pathway to Independence Award 1K99HL127295-01A1.

References

1. Guven S et al (2015) Multiscale assembly for tissue engineering and regenerative medicine. Trends Biotechnol 33(5):269–279
2. Esch EW, Bahinski A, Huh D (2015) Organs-on-chips at the frontiers of drug discovery. Nat Rev Drug Discov 14(4):248–260
3. Asghar W et al (2013) In vitro three-dimensional cancer culture models. In: Bae YH, Mrsny RJ, Park K (eds) Cancer targeted drug delivery: an elusive dream. Springer, New York, NY, pp 635–665
4. Arslan-Yildiz A et al (2016) Towards artificial tissue models: past, present, and future of 3D bioprinting. Biofabrication 8(1):014103
5. Serpooshan V, Mahmoudi M, Hu DA, Hu JB, Wu SM (2017) Bioengineering cardiac constructs using 3D printing. J 3D Printing Med 1(2):123–139
6. Chen P, Guven S, Usta OB, Yarmush ML, Demirci U (2015) Biotunable acoustic node assembly of organoids. Adv Healthc Mater 4(13):1937–1943
7. Chen P et al (2014) Microscale assembly directed by liquid-based template. Adv Mater 26(34):5936–5941
8. Serpooshan V et al (2017) Bioacoustic-enabled patterning of human iPSC-derived cardiomyocytes into 3D cardiac tissue. Biomaterials 131:47–57
9. Lee S et al (2017) Contractile force generation by 3D hiPSC-derived cardiac tissues is enhanced by rapid establishment of cellular interconnection in matrix with muscle-mimicking stiffness. Biomaterials 131:111–120
10. Burridge PW et al (2014) Chemically defined generation of human cardiomyocytes. Nat Methods 11(8):855–860
11. Chase LCJDP (2015) Epigenetic regulation of phosphodiesterases 2A and 3A underlies compromised b-adrenergic signaling in an iPSC model of dilated cardiomyopathy. Front Physiol 7

Cell Microencapsulation in Polyethylene Glycol Hydrogel Microspheres Using Electrohydrodynamic Spraying

Mozhdeh Imaninezhad, Era Jain, and Silviya Petrova Zustiak

Abstract

Microencapsulation of cells is beneficial for various biomedical applications, such as tissue regeneration and cell delivery. While a variety of techniques can be used to produce microspheres, electrohydrodynamic spraying (EHS) has shown promising results for the fabrication of cell-laden hydrogel microspheres in a wide range of sizes and in a relatively high-throughput manner. Here we describe an EHS technique for the fabrication of cell-laden polyethylene glycol (PEG) microspheres. We utilize mild hydrogel gelation chemistry and a combination of EHS parameters to allow for cell microencapsulation with high efficiency and viability. We also give examples on the effect of different EHS parameters such as inner diameter of the needle, voltage and flow rate on microsphere size and encapsulated cell viability.

Keywords: Cell microencapsulation, Electrospraying, Hydrogels, Microspheres, Polyethylene glycol (PEG)

1 Introduction

Microencapsulation of cells within hydrogels is gaining momentum in cell delivery and tissue regeneration applications as well as for the development of in vitro tissue models [1–5]. Cells microencapsulated in hydrogel microspheres can be delivered locally *via* an injection and the high surface to volume ratio of the multiple microspheres, as opposed to a single bulk hydrogel, enable efficient transport of nutrients and secretory products [1, 6, 7]. In particular, polyethylene glycol (PEG) is an excellent choice for cell microencapsulation due to its inert and biocompatible nature, versatile gelation chemistries, immuno-protective properties, and exceptional control over the resulting hydrogel physical, mechanical, and biochemical properties [8, 9].

A variety of techniques can be employed to fabricate microspheres of controlled sizes, including emulsification [10], dispersion or precipitation [11], solvent evaporation [11], and microfluidics [12]. Most such techniques involve the use of organic solvents or high speed vortexing to prepare hydrogel microspheres. The cytotoxicity of the solvents or the high shear forces imposed by vortexing, make such techniques incompatible with cell encapsulation [13]. While microfluidic techniques have been successfully used for cell microencapsulation [14], they are relatively low-throughput and require specialized equipment as well as repeated manufacturing of the microfluidic device [15]. Electrohydrodynamic spraying (EHS) has been recently adopted for the fabrication of cell-laden polymeric microspheres [16]. EHS enables a relatively high-throughput fabrication of hydrogel microspheres in a wide range of sizes and is also compatible with cell encapsulation [17]. In EHS a high voltage is applied to a fine jet of flowing polymer solution, breaking it into numerous small droplets, which are then collected in a grounded collector bath. At a certain applied voltage, namely critical applied voltage, V_{cr}, the flowing liquid overcomes the surface tension forces by accumulating charges on the surface of the liquid droplets. At or above V_{cr}, a Taylor cone formation is observed, which is a conical shape liquid meniscus formed at the needle tip [14, 16].

Here, we describe a method for the microencapsulation of cells in PEG hydrogel microspheres produced via EHS. The PEG hydrogels were crosslinked using a Michael addition reaction between an acrylate and a thiol, which is a mild and highly specific gelation chemistry [18]. We have previously demonstrated that by varying critical EHS parameters, namely voltage, flow rate, tip-to-collector distance and inner needle diameter, PEG microspheres in a wide range of sizes could be obtained [19]. From these EHS parameters, only applied voltage ≥15 kV was found to significantly impact cell viability [20]. Overall, our previous work has shown that the described EHS technique results in a wide range of PEG microsphere sizes, namely 70–700 μm, with a low coefficient of variance of 6–23% [19]. We were also able to encapsulate various cell types with high cell viability and an encapsulation efficiency >90% [20]. Thus, the technique described here can be used to successfully produce cell-laden hydrogel microspheres for use in various biomedical applications.

2 Materials

Use analytical grade reagents and de-ionized water for the preparation of all solutions. Follow applicable disposal regulations for waste products and Material Safety Data Sheets available from vendors for reagent storage, handling, and disposal.

2.1 PEG Hydrogel

1. Triethanolamine (TEA) buffer: To prepare 0.3 M TEA (Sigma Aldrich, St. Louis, MO) buffer, add 1 ml of TEA to 20 ml of phosphate buffer saline (PBS; 1×, pH 7.4). Vortex the solution for 30 s and adjust the pH to the desired value (e.g., 7.4) by adding 1 N HCl drop-wise with constant stirring. When pH is

adjusted, add more PBS to reach a final volume of 25 ml. Sterilize the buffer by filtering through a 0.22 μm filter. Store at 4 °C.

2. Four-arm polyethylene glycol-acrylate (4-arm PEGAc; MW 10,000 Da, JenKem Technology, Plano, TX): Store at −20 °C under an inert gas (e.g., argon, nitrogen) in a desiccated container. Before opening, let the container equilibrate to room temperature. Purge with an inert gas after each opening (*see* **Note 1**).

3. PEG-diester-dithiol was synthesized in-house following a previously developed protocol [21] (*see* **Note 2**): Store at −20 °C under an inert gas (e.g., argon, nitrogen) in a desiccated container. Before opening, let the container equilibrate to room temperature. Purge with an inert gas after each opening (*see* **Note 1**).

4. Arg-Gly-Asp-Ser (RGDS)-modified 4-arm PEGAc: To prepare RGDS-4-arm PEGAc [21], first dissolve 8% w/v GRCD-RGDS (Peptide 2.0, Chantilly, VA) in 5% acetic acid in de-ionized water, immediately mix with 11.5% w/v 4-arm PEGAc (in 0.3 M TEA buffer) at a ratio of 1:6.6 and leave to react for 30 min at room temperature. Upon reaction, lyophilize the product overnight (VirTis Sentry 2.0 Lyophilizer), purge with inert gas, and store in a desiccated container at −20 °C until use.

2.2 PEG Microsphere Fabrication

1. Olive oil (local grocery store) (*see* **Note 3**).

2. Programmable syringe pump (Harvard Apparatus 22, Biosurplus, San Diego, CA).

3. High voltage source (Spellman High Voltage Electronics Corporation, Hauppauge, NY). The high voltage source should cover the range of at least 1.5–20 kV applied voltage.

4. Blunt stainless steel syringe needles, such as Hamilton™ metal hub blunt point needles with luer lock, 1.5″ length.

5. Teflon-coated vessel for the olive oil bath for microsphere collection (*see* **Note 4**).

6. Lab jack to support the syringe pump and to adjust the tip-to-collector distance (*see* **Note 5**).

2.3 Cell Culture and Evaluation of Cell Viability

1. Complete cell culture medium: RPMI-1640 medium (Hyclone, Logan, UT) with 10% fetal bovine serum (FBS; Hyclone, Logan, UT) and 1% penicillin/streptomycin (MP Biomedicals, Santa Ana, CA).

2. Trypsin for cell harvesting: harvest cells by a 5 min exposure to 0.1 M trypsin/EDTA (Thermo Fisher Scientific, Waltham, MA).

3. Live/dead staining: Acridine orange (AO; EMD Millipore Corp, Billerica, MA), which is a green membrane permeable

nuclear stain (*see* **Note 6**), can be used to stain all cells and propidium iodide (PI; Fisher Scientific, Hampton, NH), which is a red membrane impermeable nuclear stain, can be used to stain dead cells.

4. U87 human glioblastoma cell line (ATCC, Manassas, VA) (*see* **Note 7**).

5. Humidified incubator with 5% CO_2 and 37 °C for mammalian cell culture.

6. Biosafety tissue culture cabinet with a UV germicidal lamp for conducting cell culture experiments in a sterile environment.

3 Methods

3.1 Preparing Cells for Microencapsulation

All steps should be performed under sterile conditions and in a tissue culture biosafety hood.

1. Culture and maintain U87 cells in complete medium in a humidified incubator at 5% CO_2 and 37 °C. Change medium every 2–3 days.

2. For microencapsulation, harvest the cells with Trypsin/EDTA at 80% confluency. Re-suspend the cells in complete medium at 2×10^6 cells ml^{-1} (*see* **Notes 8** and **9**).

3.2 Cell-Laden Polyethylene Glycol Hydrogel Precursor Solution Preparation

All steps should be performed in a biological safety cabinet unless otherwise specified. Refer to Fig. 1 for a reaction schematic of 4-arm PEGAc with a PEG-diester-dithiol crosslinker. Example calculations are provided for 100 μl solution volumes.

1. Prepare a sterile 20% w/v stock solution of 4-arm PEGAc (*see* **Notes 10** and **11**). For example, to prepare a 100 μl stock solution, weigh 20 mg of 4-arm PEGAc and transfer into a sterile 1.5 ml microfuge tube in a laminar hood. Add approximately 100 μl of sterile 0.3 M TEA buffer of pH 7.4 (*see* **Note 12**). Vortex for 1 min to dissolve completely (*see* **Note 13**).

2. Similarly prepare a sterile 20% w/v stock PEG-diester-dithiol (*see* **Notes 14** and **15**). For example, to prepare a 100 μl stock solution weigh 20 mg of the PEG-diester-dithiol and transfer into a sterile 1.5 ml microfuge tube in a laminar hood. Add approximately 100 μl of sterile 0.3 M TEA buffer of pH 7.4 (*see* **Note 11**). Vortex for 1 min to dissolve completely (*see* **Note 10**).

3. To prepare a 10% w/v hydrogel precursor solution, combine 4-arm PEGAc and PEG-diester-dithiol at a 1:1 M ratio of acrylate to thiol moieties (*see* **Note 16**). For example, to prepare a 100 μl solution, pipette 28.6 μl of 20% w/v 4-arm PEGAc stock solution and 21.4 μl of the 20% w/v PEG-diester-dithiol stock

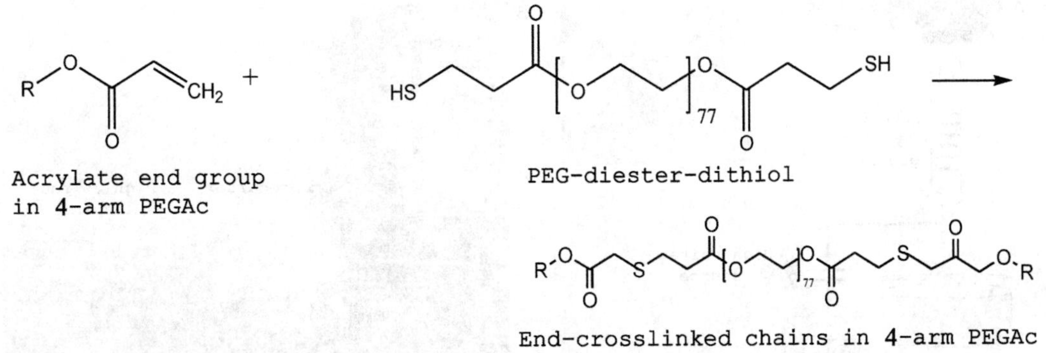

Fig. 1 Chemical structure of the end group of 4-arm polyethylene glycol-acrylate (4-arm PEGAc), PEG-diester-dithiol, and the crosslinked chain ends of the final hydrogel. Image reproduced with permission from Qauyym et al. [20]

solution in a sterile 1.5 ml microfuge tube. Add 50 μl of the cell suspension prepared earlier and mix gently but thoroughly by pipetting up and down (*see* **Note 17**).

4. Load the cell-laden hydrogel precursor solution in a 1 ml sterile syringe (*see* **Note 18**).

3.3 Electrohydrodynamic Spraying Set-Up and Parameters

The electrohydrodynamic spraying set-up (Fig. 2) should be prepared in advance, prior to the cell-laden hydrogel precursor solution preparation. Refer to Table 1 for selecting electrohydrodynamic spraying parameters.

1. Obtain a Teflon-coated collection vessel with ~15 cm inner diameter (*see* **Note 4**). Clean it with 70% ethanol and wipe with a KimWipe®. Place in a tissue culture biosafety hood, turn the UV light on, and leave for ~1 h to sterilize.

2. Sterilize olive oil by autoclaving at 132 °C for 30 min (*see* **Note 19**).

3. Measure ~25 ml of the sterile olive oil using a 50 ml pipette and place the oil in the sterilized Teflon-coated collection vessel (*see* **Note 20**).

4. Mount a syringe pump vertically on a lab jack with an adjustable height, so that the loaded syringe faces toward the collection vessel.

5. Set the height of the lab jack so that the metal needle is at a fixed desired distance from the Teflon-coated collection vessel [tip-to-collector-distance (TTCD)] (*see* **Note 21**).

6. Set the syringe pump to a desired flow rate (*see* **Note 22**).

7. Set the high voltage power supply unit to a desired voltage setting. Applied voltage has the most pronounced effect on microsphere size and cell viability (*see* **Note 23**).

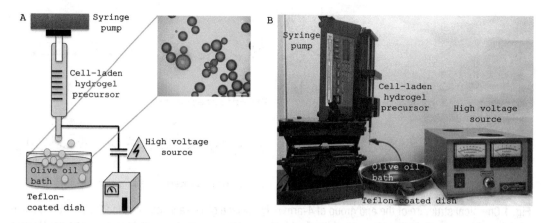

Fig. 2 (**a**) Schematic of the electrohydrodynamic spraying set-up for producing cell-laden PEG hydrogel microspheres. An example phase contrast image of PEG microspheres is shown; scale bar represents 100 μm. (**b**) Digital image of electrohydrodynamic spraying set-up for producing cell-laden PEG hydrogel microspheres. The syringe contains cell-laden PEG hydrogel precursor solution. A pre-set force is applied by the syringe pump to the syringe, resulting in the dispersion of the solution. A blunt needle is connected to a high voltage source. The microspheres are collected into a grounded Teflon-coated collection vessel containing an olive oil bath

Table 1
Effect of different EHS parameters and cell density on microsphere size, polydispersity, and U87 cell viability. Image reproduced with permission from Qauyym et al. [20]

Voltage (kV)	Flow rate (ml/h)	Inner diameter of needle (mm)	Cell density (cell/ml)	Microsphere size (μm)	Microsphere distribution (% CV)	Cell viability (%)
5	2	0.84	10^6	249.5 ± 17.3	6.8 ± 1.3	95.8 ± 1.8
10	2	0.84	10^6	132.2 ± 20.5	15.5 ± 2.1	95.2 ± 2.7
15	2	0.84	10^6	123.9 ± 32.2	23.1 ± 4.2	68.1 ± 3.1
10	0.5	0.84	10^6	133.9 ± 11.4	8.7 ± 2.7	93.5 ± 5.1
10	5	0.84	10^6	172.6 ± 32.9	18.7 ± 6.6	91.4 ± 7.1
10	2	0.26	10^6	141.7 ± 25.0	17.5 ± 2.9	92.7 ± 6.1
10	2	0.13	10^6	77.1 ± 10.6	13.7 ± 4.6	91.2 ± 2.7
10	2	0.84	10^8	142.5 ± 20.1	14.1 ± 3.9	94.4 ± 4.7
10	2	0.84	10^9	168.1 ± 36.4	21.1 ± 8.8	83.3 ± 9.5

8. Position the Teflon-coated collection vessel with olive oil directly below the vertically mounted syringe pump.

9. Position the syringe loaded with the cell-laden hydrogel precursor solution onto the syringe pump.

10. Attach a blunt stainless steel needle of a desired needle gauge to the syringe (*see* **Note 24**).

11. Connect the stainless steel needle to the high voltage power supply unit. Use an Alden wire connector to connect the power supply unit to the needle end (*see* **Note 25**: *potential fire hazard*). To connect the needle to the voltage source, attach an alligator clip to one end of the wire and clip it to the needle (*see* **Note 26**).

3.4 Electrohydrodynamic Spraying for Cell-Laden PEG Microspheres Fabrication

1. Once the set-up is assembled, turn on the syringe pump. Let the gel precursor solution flow through the needle until it reaches the tip (~1 min). When the first hydrogel precursor solution drop appears at the needle tip, turn on the high voltage power supply. This will create an electric field between the needle tip and the portion of the Teflon-coated collection vessel directly below the needle tip, resulting in a Taylor cone (Fig. 3) at the needle tip (if the applied voltage is above the critical applied voltage, V_{cr}) (*see* **Note 27**).

2. Allow the process to run until all of the hydrogel precursor solution is electrosprayed and droplets are collected in the olive oil bath (~10–15 min depending on the flow rate and other EHS parameters).

3. Once the process is complete, allow the obtained droplets to complete gelation to form cell-laden hydrogel microspheres while still in the olive oil bath (~30 min) (*see* **Note 28**).

4. Collect the olive oil containing the suspended microspheres in a conical tube. Centrifuge at 1000 RPM for 5 min and discard the oil supernatant while being careful not to discard the microspheres.

5. Re-suspend the microspheres in 2 ml of sterile cell culture medium (*see* **Note 29**) and transfer to a sterile 5 ml microfuge tube. Centrifuge again at 1000 RPM for 5 min and discard the medium supernatant while being careful not to discard the microspheres. Repeat the cell culture medium washes four times or until oil has been removed (Fig. 4) (*see* **Notes 30** and **31**).

6. After the final wash, culture the cell-laden microspheres in complete medium and use for the desired application.

7. It is recommended to use live/dead staining to monitor cell viability (Fig. 5) [20].

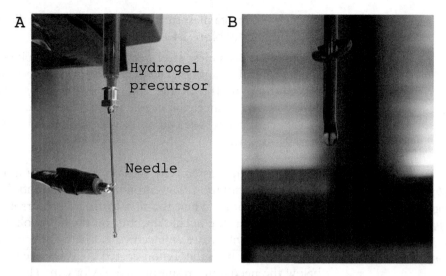

Fig. 3 (**a**) A high magnification image of the needle containing hydrogel precursor solution connected to a high voltage source, resulting in the formation of a Taylor cone. (**b**) An image of the Taylor cone at the tip of the needle. The hydrogel was mixed with 1% v/v of food coloring for better visualization

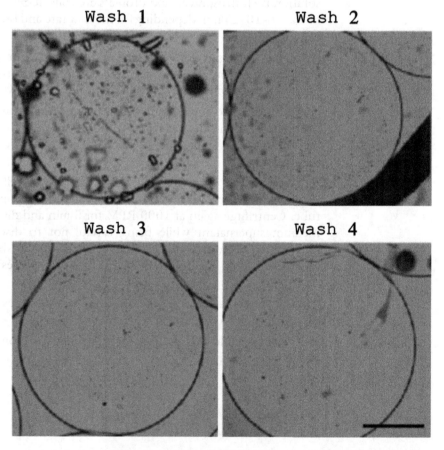

Fig. 4 Phase contrast images of PEG microspheres after four washes in phosphate buffered saline (PBS). Halo or oil droplets are no longer visible after four washes. Scale bar is 100 μm

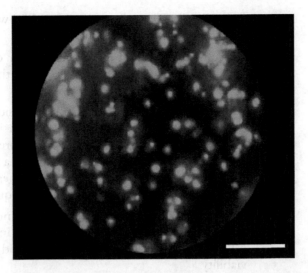

Fig. 5 Live/dead staining of encapsulated cells in a microsphere produced with an applied voltage of 10 kV. All cells were stained *green* (AO), while dead cells were stained *red* (PI). Scale bar is 200 μm

4 Notes

1. PEG polymers are extremely hygroscopic; thus, it is important to carefully follow the storage procedures. Further, thiol moieties in dithiol crosslinkers could easily deprotonate and form disulfide bonds in the presence of water. Hence, similarly stringent storage protocols are recommended for all dithiol crosslinkers.

2. Alternatively, other dithiol crosslinkers such as PEG-dithiol, dithiotreitol (DTT), or cysteine-terminated peptides can be used (refer to the following references for a more complete list of alternative dithiol crosslinkers [21–23]).

3. Olive oil could be substituted with other biocompatible water-immiscible fluids such as mineral oil or other cooking oils. However, change in oil viscosity might require additional system optimization.

4. Alternatively, other collection vessels that have a hydrophobic-coated inside surface can be used or a Parafilm lining can be applied to create such surface. The hydrophobic surface prevents the hydrogel microspheres from sticking to it and distorting their spherical shape.

5. It is highly recommended to use plastic lab jack or cover the lab jack with non-conductive materials.

6. Alternatively, 3,3′-dioctafecyloxacarbocyanine perchlorate (DiOC), which is a membrane-permeable dye, can be used to

stain all cells. The cells can be incubated with 0.02 μg ml^{-1} of DiOC for 24 h prior to experiments.

7. While here we describe U87 cells, other cell types have been tested and shown compatible with the EHS technique described here [20].

8. The cell concentration of the cell suspension will depend upon the final desired concentration of cells in the PEG microspheres. Ideally, 50% of the final gel precursor volume will constitute the cell suspension. Thus, for a final cell concentration of 1×10^6 cells ml^{-1} in the microspheres, a cell suspension concentration of 2×10^6 cells ml^{-1} is required.

9. Prepare the cell suspension only after most of the hydrogel precursor solutions have been prepared in order to minimize cell exposure to ambient conditions and maintain high cell viability.

10. Always use a fresh stock solution of 4-arm PEGAc or store protected from light at 4 °C for no longer than 3 days.

11. The 4-arm PEGAc solution can also be ultraviolet (UV) or filter sterilized. To sterilize by UV, place the uncapped stock solution in a biosafety hood with the UV germicidal light on for 20 min. The solution could also be run through a syringe filter; however, larger solution volumes (e.g., 500 μl) would be needed as a certain volume is retained in the filter during sterilization.

12. Add 85–100 μl so that the final volume of the stock solution is 100 μl. For example, add 85 μl first, measure the volume, and then add TEA buffer as needed to bring the final volume to 100 μl.

13. To avoid waste, centrifuge the stock solution at a low G force for 5 min (e.g., $1350 \times g$ on a Marathon Micro A, 04-977-AR, Fisher Scientific micro-centrifuge) to collect any solution sticking to the wall or the cap of the tube.

14. PEG-diester-dithiol stock solution should be prepared immediately prior to use because thiols quickly deprotonate in water to form disulfide bonds.

15. The PEG-diester-dithiol stock solution can also be filter sterilized but not UV sterilized. To sterilize, the solution could be run through a syringe filter; however, larger solution volumes (e.g., 500 μl) would be needed as a certain volume is retained in the filter during sterilization.

16. Prepare the cell-laden hydrogel precursor solution only after the electrospraying set-up is ready. Once the crosslinker is added, gelation will be initiated and will be complete in ~32 min.

17. The cell suspension could be added to the gel solution at a volume ratio of 1:1. The cell suspension volume should not exceed 50% of the final volume [18]. The final cell concentration in the hydrogels is 1×10^6 cell ml^{-1} unless otherwise stated. The final cell concentration is also an important factor in determining cell viability; we observed significant decrease in encapsulated cell viability at cell concentration of 10^9 cells ml^{-1} [20].

18. Always maintain an air bubble of ~100 µl in the syringe before loading the hydrogel precursor solution into the syringe. This helps to minimize dead volume hold-up in the syringe and pushes out all of the hydrogel precursor solution during electrohydrodynamic spraying.

19. Alternatively, olive oil can be sterilized by filtration through a sterile 0.22 µm filter. However, the high oil viscosity makes such sterilization difficult and slow.

20. The olive oil volume should be enough to cover the bottom of the dish and have a depth of at least 3 mm.

21. TTCD is an important parameter, which determines microsphere size. We have previously shown that TTCD can be varied between 1 and 8.5 in. to modulate microsphere size [19].

22. The syringe pump applies a pre-set force on the plunger to push the solution through the needle at a specific flow rate. Flow rate significantly affects the size and polydispersity of the microspheres, but not cell viability (Table 1) [20]. For example, we have previously observed that at lower flow rates (e.g., 0.5 ml h^{-1}) smaller microspheres with low polydispersity could be obtained, while higher flow rates (e.g., 10 ml h^{-1}) resulted in increased microsphere size and polydispersity [19, 20].

23. Voltage has the most pronounced effect on microsphere size and encapsulated cell viability, where increase in applied voltage typically leads to decrease in microsphere size and cell viability [19, 20]. In particular, a voltage $\geq$15 kV could lead to decreased cell viability possibly due to thermal or electrical influences on the cells [20].

24. Syringe needles with an inner diameter of 0.84 and 0.26 mm produced similar size microspheres, but syringe needles with an inner diameter of 0.13 mm produced significantly smaller microspheres [19, 20].

25. It is imperative to use an appropriate wire that is recommended for high voltage devices in order to prevent a fire.

26. It is best to place the alligator clip on the needle and close to the needle tip for reproducible results.

27. A Taylor cone forms at the needle tip as the solution dispenses through the needle above V_{cr} (Fig. 3) and continuous jetting is observed [24]. At higher voltages, split jetting is observed and at voltages lower than V_{cr}, dripping mode is observed. Microsphere polydispersity is increased when microspheres are made at applied voltages leading to split jetting. Under continuous jetting or dripping mode, a greater uniformity in microsphere size and dispersity is observed.

28. To ascertain gelation, poke the microspheres with a pipette tip very gently to see how they deform in response to the applied pressure.

29. Alternatively, a sterile PBS can be used, but the cell culture medium is preferable to assure high cell viability.

30. Use a 1000 μl pipette tip to change the cell culture medium with fresh medium. Be cautious when removing the supernatant medium as the microspheres may stick to the pipette tip.

31. Alternatively, microspheres can be washed with a cell strainer.

References

1. Murua A, Portero A, Orive G, Hernández RM, de Castro M, Pedraz JL (2008) Cell microencapsulation technology: towards clinical application. J Control Release 132(2):76–83. doi: https://doi.org/10.1016/j.jconrel.2008.08.010

2. Naqvi SM, Vedicherla S, Gansau J, McIntyre T, Doherty M, Buckley CT (2016) Living cell factories-electrosprayed microcapsules and microcarriers for minimally invasive delivery. Adv Mater 28(27):5662–5671

3. Xie J, Wang C-H (2007) Electrospray in the dripping mode for cell microencapsulation. J Colloid Interface Sci 312(2):247–255. doi: https://doi.org/10.1016/j.jcis.2007.04.023

4. Chen MC, Gupta M, Cheung KC (2010) Alginate-based microfluidic system for tumor spheroid formation and anticancer agent screening. Biomed Microdevices 12(4):647–654

5. Zanoni M, Piccinini F, Arienti C, Zamagni A, Santi S, Polico R, Bevilacqua A, Tesei A (2016) 3D tumor spheroid models for in vitro therapeutic screening: a systematic approach to enhance the biological relevance of data obtained. Sci Rep 6:19103

6. Wilson JL, McDevitt TC (2013) Stem cell microencapsulation for phenotypic control, bioprocessing, and transplantation. Biotechnol Bioeng 110(3):667–682

7. Orive G, Gascón AR, Hernández RM, Igartua M, Pedraz JL (2003) Cell microencapsulation technology for biomedical purposes: novel insights and challenges. Trends Pharmacol Sci 24(5):207–210

8. Leach JB, Schmidt CE (2005) Characterization of protein release from photocrosslinkable hyaluronic acid-polyethylene glycol hydrogel tissue engineering scaffolds. Biomaterials 26(2):125–135. doi: https://doi.org/10.1016/j.biomaterials.2004.02.018

9. Mahoney MJ, Anseth KS (2006) Three-dimensional growth and function of neural tissue in degradable polyethylene glycol hydrogels. Biomaterials 27(10):2265–2274

10. Yang Y-Y, Chung T-S, Ping Ng N (2001) Morphology, drug distribution, and in vitro release profiles of biodegradable polymeric microspheres containing protein fabricated by double-emulsion solvent extraction/evaporation method. Biomaterials 22(3):231–241. doi: http://dx.doi.org/10.1016/S0142-9612(00)00178-2

11. Freiberg S, Zhu XX (2004) Polymer microspheres for controlled drug release. Int J Pharm 282(1-2):1–18. doi: https://doi.org/10.1016/j.ijpharm.2004.04.013

12. Liu K, Ding H-J, Liu J, Chen Y, Zhao X-Z (2006) Shape-controlled production of biodegradable calcium alginate gel microparticles using a novel microfluidic device. Langmuir 22(22):9453–9457. doi:10.1021/la061729+

13. Cohen S, Bano MC, Visscher KB, Chow M, Allcock HR, Langer R (1990) Ionically

crosslinkable polyphosphazene: a novel polymer for microencapsulation. J Am Chem Soc 112(21):7832–7833

14. Shintaku H, Kuwabara T, Kawano S, Suzuki T, Kanno I, Kotera H (2007) Micro cell encapsulation and its hydrogel-beads production using microfluidic device. Microsyst Technol 13(8-10):951–958

15. Velve-Casquillas G, Le Berre M, Piel M, Tran PT (2010) Microfluidic tools for cell biological research. Nano Today 5(1):28–47. doi:10.1016/j.nantod.2009.12.001

16. Gasperini L, Maniglio D, Migliaresi C (2013) Microencapsulation of cells in alginate through an electrohydrodynamic process. J Bioact Compat Polym 28(5):413–425

17. Young CJ, Poole-Warren LA, Martens PJ (2012) Combining submerged electrospray and UV photopolymerization for production of synthetic hydrogel microspheres for cell encapsulation. Biotechnol Bioeng 109(6):1561–1570

18. Zustiak SP, Pubill S, Ribeiro A, Leach JB (2013) Hydrolytically degradable poly(ethylene glycol) hydrogel scaffolds as a cell delivery vehicle: characterization of PC12 cell response. Biotechnol Prog 29(5):1255–1264. doi:10.1002/btpr.1761

19. Jain E, Scott KM, Zustiak SP, Sell SA (2015) Fabrication of polyethylene glycol-based hydrogel microspheres through electrospraying. Macromol Mater Eng 300(8):823–835. doi:10.1002/mame.201500058

20. Qayyum AS, Jain E, Kolar G, Kim Y, Sell SA, Zustiak SP (2017) Design of electrohydrodynamic sprayed polyethylene glycol hydrogel microspheres for cell encapsulation. Biofabrication 9(2):025019

21. Zustiak SP, Leach JB (2010) Hydrolytically degradable poly(ethylene glycol) hydrogel scaffolds with tunable degradation and mechanical properties. Biomacromolecules 11(5):1348–1357. doi:10.1021/bm100137q

22. Jain E, Hill L, Canning E, Sell SA, Zustiak SP (2017) Control of gelation, degradation and physical properties of polyethylene glycol hydrogels through the chemical and physical identity of the crosslinker. J Mater Chem B 5(14):2679–2691. doi:10.1039/c6tb03050e

23. Lutolf M, Hubbell J (2003) Synthesis and physicochemical characterization of end-linked poly (ethylene glycol)-co-peptide hydrogels formed by Michael-type addition. Biomacromolecules 4(3):713–722

24. López-Herrera J, Barrero A, Boucard A, Loscertales I, Márquez M (2004) An experimental study of the electrospraying of water in air at atmospheric pressure. J Am Soc Mass Spectrom 15(2):253–259

Gastrointestinal Epithelial Organoid Cultures from Postsurgical Tissues

Soojung Hahn and Jongman Yoo

Abstract

An organoid is a cellular structure three-dimensionally (3D) cultured from self-organizing stem cells in vitro, which has a cell population, architectures, and organ specific functions like the originating organs. Recent advances in the 3D culture of isolated intestinal crypts or gastric glands have enabled the generation of human gastrointestinal epithelial organoids. Gastrointestinal organoids recapitulate the human in vivo physiology because of all the intestinal epithelial cell types that differentiated and proliferated from tissue resident stem cells. Thus far, gastrointestinal organoids have been extensively used for generating gastrointestinal disease models. This protocol describes the method of isolating a gland or crypt using stomach or colon tissue after surgery and establishing them into gastroids or colonoids.

Keywords Colonoid, Gastroid, Gastrointestinal epithelium, Organoid, Postsurgical tissues

1 Introduction

An organoid is a cellular structure three-dimensionally (3D) cultured from self-organizing stem cells in vitro, which has a cell population, architectures, and organ specific functions like originating organs [1]. Recent advances in long-term 3D culture of isolated intestinal crypts or intestinal stem cells have enabled the generation of intestinal epithelial organoids. These produce the human in vivo physiology because they contain all the tissue-specific cell types derived from the epithelial stem cells [2]. These models can provide an in vitro platform for studying pathophysiology, screening drug efficacy, and testing drug toxicity [3]. In addition, the organoid can restore damaged intestinal epithelium when injected into an animal model of inflammatory bowel disease. The organoid could also be used to develop therapeutic agents to regenerate damaged epithelium [4, 5].

The method of culturing the organoid can be classified as starting from pluripotent stem cells or tissue resident stem cells [2]. Current techniques can produce organoids from tissue resident stem cells derived from organs belonging to the digestive system including the gastrointestinal tract, such as the stomach [6], small intestine [7], and colon [8] and accessory organs such as the liver [9] and pancreas [10]. The tissue-resident stem cells that form the organoids originating from digestive organs commonly express leucine-rich repeat-containing G-protein-coupled receptor5 (Lgr5), and the activation of Lgr5 mediating Wnt signaling by the R-spondin protein, which is a ligand for Lgr5, is essential for the self-renewal of stem cells [11, 12].

In 2009, the Hans Clever group developed a method to establish organoids from crypts or a single Lgr5 stem cell isolated from a mouse colon or small intestine [7]. To do this, the crypts were suspended in Matrigel and cultured in a growth medium containing R-spondin, EGF, and Noggin to form an intestinal organoid with crypt-villus structures. Organoids from the gland of the stomach or the ducts of the pancreas and liver were cultured in a similar manner, and organoids from human-derived tissues were also possible [9, 10].

In this protocol, we sequentially describe the method of isolating glands or crypts using the stomach or colon tissue after surgery and establishing them as organoids.

2 Materials

2.1 Human Colonoid Culture

1. Chelating buffer: 2% w/v D-sorbitol (Sigma-Aldrich, cat no. S6021), 1% w/v sucrose (Sigma-Aldrich, cat no. S3089), 1% w/v bovine serum albumin (Sigma-Aldrich, cat no. A2153), and 1× gentamicin/amphotericin B (Life technologies, cat no. R-015-10) in DPBS. Mix and filter through a 0.2 µM syringe filter.

2. Human colon dissociation buffer: 2 mM EDTA (Sigma-Aldrich, cat no. E9884) in chelating buffer.

3. Human colonoid medium: 50% v/v Wnt-3A-conditioned medium, 10% v/v R-Spondin-1-conditioned medium, 100 ng/ml recombinant human Noggin (Peprotech, cat no. 120-10C), 50 ng/ml recombinant murine EGF (Peprotech, cat no. 315-09), 500 nM A-83-01 (Sigma-Aldrich, cat no. SML0788), 10 µM SB202190 (Sigma-Aldrich, cat no. S7067), 10 nM [Leu]15-Gastrin-1 human, SB202190 (Sigma-Aldrich, cat no. G9145), 10 mM nicotinamide (Sigma-Aldrich, cat no. N0636), 1 mM N-acetylcysteine (Sigma-Aldrich, cat no. A9165), 2 mM GlutaMAX (Life technologies, cat no. 35050-061), 10 mM HEPES (Life technologies, cat no. 15630-080), 1× penicillin–streptomycin (Life technologies, cat no. 15140-148), 1× N2 supplement (Life technologies, cat no. 17502-048), 1× B27 supplement (Life technologies, cat no. 17504-044), and 1% w/v BSA in advanced DMEM/F12 medium (Life technologies, cat no.

12634-028). Mix immediately before use and only store for 2 days at 4 °C.

4. Wnt-3a-conditioned medium: L-Wnt-3A cell line (ATCC, cat no. CRL-2647) was used for the Wnt-3A-conditioned medium. L-Wnt-3A-unconditioned medium consists of 2 mM glutamine, 10 mM HEPES, 1× penicillin–streptomycin, 1× N2 supplement, 1× B27 supplement, 1% w/v BSA, and 10% fetal bovine serum (Sigma-Aldrich, cat no. 12003c) in DMEM. To harvest the Wnt-3A conditioned medium, the L-Wnt-3A cell line is maintained for 4 days in L-Wnt-3A-unconditioned medium, collected in L-Wnt-3A-conditioned medium, and then incubated for 3 days in the L-Wnt-3A unconditioned medium and then collected in the L-Wnt-3A-condtioned medium. Filter through a 0.2 µM syringe filter. Long-term storage should be at −20 °C (see Notes 1 and 2).

5. R-spondin-1-conditioned medium: HA-R-spondin 1-Fc 293T cells (Trevigen, cat no. 3710-001-K) are used for the R-spondin-1-conditioned medium. R-spondin-1-unconditioned medium consists of 2 mM glutamine, 10 mM HEPES, 100 U/ml penicillin, 100 g/ml streptomycin, 1× N2 supplement, 1× B27 supplement, 1% BSA, and 10% FBS in advanced DMEM/F12. To harvest the R-spondin-1 conditioned medium, the L-Wnt-3A cell line is maintained for 4 days in R-spondin-1-unconditioned medium, collected in R-spondin-1-conditioned medium, and then incubated for 3 days in L-Wnt-3A-unconditioned medium and then collected in R-spondin-1-condtioned medium. Filter through a 0.2 µM syringe filter. Long-term storage should be at −20 °C (see Notes 3 and 4).

6. Matrigel, GFR, phenol free (Corning, cat no. 356231) (see Note 5).

7. Dulbecco's phosphate buffered saline, Ca^{2+} and $Mg2^+$ free (Life technologies, cat no. 14190-144).

8. Y-27632 (Sigma-Aldrich, cat no. Y0503).

9. CHIR99021 (Sigma-Aldrich, cat no. SML1046).

10. Thiazovivin (Sigma-Aldrich, cat no. SML1045).

11. CO_2 incubator.

12. Curved forceps.

13. Petri dish.

14. 26G, 1 ml syringe (Becton Dickinson, cat no. 320320).

15. Refrigerated centrifuge.

16. 15 and 50 ml conical tubes.

17. Cryogenic vials (Corning, cat no.430488).

2.2 Human Gastroid Culture

Only the following two materials are different from the human colonoid culture.

1. Human stomach dissociation buffer: 10 mM EDTA in chelating buffer.

2. Human gastroid medium: 50% Wnt-3A-conditioned medium, 10% R-spondin-1-conditioned medium, 100 ng/ml noggin, 50 ng/ml EGF, 200 ng/ml FGF10, 1 nM [Leu]15-Gastrin-1, 2 µM A-83-01, 10 mM nicotinamide, 1 mM N-acetylcysteine, 10 mM HEPES, 100 U/ml penicillin, 100 g/ml streptomycin, 1× B27 supplement, and 1% BSA in advanced DMEM/F12 medium. Mix immediately before use and only store for 2 days at 4 °C.

3 Methods

3.1 Human Gastric Glands or Colon Crypts or Isolation from Postsurgical Stomach or Colon Tissue

Tissues must be immersed in ice-cold saline. Preparation for crypt or gland isolation should be ready as soon as possible (see Note 6).

1. Thaw Matrigel matrix in ice and pre-warm the culture plate at 37 °C in a CO_2 incubator.

2. Transfer the tissue in a 50 ml conical tube with 30 ml ice-cold DPBS and gently invert the tube 4–6 times. Stand the conical tube for 1 min and aspirate the supernatant as the tissue sinks. Repeat this step until the supernatant becomes clear enough.

3. Transfer the tissue into a silicone-coated glass Petri dish with cold DPBS. Stretch the tissue and fix with the mucosa side facing up using needles.

4. Gently scratch the mucosa surface with curved forceps to remove the mucous and any debris.

5. Separate the mucosa from the submucosa and connective tissue using curved forceps and scissors. Discard the remaining submucosa and connective tissue and transfer the mucosa to a 50 ml conical tube.

6. Add 30 ml ice-cold chelating buffer. Gently invert the tube 4–6 times. Stand the conical tube for 1 min and aspirate the supernatant as the tissue sinks. Repeat this step 4 times.

7. Add 30 ml ice-cold dissociation buffer. Place on ice and incubate on a rocker for 30 min for a colon or 10 min for a stomach. Aspirate the supernatant as the tissue sinks.

8. Add 30 ml ice-cold chelating buffer and gently invert the tube 4–6 times. Stand the conical tube for 1 min and aspirate the supernatant as the tissue sinks. Repeat this step 4 times.

9. Transfer the mucosa to a Petri dish with ice-cold chelation buffer. Scratch the mucosa side using curved forceps. Check

microscopically that the crypts or glands are isolated and go to the next step when almost all of the crypts or glands of the mucosa are dissociated.

10. Filter the crypts or glands suspension through a 150 μm filter mesh into a 50 ml conical tube and collect the crypt or gland fraction.

11. Centrifuge the crypt or gland fraction at 150 × g for 5 min at 4 °C and aspirate the supernatant.

3.2 Establishment of Human Colonoids or Gastroids

1. Add 5 ml ice-cold chelating buffer and resuspend a crypt or gland pellet by pipetting up and down several times.

2. Transfer 10 μl of the crypt or gland suspension into a Petri dish and count the number of crypts or glands. Calculate the volume for 100 crypts or glands per well of 48-multiwell plates and transfer the required volume to a new 15 ml conical tube.

3. Centrifuge the crypt or gland suspension at 150 × g for 5 min at 4 °C and aspirate the supernatant.

4. Add the required total amount of ice-cold human colonoid or gastroid medium, calculated as 10 μl per well of a 48-multiwell plate.

5. Add an amount of ice-cold Matrigel equivalent to the amount of human colonoid or gastroid medium and mix by pipetting up and down several times with p-200 tips.

6. Make a dome shape by placing 20 μl of the Matrigel mixture on a pre-warmed 48-multiwell plate (see Notes 7 and 8).

7. Solidify the Matrigel mixture at 37 °C in a CO_2 incubator for 10 min.

8. Overlay 300 μl of human colonoid or gastroid medium supplemented with 2.5 μM CHIR99021 and 2.5 μM thiazovivin (see Note 9).

9. Change completely with fresh human colonoid or gastroid medium every 2 days (Figs. 1 and 2).

3.3 Passaging of Human Colonoids or Gastroids

The organoids are passaged at a ratio of 1: 3–4 at 6–8 days after seeding.

1. Remove the old human colonoid or gastroid medium. To depolymerize the Matrigel, add 1 ml of ice-cold DPBS, and incubate for 1 min. Transfer to a new 15 ml conical tube after pipetting up and down several times.

2. To dissociate mechanically, transfer the organoid suspension to 1 ml syringe with a 26G needle and push the plunger to pass the organoid suspension through the needle (see Note 10).

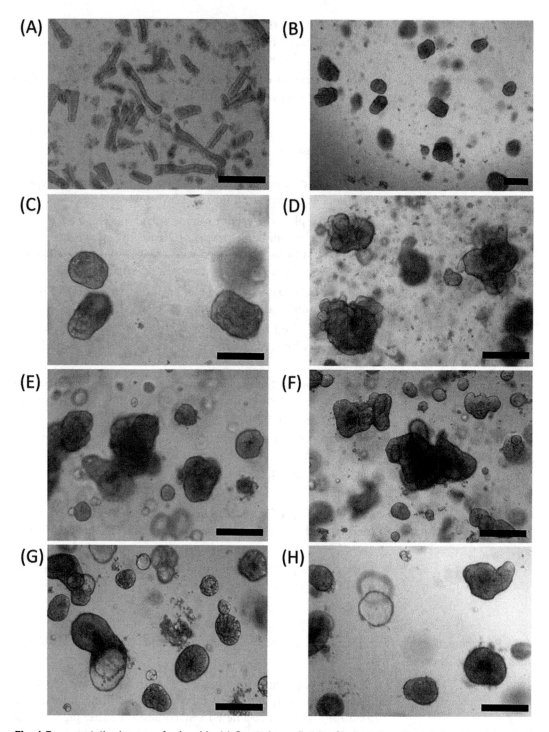

Fig. 1 Representative images of colonoids. (**a**) Crypts immediately after isolation, (**b**, **c**) Colonoids at day 5, (**d**) Passage 1, (**e**) Passage 2, (**f**) Passage 3, (**g**) Passage 5, (**h**) Passage 7

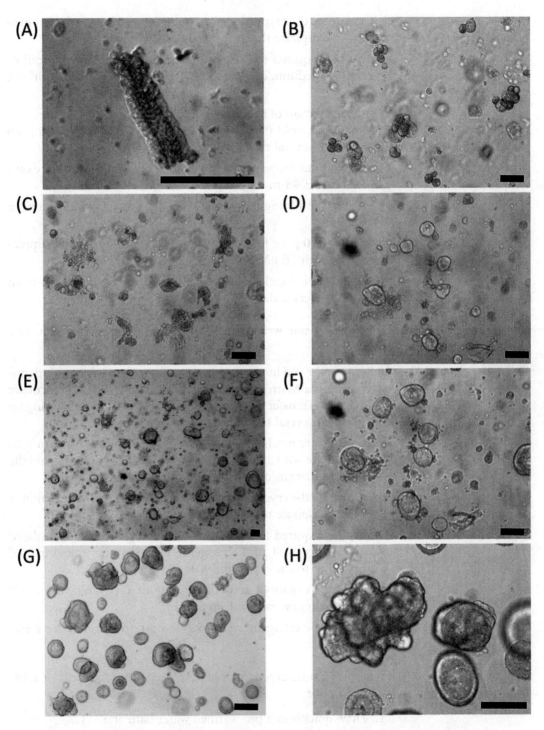

Fig. 2 Representative images of gastroids. (**a**) Glands immediately after isolation, (**b**) closed glands immediately after Matrigel embedding, (**c**) gastroids at day 1, (**d**) day 4, (**e, f**) day 6, (**g, h**) Passage 1

3. Centrifuge the crypt or gland suspension at 150 × g for 5 min at 4 °C and aspirate the supernatant.

4. Add the required total amount of ice-cold human colonoid or gastroid medium, calculated as 10 μl per well of a 48-multiwell plate.

5. Add an amount of ice-cold Matrigel equivalent to the amount of human colonoid or gastroid medium and mix by pipetting up and down several times with p-200 tips.

6. Make a dome shape by placing 20 μl of the Matrigel mixture on a pre-warmed 48-multiwell plate.

7. Solidify the Matrigel mixture at 37 °C in a CO_2 incubator for 10 min.

8. Overlay 300 μl of human colonoid or gastroid medium supplemented with 10 μM Y-27632 (see Note 11).

9. Change completely with fresh human colonoid or gastroid medium every 2 days.

3.4 Freezing of Human Colonoids or Gastroids

Organoids in one well of the 48-multiwell plate is frozen in one cryovial.

1. Remove the old human colonoid or gastroid medium. To depolymerize the Matrigel, add 1 ml of ice-cold DPBS, and incubate for 1 min. Transfer to a new 15 ml conical tube after pipetting up and down several times.

2. To dissociate mechanically, transfer the organoid suspension to 1 ml syringe with a 26G needle and push the plunger to pass the organoid suspension through the needle.

3. Centrifuge the crypt or gland suspension at 150 × g for 5 min at 4 °C and aspirate the supernatant.

4. Add the required total amount of freezing medium, calculated as 1 ml per well of 1 cryogenic vial. Transfer the organoid suspension to the cryogenic vials.

5. Freeze the organoid at a cooling rate of 1 °C/min, using cell freezing containers in a deep freezer.

6. Transfer the cryogenic vials to a liquid nitrogen tank for long-term storage.

3.5 Thawing of Human Colonoids or Gastroids

Organoids in one cryogenic vial is seeded into three wells of a 48-multiwell plate.

1. Thaw quickly in a pre-warmed water bath at 37 °C.

2. Centrifuge the organoid suspension at 150 × g for 5 min at 4 °C and aspirate the supernatant.

3. Add 2 ml of human organoid basal medium.

4. Centrifuge the organoid suspension at $150 \times g$ for 5 min at 4 °C and aspirate the supernatant.
5. Add the required total amount of ice-cold human colonoid or gastroid medium, calculated as 30 μl per well for 1 cryogenic vial.
6. Add an amount of ice-cold Matrigel equivalent to the amount of human colonoid or gastroid medium and mix by pipetting up and down several times with p-200 tips.
7. Make a dome shape by placing 20 μl of a Matrigel mixture on a pre-warmed 48-multiwell plate.
8. Solidify the Matrigel mixture at 37 °C in a CO_2 incubator for 10 min.
9. Overlay 300 μl of human colonoid or gastroid medium supplemented with 10 μM Y-27632.
10. Change completely with fresh human colonoid or gastroid medium every 2 days.

4 Notes

1. L-Wnt-3a stable cell line is selected with 0.4 mg/ml G-418 and split 1:4. When 70–80% confluency is observed, replace the medium in the Wnt3a-unconditioned medium to make Wnt3a-conditioned medium.
2. It is also possible to use human recombinant Wnt-3a (R&D systems, cat no. 5036-WN) instead of Wnt3a-conditioned medium.
3. R-spondin-1 stable cell line is selected with 0.3 mg/ml Zeocin and split 1:4. When 70–80% confluency is observed, replace the medium in the R-spondin-1-unconditioned medium to make R-spondin-1-conditioned medium.
4. It is also possible to use human recombinant R-spondin-1 (PeproTech, cat no. 120-38) instead of R-spondin-1-conditioned medium.
5. Be careful when handling Matrigel. Matrigel rapidly polymerizes at room temperature and changes to gel. The frozen Matrigel is slowly dissolved at 4 °C overnight. Do not freeze-thaw repeatedly.
6. It is also possible to use isotonic buffers such as PBS, DMEM, and RPMI in addition to saline for short-term storage of tissues.
7. Scale-up or -down for crypt seeding is possible. Make a dome of Matrigel mixture with 10 μl for a 96-multiwell plate and 50 μl for a 24-multiwell plate. Because the growth rate of the

8. When seeding organoids with the basement matrix, perform seeding as quickly as possible before the basement matrix and organoids are aggregated in a vial.

9. The addition of CHIR99021 and Thiazovivin greatly increases the rate at which a crypt becomes an organoid [13]. However, they alter the characteristics of the organoids; thus, they should be removed from day 2 after organoid seeding.

10. Because the organoid is sticky, there is loss on the wall of the cultureware. To avoid this, coat the walls with chelating buffer or 1% BSA in PBS.

11. Y-27632 reduces cell apoptosis and increases the efficiency of organoid growth during passaging. However, it alters characteristics of the organoids, so they should be removed from day 2 after passage [7].

(Note: The organoid located in the middle of the dome is slow, it does not make a dome with over 50 μl of the Matrigel mixture — this appears at top of page as continuation.)

Acknowledgments

This work was supported by the Basic Science Research Program through the National Research Foundation of Korea funded by the Ministry of Science, ICT & future Planning, Republic of Korea (NRF-2016M3A9D9945475) and by a grant of the Korea Health Technology R&D Project through the Korea Health Industry Development Institute, funded by the Ministry of Health & Welfare, Republic of Korea (HI16C1634, HI17C2094).

References

1. Willyard C (2015) The boom in mini stomachs, brains, breasts, kidneys and more. Nature 523(7562):520–522. doi:10.1038/523520a
2. Clevers H (2016) Modeling development and disease with organoids. Cell 165(7):1586–1597. doi:10.1016/j.cell.2016.05.082
3. Bredenoord AL, Clevers H, Knoblich JA (2017) Human tissues in a dish: the research and ethical implications of organoid technology. Science 355(6322). doi:10.1126/science.aaf9414
4. Yui S, Nakamura T, Sato T, Nemoto Y, Mizutani T, Zheng X, Ichinose S, Nagaishi T, Okamoto R, Tsuchiya K, Clevers H, Watanabe M (2012) Functional engraftment of colon epithelium expanded in vitro from a single adult Lgr5(+) stem cell. Nat Med 18(4):618–623. doi:10.1038/nm.2695
5. Fordham RP, Yui S, Hannan NR, Soendergaard C, Madgwick A, Schweiger PJ, Nielsen OH, Vallier L, Pedersen RA, Nakamura T, Watanabe M, Jensen KB (2013) Transplantation of expanded fetal intestinal progenitors contributes to colon regeneration after injury. Cell Stem Cell 13(6):734–744. doi:10.1016/j.stem.2013.09.015
6. Stange DE, Koo BK, Huch M, Sibbel G, Basak O, Lyubimova A, Kujala P, Bartfeld S, Koster J, Geahlen JH, Peters PJ, van Es JH, van de Wetering M, Mills JC, Clevers H (2013) Differentiated troy+ chief cells act as reserve stem cells to generate all lineages of the stomach epithelium. Cell 155(2):357–368. doi:10.1016/j.cell.2013.09.008
7. Sato T, Vries RG, Snippert HJ, van de Wetering M, Barker N, Stange DE, van Es JH, Abo A, Kujala P, Peters PJ, Clevers H (2009) Single Lgr5 stem cells build crypt-villus structures

in vitro without a mesenchymal niche. Nature 459(7244):262–265. doi:10.1038/nature07935

8. Jung P, Sato T, Merlos-Suarez A, Barriga FM, Iglesias M, Rossell D, Auer H, Gallardo M, Blasco MA, Sancho E, Clevers H, Batlle E (2011) Isolation and in vitro expansion of human colonic stem cells. Nat Med 17 (10):1225–1227. doi:10.1038/nm.2470

9. Huch M, Dorrell C, Boj SF, van Es JH, Li VS, van de Wetering M, Sato T, Hamer K, Sasaki N, Finegold MJ, Haft A, Vries RG, Grompe M, Clevers H (2013) In vitro expansion of single Lgr5+ liver stem cells induced by Wnt-driven regeneration. Nature 494(7436):247–250. doi:10.1038/nature11826

10. Huch M, Bonfanti P, Boj SF, Sato T, Loomans CJ, van de Wetering M, Sojoodi M, Li VS, Schuijers J, Gracanin A, Ringnalda F, Begthel H, Hamer K, Mulder J, van Es JH, de Koning E, Vries RG, Heimberg H, Clevers H (2013) Unlimited in vitro expansion of adult bi-potent pancreas progenitors through the Lgr5/R-spondin axis. EMBO J 32(20):2708–2721. doi:10.1038/emboj.2013.204

11. Koo BK, Clevers H (2014) Stem cells marked by the R-spondin receptor LGR5. Gastroenterology 147(2):289–302. doi:10.1053/j.gastro.2014.05.007

12. de Lau W, Barker N, Low TY, Koo BK, Li VS, Teunissen H, Kujala P, Haegebarth A, Peters PJ, van de Wetering M, Stange DE, van Es JE, Guardavaccaro D, Schasfoort RB, Mohri Y, Nishimori K, Mohammed S, Heck AJ, Clevers H (2011) Lgr5 homologues associate with Wnt receptors and mediate R-spondin signalling. Nature 476(7360):293–297. doi:10.1038/nature10337

13. Wang F, Scoville D, He XC, Mahe MM, Box A, Perry JM, Smith NR, Lei NY, Davies PS, Fuller MK, Haug JS, McClain M, Gracz AD, Ding S, Stelzner M, Dunn JC, Magness ST, Wong MH, Martin MG, Helmrath M, Li L (2013) Isolation and characterization of intestinal stem cells based on surface marker combinations and colony-formation assay. Gastroenterology 145 (2):383–395. e381-321. doi:10.1053/j.gastro.2013.04.050

Drug Sensitivity Assays of Human Cancer Organoid Cultures

Hayley E. Francies, Andrew Barthorpe, Anne McLaren-Douglas, William J. Barendt, and Mathew J. Garnett

Abstract

Drug sensitivity testing utilizing preclinical disease models such as cancer cell lines is an important and widely used tool for drug development. Importantly, when combined with molecular data such as gene copy number variation or somatic coding mutations, associations between drug sensitivity and molecular data can be used to develop markers to guide patient therapies. The use of organoids as a preclinical cancer model has become possible following recent work demonstrating that organoid cultures can be derived from patient tumors with a high rate of success. A genetic analysis of colon cancer organoids found that these models encompassed the majority of the somatic variants present within the tumor from which it was derived, and capture much of the genetic diversity of colon cancer observed in patients. Importantly, the systematic sensitivity testing of organoid cultures to anticancer drugs identified clinical gene–drug interactions, suggestive of their potential as preclinical models for testing anticancer drug sensitivity. In this chapter, we describe how to perform medium/high-throughput drug sensitivity screens using 3D organoid cell cultures.

Keywords: Drug screening, Organoids, Cancer, Cell lines, Cancer models, Drugs, Targeted therapy, Preclinical

1 Introduction

The research community currently has access to approximately 1000 experimentally tractable, genetically unique cancer cell lines. These cell lines have been instrumental in the advances made in cancer research and are an important tool during drug development, in part due to their ability to model patient responses to therapies. For example, cell lines were used for the preclinical development of the small-molecule EGFR inhibitor gefitinib, which is approved for the treatment of *EGFR*-mutated non-small-cell lung cancer [1, 2]. Drug sensitivity screening of cancer cell lines is routinely performed and large-scale datasets are publically available [3–6].

Despite their utility, there are shortcomings associated with traditional 2D cell lines which include (but are not limited to) their failure to reflect tissue/tumor architecture, a poor success

---

The original version of this chapter was revised. A correction to this chapter can be found at DOI 10.1007/7651_2018_138

rate of derivation, and a lack of associated patient pathological and clinical data. In addition, they likely reflect a subset of aggressive tumors emendable to growth in 2D culture, and fail to encompass the histological and molecular diversity of cancer [7–9]. These deficiencies hamper our ability to identify and successfully translate novel treatment strategies for patients into the clinic.

New cell line derivation technologies such as organoids [10, 11] and conditional reprogramming [12] have provided an opportunity to improve the set of available cancer models. Organoids are a long-term culture system which allow cells, both normal and diseased, to be grown in specialized 3D culture conditions that maintain the stem cell population, as well as much of the cell-type composition and tissue architecture found in vivo [10, 11, 13–16]. Cancer organoid cultures recapitulate features of the original tissue/tumor and can model clinically relevant drug responses [13, 14, 16]. Thus, it should be possible to more effectively develop new therapeutic modalities by performing drug sensitivity assays in a larger and more representative collection of cancer organoid cultures. In the future, it may be possible to individualize patient care through prospective modeling of drug sensitivity in patient-derived cancer organoids.

The utilization of organoid cultures derived from matched healthy tissue such as the liver could be useful in identifying drug toxicity. Additionally, organoids have been generated from diseased tissues other than cancer including cystic fibrosis and liver disorders such as α1-antitrypsin deficiency and Alagille syndrome, and shown to mimic disease pathology [15].

Here we describe a protocol for medium/high-throughput drug screening of human organoid cultures in 384-well plates. This is based on our experience of performing drug sensitivity screens using colorectal and esophageal cancer organoids. We provide details on assay design, protocols for the manipulation of organoid cultures, and parameters for evaluating assay quality.

2 Materials

2.1 Preparation of Organoid Media and BME-2

Prepare the following solutions.

1. 7.5 mg/mL BME-2 (Cultrex® Reduced Growth Factor Basement Membrane Extract, Type 2, PathClear® (Amsbio)): For each 384-well plate approximately 3 mL of 7.5 mg/mL BME-2 is required. An additional 10 mL should be added to the total volume to account for dispensing dead volume. BME-2 (~15 mg/mL) is diluted 1:1 using organoid media to ~7.5 mg/mL (*see* **Note 1**). Determine the number of 384-well plates required for your drug screen and ensure sufficient volumes of BME-2 are chilled overnight on ice (*see* **Note 2**).

2. Organoid media: Prepare organoid media as described for esophagus [11], pancreas [13], and colon [11, 14]. We have

Table 1
Media components for colon and esophageal organoids

Organoid media component	Colon	Esophagus
Advanced DMEM F12 (1× HEPES, 1× Glutamax)	+	+
WNT3A cond. media		50 %
R-Spondin 1 cond. media	20 %	20 %
Noggin cond. media	10 %	10 %
Nicotinamide	10 mM	10 mM
N-acetyl cysteine	1 mM	1 mM
B27-supplement	1×	1×
Recombinant human EGF	50 ng/mL	50 ng/mL
A83-01	500 nM	500 nM
SB202190	10 μM	10 μM
Y-27632	10 μM	10 μM
FGF-10		100 ng/mL
Gastrin		10 nM

The WNT3A and noggin conditioned media (cond. media) producing cell lines were provided by the laboratory of Hans Clevers. ATCC have an alternative WNT3A producing line, L Wnt-3A (ATCC® CRL-2647™). The R-spondin 1 producing cell line was provided by laboratory of Calvin Kuo. An alternative cell line is available from AMSBIO, Cultrex R-spondin1 (RSPO1) Cells

provided details in Table 1 of the media used for colorectal and esophageal cancer organoid drug screening (*see* **Note 3**). The media can be stored for up to 4 weeks at 4 °C until required. Approximately 15 mL of organoid/media suspension is required per 384-well plate. An additional 10 mL should be added to the total volume to account for dispensing dead volume.

3. Drug screening media: Organoid media in point 2 is supplemented with 2 % BME-2 (~0.3 mg/mL). For each 100 mL, add 96 mL of media and 4 mL of 15 mg/mL BME-2.

2.2 Drug Screening Assay

1. 384-well flat clear bottom, opaque walls, tissue culture treated polystyrene plates (assay plate).
2. 384-well polypropylene Echo qualified plates (stock plate).
3. 384 low dead volume (LDV) Echo qualified cyclic olefin copolymer plate (source plate).
4. Microscope slide.
5. Phase contrast microscope.

6. Cell scraper(s).
7. Centrifuge.
8. 15 mL conical centrifuge tube(s).
9. FluidX XRD-384 reagent dispenser, or alternative reagent dispenser with the ability to accurately dispense between 8 and 40 μL volume.
10. Echo 555 acoustic dispenser (Labcyte), or similar liquid handling platform.
11. CellTiter-Glo® 2.0 Luminescent Cell Viability Assay (Promega).
12. Molecular Devices Paradigm plate reader fitted with a luminescence cartridge, or another luminescence plate reader.
13. Drugs to be included in the assay, including positive control drugs.

3 Methods

Figure 1 provides an overview of the organoid drug screening process. We utilize liquid-handling robotics for our drug screening experiments to improve accuracy and reproducibility, and increase throughput. In this protocol we indicate the use of specific

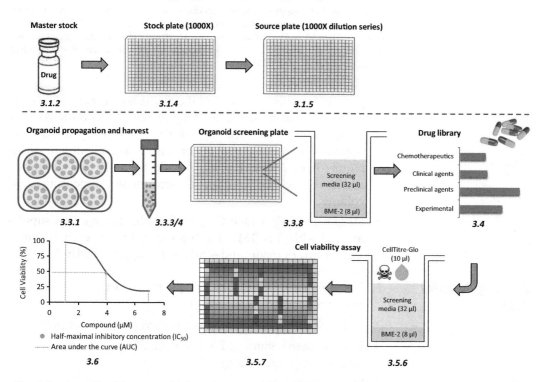

Fig. 1 A schematic of the organoid drug screen workflow. Drug preparation is shown in the *upper panel* and the screening workflow in the *lower panel*, together with associated protocol sections (*italics*)

liquid-handling robotics but, with only minor modifications to these protocols, other equipment could be utilized. Similarly, with some modifications, these protocols could be performed using manual pipetting. Perform all procedures at room temperature unless otherwise specified.

3.1 Preparation of Drug Stocks and Drug Plates

1. Design the layout of your drug plate (*see* **Note 4** and Fig. 2).
2. Drugs are reconstituted in 100 % DMSO and stored frozen at −80 °C or ideally in StoragePods® (Roylan Developments) kept at room temperature, providing a moisture-free, low-oxygen environment and protection from UV damage (*see* **Note 5**).
3. Master stocks of drugs are reconstituted at 1000–10,000× of the final desired maximum screening concentration.

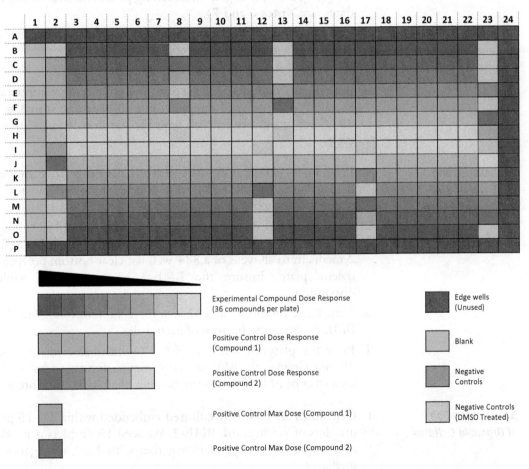

Fig. 2 An illustration of an optimized screening plate layout. Positive and negative control wells are distributed across the plate and drug concentration curves are arranged vertically from the exterior to the interior of the plate (also see Note 4). The edge wells are not used due to edge effects

4. Following the reconstitution of the drug(s) in DMSO, the master stock drugs are cherry-picked to a 384-well polypropylene plate using a liquid-handling robot Biomek FX (Beckman Coulter). This plate is called the stock plate. If the master stock is at 1000× of the final screening concentration, a direct transfer to the stock plate can be performed. If at greater than 1000× concentration, a dilution in DMSO is performed during the transfer to the stock plate, resulting in all drugs on the stock plate being at 1000×.

5. Using the stock plate we generate a source plate (384-well low dead volume (LDV) plate). The source plate contains all the drugs to be screened along with the dilution series of the drugs. Half-log dilutions are performed in DMSO and all wells are 1000× of the final desired concentration, typically containing 10 μL per well. A single source plate can be used for drugging multiple organoid screening plates and over multiple dates (*see* **Notes 6** and **7**).

6. The generation of the source plate is performed using an Echo 555 acoustic dispenser.

7. For storage, seal plates using adhesive plate foil seals.

8. Both the stock and source drug plates can be stored in compound StoragePods® and used multiple times. If plates are stored at -80 °C we recommended you limit the number of freeze–thaw cycles.

3.2 Dispensing the Layer of BME-2 to 384-well Plates

1. Program a FluidX XRD-384 reagent dispenser to dispense 8 μL. Flush the XRD-384 with ethanol, followed by sterile water and lastly organoid media.

2. Prime the XRD-384 with 7.5 mg/mL BME-2 until all tubing is loaded. Once primed, dispense 8 μL of 7.5 mg/mL BME-2 media in to all wells of a 384-well flat clear bottom polypropylene plate. Ensure the BME-2 is kept chilled while dispensing.

3. Centrifuge the plates at 182 × g for 1 minute to ensure the BME-2 covers the bottom of each well.

4. Place the plates in a 37 °C, 5 % CO_2 incubator for at least 20 min allowing the BME-2 to polymerize, forming a polymerized layer of BME-2 across the bottom of all wells (*see* **Note 8**).

3.3 Preparation of Organoid Cultures

1. Organoids are routinely cultured embedded within 10–15 μL droplets of 7.5 mg/mL BME-2. We seed 15 droplets per well of a 6-well cell culture plate together with 2 mL of organoid media.

2. To harvest organoids from a 6-well plate, use a cell scraper to disrupt the 7.5 mg/mL BME-2 droplets containing organoids from the bottom of the well.

3. Using a 1 mL pipette, mix the disrupted organoids along with the media to initiate release of the organoids from the BME-2. This should physically disrupt the organoids but avoid breaking them down to single cells, the goal being to plate formed organoids, not single cells.

4. Transfer the organoid suspension to a 15 mL centrifuge tube and centrifuge at 364 × g for 2 min.

5. Aspirate the supernatant from the 15 mL falcon tube and resuspend the organoid pellet in 5 mL of drug screening media.

6. As organoids are too large to be counted by an automated cell counter or a hemocytometer, dispense three 10 µL droplets of the organoid suspension on to a microscope slide and count the number of organoids in each 10 µL droplet (*see* **Note 9**). Multiply the average number of organoids per droplet by 100 to calculate the number of organoids per mL (*see* **Note 10**).

7. A suspension of 15,000–20,000 organoids per mL is required for drug screening. To dispense two 384-well plates requires 35 mL of organoids prepared in drug screening media at 15,000 organoids per mL. Approximately three confluent 6-well plates are sufficient to generate the 525,000 organoids necessary.

8. Program the XRD-384 to dispense 32 µL. Flush the XRD-384 as previously described in Section 3.2 with ethanol, sterile water and organoid media. Ensure the XRD-384 is primed, then dispense 32 µL of drug screening media into all wells of column 1 (media only, no organoids).

9. Once completed, dispense 32 µL of the organoid suspension in to all wells with the exception of column 1 (*see* **Note 11**). This gives a total well volume of 40 µL.

10. Once completed, place the plates in a 37 °C, 5 % CO_2 incubator until the following day allowing the organoids to settle (*see* **Note 12**).

3.4 Drugging of Screening Plates

1. Following overnight incubation, organoid plates are manually inspected using a phase-contrast microscope to check cell health prior to the addition of drugs.

2. Using an Echo 555 acoustic dispenser, 40 nL of compound previously prepared at 1000× are dispensed from a source plate (described in Section 3.1) into the cell plate, to a final concentration of 1× compound and 0.1 % DMSO.

3. Return plates to a 37 °C, 5 % CO_2 incubator.

3.5 Cell Viability Assay

1. Organoid plates can be incubated with compound/drug for up to 6 days before ending the assay. After more than 6 days, the plates are liable to be effected by evaporation and a reduction in data quality may be observed.

2. Following incubation with drug, the plates are manually checked under the microscope to ensure the positive control drug(s) have effectively killed organoids, and organoids present in the negative control wells are healthy.

3. To quantitate the effect of drugs on cell viability, CellTiter-Glo® 2.0 Luminescent Cell Viability Assay (Promega) can be used. CellTiter-Glo® 2.0 determines the number of viable cells in a well based on the quantification of ATP present, an indicator of metabolically viable cells. Addition of CellTiter-Glo® 2.0 leads to cell lysis and generation of a luminescent signal proportional to the amount of ATP present, and thus indirectly measures cell number.

4. For each 384-well plate approximately 4 mL of CellTiter-Glo® 2.0 is required, add 10 mL to the total volume required to account for dead volume of the XRD-384.

5. Prior to the addition of CellTiter-Glo® 2.0, remove plates from the incubator and allow them to equilibrate to room temperature for 10 min.

6. Program the XRD-384 to dispense 10 μL. Following the flushing of the machine as described in Section 3.2, prime the machine with CellTiter-Glo®2.0 and dispense 10 μL to all wells of the 384-well plate(s) (*see* **Note 13**).

7. Incubate the plates at room temperature for 20 min before reading plate luminescence using a Paradigm plate reader (Molecular Devices) (*see* **Note 14**).

3.6 Data Quality Checks and Downstream Analysis

1. Quality control metrics are calculated for each plate prior to further analysis:

    (a) Coefficient of variation (CV): The CV of the negative control wells determines the variation in the luminescence of the negative control wells on each plate. It is calculated by dividing the standard deviation of the luminescence of the negative controls wells by their mean luminescence. For an organoid drug screen, CV values of less than 0.22 are acceptable.

    (b) Z-factor: The Z-factor (also known as Z-prime) is commonly used in high-throughput screens and measures the assay dynamic range and data variation using both the positive and negative controls [17]. The calculation determines if the effect observed between the positive and negative controls is large enough to make comparisons with drugs of unknown effect. The Z-factor can be calculated using each of the positive control drugs present on the drug plate (*see* **Note 4d**). Organoid plates with a Z-factor of 0.4 or greater are generally of good quality. This threshold is slightly lower than we would typically use for a cancer cell line (Z-factor >0.5) due to the increased technical complexity of screening 3D organoid cultures.

(c) DMSO effect: The ratio between the negative control wells containing DMSO and those without DMSO is calculated. This is to ensure that the concentration of DMSO present in the drugged wells (0.1 % DMSO) does not have a significant effect on cell viability. A ratio of between 0.8 and 1.2 is required.

2. Following completion of the above quality metrics, the data can be normalized or the raw intensity data can be used for curve-fitting.

(a) Data normalization.
Normalization is completed using the following calculation:

(Raw intensity signal − Mean of positive control)/
(Mean of negative control − Mean positive control)

For drugs where the concentrations selected have generated a dose–response curve, measurements such as the IC_{50} (half-maximal inhibitory concentrations) and AUC (area under the curve) can be calculated to assess and compare sensitivity.

(b) Curve-fitting.
Commercial software packages such as GraphPad Prism and Microsoft Office Excel can also be used to analyze the data generated from the drug screen assay. Curve-fitting algorithms for modeling drug response are also available [18].

4 Notes

1. Basement membranes are specialized extracellular matrixes in tissues that form an interface between a number of cell types and the adjacent stroma. The basement membrane supports cells and also plays a role in tissue organization. BME-2 is a soluble basement membrane purified from Engelbreth-Holm-Swarm (EHS) tumor. The major components of BME-2 are laminin, collagen IV, entactin, and heparin sulfate proteoglycan.

2. Always keep BME-2 on ice, even when diluted with media, it will begin to polymerize at temperatures above 10 °C and will rapidly polymerize between 22 °C and 35 °C.

3. The Y-27632 compound is not required to be added to the media to prevent anoikis, as the organoids are dispensed fully formed and not as single cells for drug screening.

4. When designing a plate layout to assess the drug sensitivity of multiple drugs, adhering to the following guidelines will increase the likelihood of generating usable, quality data, as well as providing sufficient data to calculate quality control metrics (discussed in Section 3.6).

   Guidelines and decisions to be made:

   (a) Avoid using the outside wells of a 384 well-plate. These wells are liable to edge effects and thus the data generated in these wells is often unreliable.

   (b) Distribute the negative control wells across the 384 well-plate. This allows you to detect areas of the plate that may have been affected by edge effects, thermal gradients, or other artifacts such as dispensing errors, and which may affect downstream interpretation of results. Dedicate approximately 25 wells as negative control wells.

   (c) Decide on the number of drug concentrations per dose-response. Using a greater number of concentrations per dose-response generally results in a more accurate measurement of drug sensitivity. However, there is a compromise between the number of concentrations assayed and the number of drugs which can be screened on a plate. We find that a 7-point dose–response curve using a half-log fold dilution, covering a 1000-fold concentration range, allows for the generation of an informative dose–response curve with sufficient data from which the IC_{50} and AUC can be calculated.

   (d) Distribute the positive control wells across the 384 well-plate. Dedicate at least six wells as positive control wells. If possible include more than one positive control drug because it is our experience that some cancer organoid cultures are resistant to one or more of the positive control drugs. It is also good practice to include a dose-response of a positive control drug to ensure a dose–response effect is observed. We use MG-132 at 4 μM and staurosporine at 2 μM as positive controls. If an organoid culture is insensitive to the positive control drug(s), the Z-factor can be calculated using the blank wells.

5. If drugs are stored frozen, avoid more than 3–4 freeze–thaw cycles as drug activity can be affected from multiple cycles. Drugs can be kept long-term in StoragePods® (Roylan Developments).

6. Drug dilutions are prepared in a source plate at a concentration of 1000× of the desired final screening concentration, rather

than directly adding drugs from the stock to the screening plate. This is to ensure the volumes dispensed are accurate and to maintain a final DMSO concentration of 0.1 %. For example, the highest drug concentration at 1000× requires transfer of 40 nL to a well of a screening plate containing 40 μL total volume. However, less than 1 nL of this stock would be required at lower drug concentrations. Because such low volumes cannot be accurately dispensed using our robotics, we generate the dose–response curves in a source plate but using larger volumes (typically 10 μL). The source plate can then be used for drugging multiple screening plates.

7. Drug concentrations are selected to fully inhibit the target of interest based on prior knowledge and reading of the literature. We generally avoid exceeding concentrations of 10 μM due to the increased likelihood of off-target drug activity at higher concentrations.

8. The plates containing a layer of 7.5 mg/mL of BME-2 can be stored in a 37 °C, 5 % CO_2 incubator for a number of hours but we do not recommend storing them overnight as the consistency of the BME-2 can be affected.

9. The organoid suspension should be as homogeneous in size as possible to allow for even dispensing of organoids in to screening plates. During the counting of the organoids, if it appears that the organoid suspension is not homogenous in size, the suspension can be filtered using a 20 μM cell strainer prior to recounting.

10. Cultures are plated as formed multicellular organoids. During the harvesting procedure, some organoids may break into smaller clumps of cells and these will reform into organoids following the overnight incubation prior to drugs being added. As there are multiple cells per organoid, we are unable to give a precise cell number per well but our work has found that 15,000–20,000 organoids per mL generates sufficient signal to measure drug response.

11. Keep the organoid suspension moving when dispensing to prevent the organoids falling to the bottom of the collection tube, leading to uneven dispensing.

12. For the drug screen, the organoids are dispensed on top of a layer of 7.5 mg/mL BME-2 rather than embedded in the BME-2. The end-point of the assay requires the cells to be lysed in order to measure cell viability and we have found improved cell lysis when the organoids are not embedded in BME-2, which subsequently improved data quality and consistency. We have not observed any effects on organoid growth or morphology when overlaying rather than embedding cells.

13. A 1:1 dilution of CellTiter-Glo® 2.0 is recommended by the manufacturer. We use a 4:1 dilution (4 media:1 CellTiter-Glo® 2.0) and have found no improvement in data quality when using the recommended volume.

14. An increased incubation time is used when assaying organoids with CellTiter-Glo® 2.0 as we have found improved cell lysis and signal.

Acknowledgments

We would like to thank the Wellcome Trust Sanger Institute cell line drug screening team for their help in developing these protocols and Stacey Price for critical reading of this manuscript. This work was funded with Awards from the Wellcome Trust (102696), Stand Up To Cancer (SU2C-AACR-DT1213), The Dutch Cancer Society (H1/2014-6919), and Cancer Research UK (C44943/A22536).

References

1. Lynch TJ, Bell DW, Sordella R et al (2004) Activating mutations in the epidermal growth factor receptor underlying responsiveness of non-small-cell lung cancer to gefitinib. N Engl J Med 350:2129–2139
2. Paez JG, Jänne PA, Lee JC et al (2004) EGFR mutations in lung cancer: correlation with clinical response to gefitinib therapy. Science 304:1497–1500
3. Garnett MJ, Edelman EJ, Heidorn SJ et al (2012) Systematic identification of genomic markers of drug sensitivity in cancer cells. Nature 483:570–575
4. Barretina J, Caponigro G, Stransky N et al (2012) The cancer cell line encyclopedia enables predictive modelling of anticancer drug sensitivity. Nature 483:603–607
5. Basu A, Bodycombe NE, Cheah JH et al (2013) An interactive resource to identify cancer genetic and lineage dependencies targeted by small molecules. Cell 154:1151–1161
6. Seashore-Ludlow B, Rees MG, Cheah JH et al (2015) Harnessing connectivity in a large-scale small-molecule sensitivity dataset. Cancer Discov 5:1210–1223
7. Holliday DL, Speirs V (2011) Choosing the right cell line for breast cancer research. Breast Cancer Res 13:215
8. Wistuba II, Behrens C, Milchgrub S et al (1998) Comparison of features of human breast cancer cell lines and their corresponding tumors. Clin Cancer Res 4:2931–2938
9. Burdall SE, Hanby AM, Lansdown MR et al (2003) Breast cancer cell lines: friend or foe? Breast Cancer Res 5:89–95
10. Sato T, Vries RG, Snippert HJ et al (2009) Single Lgr5 stem cells build crypt-villus structures in vitro without a mesenchymal niche. Nature 459:262–265
11. Sato T, Stange DE, Ferrante M et al (2011) Long-term expansion of epithelial organoids from human colon, adenoma, adenocarcinoma, and Barrett's epithelium. Gastroenterology 141:1762–1772
12. Liu X, Ory V, Chapman S (2012) ROCK inhibitor and feeder cells induce the conditional reprogramming of epithelial cells. Am J Pathol 180:599–607
13. Boj SF, Hwang CI, Baker LA et al (2015) Organoid models of human and mouse ductal pancreatic cancer. Cell 160:324–338
14. van de Wetering M, Francies HE, Francis JM et al (2015) Prospective derivation of a living organoid biobank of colorectal cancer patients. Cell 161:933–945
15. Huch M, Gehart H, van Boxtel R et al (2015) Long-term culture of genome-stable bipotent stem cells from adult human liver. Cell 160:299–312

16. Gao D, Vela I, Sbiner A et al (2014) Organoid cultures derived from patients with advanced prostate cancer. Cell 159:176–187
17. Zhang JH, Chung TD, Oldenburg KR (1999) A simple statistical parameter for use in evaluation and validation of high throughput screening assays. J Biomol Screen 4:67–73
18. Vis DJ, Bombardelli L, Lightfoot H et al (2016) Multilevel models improve precision and speed of IC50 estimates. Pharmacogenomics 7:691–700

Open Access This chapter is licensed under the terms of the Creative Commons Attribution 4.0 International License (http://creativecommons.org/licenses/by/4.0/), which permits use, sharing, adaptation, distribution and reproduction in any medium or format, as long as you give appropriate credit to the original author(s) and the source, provide a link to the Creative Commons license and indicate if changes were made.

The images or other third party material in this chapter are included in the chapter's Creative Commons license, unless indicated otherwise in a credit line to the material. If material is not included in the chapter's Creative Commons license and your intended use is not permitted by statutory regulation or exceeds the permitted use, you will need to obtain permission directly from the copyright holder.

Erratum to: Drug Sensitivity Assays of Human Cancer Organoid Cultures

Hayley E. Francies, Andrew Barthorpe, Anne McLaren-Douglas, William J. Barendt, and Mathew J. Garnett

Erratum to:
Methods in Molecular Biology
https://doi.org/10.1007/7651_2016_10

The protocol *Drug Sensitivity Assays of Human Cancer Organoid Cultures* has now been made available open access under a CC BY 4.0 license.

The updated online version of this protocol can be found at
https://doi.org/10.1007/7651_2016_10

Correction to: The Three-Dimensional Culture of Epithelial Organoids Derived from Embryonic Chicken Intestine

Malgorzata Pierzchalska, Malgorzata Panek, Malgorzata Czyrnek, and Maja Grabacka

Correction to:
Methods in Molecular Biology
https://doi.org/10.1007/7651_2016_15

There are two corrections for this chapter:

In Figure 4 Section A, the upper right corner should read "3d", whereas it was incorrectly printed as "4d."

Also, the caption of Figure 5 should read: "The cultures were photographed on the first, second and fifth day of culture (1d, 2d, 5d) with differential interference contrast optics." rather than the incorrect "The cultures were photographed on second, third, and fifth day of culture (2d, 3d, 5d) with differential interference contrast optics".

These issues have now been corrected.

The updated online version of this protocol can be found at
https://doi.org/10.1007/7651_2016_15

INDEX

A

AB/TL wild type (WT) 257
Acetylcholine (ACh), *see* Non-neuronal ACh
Acoustic assembly
    bioprinting modalities 301
    3D organotypic microtissues
        cardiomyocytes 309–310
        CDM3 304
        cell spheroids 302
        cell viability and proliferation assays 307
        experimental setup 302–303
        fibroblast cells culture medium 303
        fibroblast spheroids 304, 306
        HBSS 304–305
        hiPSCs 303–304, 307–309
        immunocytochemistry 307
        immunohistochemistry 305–306
        metabolic activity assay 309
        patterned constructs 309
        structure and function 302–303
Adaptive immune system 34
Adult intestinal organoids
    Cre-recombinase enzyme 131
    differentiated liver hepatocytes 131
    DNA sequence 124
    Lgr5 stem cells 124
    liver ductal tissue 130–131
    organoid culture and cre-recombinase
        induction 125–126
    3D culture 123, 124
    tissue culture
        and antral stomach gland isolation 128–130
        and intestinal epithelial crypt isolation 126–128
    tissue dissection 125
    2D culture 123–124
Adult stem cells (aSC) cells 23, 291
Advanced DMEM/F12 base medium (ADF) 125
Alagille syndrome 340
α1-antitrypsin deficiency 340
Amniotic fluid stem cells (AFSCs) 102
Anterior foregut endoderm (AFE) 56
    components 63, 64, 80
    DE medium 79–80
    FOXA2 and NKX2.1 80
    IMDM basal medium 63
    Matrigel 80
Asymmetric cell division (ACD) 273–274
5-Azacytidine (5-aza-CR) treatment 293, 295–297

B

β-napthoflavone (BNF) 126
Blebbistatin 114
Bone marrow-derived lymphocyte progenitors 34
Bovine serum albumin (BSA) 159, 294
Bürker's chamber 141

C

Cancer models, *see* Organoids
Cancer stem cell (CSC) hypothesis 24
Cardiomyocyte differentiation medium
    (CDM3) 304
Cardiomyocytes (CMs) 308
Cell suspensions 105
    dissociation and reaggregation 102
    human–mouse chimeric organoids in vitro
        construction and culture 102–103, 106–109
        implantation of 104, 109–110
    mouse organoids in vitro
        construction and culture 105–106, 108
        immunofluorescence analysis 103–104, 108
    organoids in vivo 102, 104, 111
    rudimental nephron-like structures 101
CellTiter-Glo® 2.0 346
Cerebral organoids 1
Cerebral tissue growth 3, 9
Chemically defined medium supplemented with polyvinyl
    alcohol (CDM PVA) 160–161
Cholesterol 199
Chylomicron 196
Cilia 52
Clustered regularly interspaced short palindromic repeats
    (CRISPR)/Cas9 gene 205
Coefficient of variation (CV) 346
Collagenase–dispase enzyme digestion 125
Colonic organoids
    commercial product 172
    immunohistochemistry analysis 178–179
    isolation of 173–177
    "mini-gut" model 171
    R-Spondin conditioned medium generation 173
    single cell harvesting 179
    Wnt conditioned medium generation 172–173
Colonoids
    chelating buffer 328
    crypt/gland suspension 331–333
    freezing of 334
    human colon dissociation buffer 328

Colonoids (*cont.*)
    human colonoid medium 328–329
    passaging of 331, 334
    R-spondin-1-conditioned medium 329
    thawing of 334–335
    Wnt-3a-conditioned medium 329
Conventional 2D neural cell culture 13
Cortical spheroids 1
Crypt–villus organoids 145, 146, 149–150, 153
Curve-fitting 347
Cysteine-terminated peptides 321

D

Data normalization 347
Definitive endoderm (DE) 56
Dimethyl sulfoxide (DMSO) 159, 347
3,3′-Dioctafecyloxacarbocyanine perchlorate
    (DiOC) 321–322
Disease pathogenesis 13
Dithiotreitol (DTT) 321
Drug screening 353
    cell viability assay 345–346
    overview of 342–343
    phase-contrast microscope 345
    preparation of 344–345
    quality control metrics 346–347
    stocks and plates 343–344
    384-well plates 341–342, 344
    XRD-384 344
Drug Sensitivity Assays of Human Cancer Organoid Cultures
Dulbecco's Modified Eagle Medium
    (DMEM) 66–67, 137, 293
Dulbecco's phosphate-buffered saline (DPBS) 159

E

Electrohydrodynamic spraying (EHS)
    live/dead staining 319, 321
    parameters 317–319
    phase contrast images 319–320
    Taylor cone 319–320
Embryoid bodies (EBs) 1
    feeder-dependent hPSC culture 3, 5
    feeder-independent hPSC culture 3, 5–6
Embryonic stem cells (ESCs) 23
Embryonic zebrafish forebrain 274–275
    antibody complex 274
    antibody injection 276
    antibody uptake assay 276–278
    fluorescent confocal imaging 274
    in vivo dynamics 274
    preparation of embryos 275

    radial glia 273–274, 278
    stem cells 273–274
Engelbreth-Holm-Swarm (EHS) tumor 347
Enteroids
    advantages 196–197
    animals 197
    crypt culture 199–200
    culture buffers 197–198
    fatty acid treatment 198–199
    gut immune system 195
    ISCs 196, 199–200
    lipid micelle treatment 199
    luminal face 202
    maturation and passaging 201–202
    villus and crypt 196
Epidermal growth factor (EGF) 114
Epithelial organoids 355
    adult chicken organoid cultures 142, 143
    chicken serum and PGE$_2$ influence 141, 142
    cultureware and equipment 137
    Matrigel 136
    methods 137–140
    "mini-guts" model 136
    reagents and media 136–137
    types 135

F

Feeder-dependent hPSC culture 2
    components 57–59
    embryoid bodies generation 3
    and passaging 4
    reagents 57–58
Feeder-independent hPSC culture 2
    embryoid bodies generation 3
    Essential 8™ Medium 60–61
    Matrigel™ (*see* Matrigel™)
    MTeSR™ Medium 59–60
    and passaging 4–5
    Vitronectin (*see* Vitronectin)
Fetal Bovine Serum (FBS) 293
Fetal development 13
Fetal neocortical development 1
Forebrain organoids 1

G

Gastric organoids, *see* Adult intestinal organoids
Gastroids
    crypt/gland isolation 330–331
    crypt/gland suspension 331–333
    culture medium 330
    freezing of 334
    passaging of 331, 334
    thawing of 334–335

Gastrointestinal viral infections, *see* Human intestinal enteroids (HIEs)
Germ layer differentiation 3, 6
GFAP:GFP transgenic line 257
Glial cell-derived neurotrophic factor (GDNF) 102
Goblet cells 49
Growth factor cocktails 23, 114

H

Hank's Balanced Salt Solution (HBSS) 304–305
Host–bacteria interactions
    intestinal organoids
        bacterial colonization 252–253
        isolation 250–251
        microscope observation 252
        passage of 252
    isolated crypts 249
    reagents 250
    Salmonella
    stem cell 249
Human airway basal stem cells
    environmental barrier 43
    maintenance of 48–49
    multipotent differentiation capacity 43
    pseudostratified ciliated and mucosecretory
        epithelium 43
Human airway epithelial cell culture 45–46
Human airway epithelial cell isolation 44–45
    from biopsies 47–48
    from brushings 47
Human brain organoids 13
    cast GelMA hydrogel preparation 14–15, 17–18
    clinical compliance 13–14
    high-throughput production 20
    neural differentiation medium 21
Human cancer stem cells
Human-derived fundic gastric organoids (hFGOs)
    CRISPR/Cas9 gene 205
    gastric cancer tissue 209–210
    growth medium 206–207
    incubation media 206
    isolation 208–209
    limitation 206
    nucleofection 207, 209
    orthotopic transplantation
        materials 207
        methods 210–212
    storage and washing buffer 206
    three dimensional organotypic cancer
        models 205–206
    tumor-derived gastric organoids 207
Human embryonic stem cells (hES cells) 56
Human induced pluripotent stem cell-derived
    cardiomyocytes (hiPSC-CMs) 305–306

Human induced pluripotent stem cells (hIPSCs)
    assembly of cells 307–309
    CDM-PVA culture system 162
    culture medium 303–304,
        307–309
    definitive endoderm 56
    feeder-dependent culture protocol
        components 57–59
        reagents 57–58
    feeder-free culture protocol 59
        Essential 8™ Medium 56, 60–61
        Matrigel™ (*see* Matrigel™)
        MTeSR™ Medium 56, 59–60
        Vitronectin (*see* Vitronectin)
    growth factors 159
    hES cells 56
    human lung epithelial cells 56
    iHOs 157
        differentiation 158, 162–164
        intestinal epithelium 158, 159
        microinjection 158, 160,
        165–166
    lentiviral vectors protocol
        colonies 87–88
        components 67
        cryopreservation medium 67
        DMEM medium 66–67
        Dox induction medium 67
        equipment and supplies 65
        fibroblasts 81–82
        HEK293T 82–84
        infection efficiency 86
        polybrene 68
        re-plate infected cells 86–87
        reprogramming 85–86
        required reagents 65–66
        skin biopsy 81
        virus concentration 84–85
        virus titration 85
        VPA 67
    lung progenitors
        anterior foregut endoderm differentiation
            (*see* Anterior foregut endoderm (AFE))
        definitive endoderm differentiation 62–63, 79
        equipment and supplies 61–62
        reagents 62
    passaging
        maintenance 164–165
        and organoid differentiation 161
        and stem cell growth 160–161
    reprogramming factors 55–56
    secretory and adsorptive IECs 158, 159
        (*see also* Induced pluripotent stem cells (iPSC))
Human intestinal enteroids (HIEs)

Human intestinal enteroids (HIEs) (*cont.*)
   HRV
      Differentiation Media 237–238
      disease reduction 230
      efficacy of 230
      flow cytometry analysis 236, 238–239
      infection and analysis 230–231, 233
   HuNoV.. 240
      diarrhea-related deaths...................... 229
      differentiated monolayer HIEs 234
      epithelial lining 236, 244
      flow cytometry analysis 241–242
      immunofluorescent staining 242–243
      infections 230–231
      in vitro cultivation system................. 230
      96-well plates.....................................240–241
   maintenance and differentiation
      composition of 232–233
      flow cytometry analysis 235–237
      frozen stocks from liquid nitrogen 234–235
      passage 235
   transwells and 96-well plates 233, 239–240
Human norovirus (HuNoV).........................240
   diarrhea-related deaths.......................... 229
   differentiated monolayer HIEs 234
   epithelial lining 236, 244
   flow cytometry analysis 241–242
   immunofluorescent staining 242–243
   infections 230–231
   in vitro cultivation system...................... 230
   96-well plates..240–241
Human rotavirus (HRV)
   Differentiation Media 237–238
   disease reduction 230
   efficacy of 230
   flow cytometry analysis 236, 238–239
   infection and analysis 230–231, 233
4-Hydroxytamoxifen (4OHT) 126

I

Immunofluorescence staining of
      tracheospheres 46, 49
Induced pluripotent stem cells (iPSC)................ 13, 291
   culture and passaging............................ 14, 16–17
   neural induction 15
   neural progenitor cells 15, 19–20
   surface seeding 18–19
Insulin-Transferrin-Selenium (ITS) 294
Intestinal epithelial cells (IECs) 157, 158, 165, 249
Intestinal human organoids (iHOs)............................ 157
   differentiation.........................158, 162–164
   intestinal epithelium............................ 158, 159
   microinjection 158, 160, 165–166

Intestinal organoids
   bacterial colonization 252–253
   culture 116
   hIPSCs (*see* Human induced pluripotent
      stem cells)
   imaging and immunofluorescence 115
   intestinal crypt isolation......................... 114
   isolation 250–251
   Lgr5+ stem cell colony formation............. 116–118
   Lgr5+ stem cell sorting............................ 114
   microbes–epithelium interactions
      (*see* Microbes–epithelium interactions)
   microscope observation 252
   passage of... 252
   self-renewal, proliferation, differentiation, and
      apoptosis 118–119
   stem cell culture 114–115
   TA cells 113, 114
Intestinal stem cells (ISCs) 196
   confocal microscopic observation 216, 220–222
   culture and passage 218–220
   digestive and absorptive functions 223
   isolation 216–218
   MetaXpress analysis software 223, 224
   regenerative medicine 215
   three-dimensional imaging 215
In vitro primary culture 171–172
In vitro 3D models of human brain................ 1
In vivo brain development............................. 2
Iscove's Modified Dulbecco's Medium (IMDM)........ 160

L

Leucine-rich-repeat-containing G-protein-coupled
      receptor 5 (Lgr5)............................ 124
Leukemia Inhibitory Factor (LIF) 294
Lgr5+ stem cells
   colony formation 116–118
   sorting.. 114
Lineage-restricted cells 23
Lingual epithelial cells (LECs)
   Bmi1-rainbow mice................................ 95, 98
   concentric cell arrangements 94
   Cre-mediated recombination 95
   culture medium 96
   fetal tongues 93–94
   Matrigel 97
   4-NQO 98
   round-shaped organoids 94
   rugged-and round-shaped organoids 94
   separation of 96–97
   stratum corneum 95, 98
   3D growth 94
Liquid Marbles (LM)....................................... 292

Listeria monocytogens
Liver organoids, *see* Adult intestinal organoids
LoxP sites 124

M

Matrigel™
    aliquots 60
    coated plates 72–73
    equipment and supplies 59, 60
    feeding method 73–74
    freezing 75–76
    passaging 74–75
    reagents 59
    thawing 73
MetaXpress analysis software 223, 224
Microbe-associated molecular patterns
    (MAMPs) 185
Microbes–epithelium interactions
    bacterial products 184
    crypt extraction 184–186
    FACS 185, 188–189
    internal bacteria delivery 185, 187–188
    Listeria monocytogens
        MAMPs/dead bacteria 185–187
        stimulation 184
        transformation 183–184
MicroRNA-expressing lentivirus 24
2-Monooleoylglycerol (2-MG) 199
Mouse embryonic feeder (MEF) medium 58
    components 58
    freezing 71–72
    preparation 68
    standard passaging 70–71
    stem cell culture medium 69–70
    thawing 69
Mouse metanephric mesenchyme (MM) 101
Mouse thymus decellularization 35–36
Muscarinic ACh receptors (mAChRs) 146

N

Neural organoids 1
Neural progenitor cells generation 15
Neural stem cells 9
Neurodevelopmental and neurological disorders 1
Neuroepithelial tissue expansion 3, 6–8
Niche-independent crypt expansion 171
4-Nitroquinoline 1-oxide (4-NQO) 98
Noggin 114
Nonmuscle myosin II (NMII) 114
Non-neuronal ACh
    biological assay 154
    cholinergic system 145
    crypt isolation and culture 147–150

crypt–villus organoid culture 149–150
crypt–villus structure 145, 146
fluorescent immunohistochemistry 149
lineages of 146
PCR 147
pharmacological assay 153
quantitative RT-PCR 151
RT-PCR 151
self-renewing stem cells 146
separation of 146
whole-mount immunohistochemistry 151–152
Notch signaling, *see* Embryonic zebrafish forebrain
N-Phenylthiourea (PTU) 257

O

Oleic acid (OA) 199
Oncogenic transformation, *see* Human-derived
    fundic gastric organoids (hFGOs)
Organoids 196, 353
    BME-2 340–341
    conditional reprogramming 340
    drug screening
        cell viability assay 345–346
        overview of 342–343
        phase-contrast microscope 345
        preparation of 344–345
        quality control metrics 346–347
        stocks and plates 343–344
        384-well plates 341–342, 344
        XRD-384 344
    of human cancer stem cells 29
    in vitro propagation, CSCs 24
    murine small intestinal epithelium 23
    pancreatic (*see* Pancreatic organoids)
    passage of organoids 26, 28
    plate preparation and seeding 25–27
    preclinical development 339
    seeding of sorted cells 26–28
    2D cell lines 339–340
    Wnt3A-conditioned medium 28

P

Pancreatic organoids
    5-aza-CR treatment 293, 295–297
    differentiation 293–294, 297
    epigenetic modifier 293
    fibroblast
        culture 293–295
        isolation 293–295
    microbioreactor preparation 293
    morphological changes 292–293
    PTFE 292
Pattern recognition receptor (PRR) 184

Index

PEG-dithiol 321
Phosphatidylcholine (PC) 199
Pluripotent embryonic stem (ES) cells 291
Pluripotent stem cells (PSCs) technology 1
Polyethylene glycol (PEG)
    cell culture medium 315–316
    cell-laden precursor solution preparation 316–317
    EHS
        live/dead staining 319, 321
        parameters 317–319
        phase contrast images 319–320
        Taylor cone 319–320
    four-arm polyethylene glycol-acrylate 315
    microencapsulation 316
    microsphere fabrication 315
    PEG-diester-dithiol 315
    RGDS-4-arm PEGAc 315
    TEA buffer 314–315
Polytetrafluoroethylene (PTFE) 292
Postsurgical tissues
    classification 327–328
    colonoids
        chelating buffer 328
        crypt/gland suspension 331–333
        freezing of 334
        human colon dissociation buffer 328
        human colonoid medium 328–329
        passaging of 331, 334
        R-spondin-1-conditioned medium 329
        thawing of 334–335
        Wnt-3a-conditioned medium 329
    gastroids
        crypt/gland isolation 330–331
        crypt/gland suspension 331–333
        culture medium 330
        freezing of 334
        passaging of 331, 334
        thawing of 334–335
Primary enterocytes 196
Primitive neuroepithelia induction 3
Progenitor cell isolation from bone marrow 38
Pseudostratified ciliated and mucosecretory
    epithelium 43, 44

Q

Quality assurance (QA) 14

R

Recombinant Human FGF basic (bFGF) stock 294
Renal organoids 102
Replenish medium (RP) 293
Rho-associated protein kinase (ROCK) inhibitor 101
R-spondin 114

S

Salmonella Typhimurium
SFEBq 1
Small intestine 196, 198, 199
Spectrum of Fates (SoFa1) 257

T

3T3-J2 feeder cell culture 45, 47
Taurocholate (TC) 199
Testicular organoids (TOs)
    scaffolds
        adult/pubertal donors 284–285
        generation of 284, 286
        immature rodent 283
        in vitro study 284
        pig testicular cells 283
        testicular tissue digestion 284, 286–287
    in vitro spermatogenesis 283–284
Three-dimensional brain organoids 1
    cerebral tissue growth 3, 9
    embryoid bodies 1, 3, 5–6
    feeder-dependent hPSC culture 2
        embryoid bodies generation 3
        and passaging 4
    feeder-independent hPSC culture 2
        embryoid bodies generation 3
        and passaging 4–5
    germ layer differentiation 3, 6
    neuroepithelial tissue expansion 3, 6–8
    primitive neuroepithelia induction 3
3D human tracheospheres (lumen-in) 46, 49
Thymic cell isolation 36–37
Thymic epithelial cells (TECs) 34
    enrichment 37
    isolation by FACS 37–38
    physical colonization 34
    T-cell development 34
Thymus involution 33
Thymus organoid construction 39
    immune self-tolerance 33
    T cell repertoire 34
    TECs (*see* Thymic epithelial cells)
Tip-to-collector-distance (TTCD) 323
Tissue engineering 34
    See also Acoustic assembly
Transient amplifying (TA) cells 113, 114
Triethanolamine (TEA) buffer 314–315

U

Ultraviolet (UV) 322

V

Valproic acid (VPA) 67
Vascular endothelial growth factor (VEGF) 102
Vitronectin
    aliquots 61
    coated plates 76
    equipment and supplies 60, 61
    freezing 78–79
    passaging 76–78
    reagents 61

W

Water for injection (WFI) quality 304

Y

Yamanaka factors 55

Z

Zebrafish retinal cells
    agarose microwell culture dishes 259
    culture of 262
    dissociation and culture medium 257
    double-distilled sterile water 257
    embryo maintenance 257
    equipment and materials 260–261
    fixation, mounting and imaging 262–264
    immunostaining 260, 264–265
    isocontour profiling 265–268
    isolation of 259, 261–262
    in vivo studies 256
    Müller glia 255
    photoreceptors 255
    re-aggregation studies 256
    sample fixation and mounting 259–260
    solutions and cell culture reagents 258–259
    sterile technique 257
Z-factor 346–347
Z-prime 346–347